INTEGRATION OF
SWARM INTELLIGENCE AND
ARTIFICIAL NEURAL NETWORK

SERIES IN MACHINE PERCEPTION AND ARTIFICIAL INTELLIGENCE*

Editors: **H. Bunke** (Univ. Bern, Switzerland)
P. S. P. Wang (Northeastern Univ., USA)

*For the complete list of titles in this series, please write to the Publisher.

Series in Machine Perception and Artificial Intelligence – Vol. 78

INTEGRATION OF SWARM INTELLIGENCE AND ARTIFICIAL NEURAL NETWORK

Satchidananda Dehuri
Fakir Mohan University, India

Susmita Ghosh
Jadavpur University, India

Sung-Bae Cho
Yonsei University, South Korea

editors

World Scientific

NEW JERSEY · LONDON · SINGAPORE · BEIJING · SHANGHAI · HONG KONG · TAIPEI · CHENNAI

Published by

World Scientific Publishing Co. Pte. Ltd.

5 Toh Tuck Link, Singapore 596224

USA office: 27 Warren Street, Suite 401-402, Hackensack, NJ 07601

UK office: 57 Shelton Street, Covent Garden, London WC2H 9HE

British Library Cataloguing-in-Publication Data
A catalogue record for this book is available from the British Library.

INTEGRATION OF SWARM INTELLIGENCE AND ARTIFICIAL
NEURAL NETWORK
Series in Machine Perception and Artificial Intelligence — Vol. 78

Copyright © 2011 by World Scientific Publishing Co. Pte. Ltd.

ISBN-13 978-981-4280-14-3
ISBN-10 981-4280-14-3

Printed in Singapore.

Preface

In the past few years a significant amount of research has been done on training Artificial Neural Networks (ANNs) with Swarm Intelligence (SI) techniques rather than the standard technique of back-propagation, evolving the topology of neural networks by genetic algorithms (GAs), and so on. However, there is still a large gap between the researchers of SI and ANNs. Hence, to bridge the gap between these two disciplines an integration of SI and ANNs is required. As a result this will provide a new medium for dissemination of knowledge in both theoretical and applied research on SI and ANNs. This medium accelerates interaction between the above two bodies of knowledge, and fosters a unified development in the next generation computational model for machine learning. Keeping this as a target this book encompasses eleven chapters of different flavors.

In Chapter 1 the recent development of swarm intelligence and neural networks are discussed. Specifically, this chapter is focusing on particle swarm optimization (PSO), ant colony optimization (ACO) and bee colony optimization (BCO) techniques of swarm intelligence. Artificial neural networks (ANNs) are known as the universal approximators and computational models with particular characteristics such as the ability to learn and adapt. In particular, it is very difficult to obtain an optimal neural network architecture for solving a complex problem by a human expert. For instance, automatic determination of the optimal number of hidden layers and hidden neurons in each (hidden) layer is the most critical task. Several earlier attempts can provide a few off-the-self solutions however, this chapter argues that the application of swarm intelligence can lead towards better solutions in terms of robustness and time. Some of the potential higher order neural networks and their possible integration with swarm intelligence is also another attracting coverage of this chapter.

Omkar and Senthilnath describes the classification task of data mining using nature inspired techniques such as neural network and swarm intelligence algorithm in Chapter 2. Neural network and swarm intelligence techniques have an advantage over conventional statistical techniques like maximum likelihood classifier, nearest neighbor classifier etc, because they are distribution free, i.e., no knowledge is required about the distribution of data. Authors reported that the extraction of knowledge from the dataset in the form of weights and rules using various neural network and swarm intelligence methods is increasing over the last few years. In this chapter, Omkar and Senthilnath present the knowledge discovery process using three types of testbeds collected from UCI repository. The performance analysis of the classification task of data mining is done based on per class accuracy as well as overall accuracy. Finally, they evaluated the computational complexity of neural network and swarm intelligence techniques.

Ant Colony Optimization (ACO) is one of the most representative metaheuristics derived from the broad concept of SI where the behavior of social insects is the main source of inspiration. Being a particular SI approach, the ACO metaheuristic is mainly characterized by its distributiveness, flexibility, capacity of interaction among simple agents, and its robustness. The ACO metaheuristic has been successfully applied to an important number of discrete and continuous single-objective optimization problems. However, this metaheuristic has shown a great potential to cope with multi-objective optimization problems as evidenced by the several proposals currently available in this regard. Chapter 3 contributed by Leguizamon and Coello is aimed at describing the most relevant and recent developments on the use of the ACO metaheuristic for solving multi-objective optimization problems. Additionally, they also derive a refined taxonomy of the types of ACO variants that have been used for multi-objective optimization. Such a taxonomy intends to serve as a quick guide for those interested in using an ACO variant for solving multi-objective optimization problems. In the last part of the chapter, they provide a some potential paths for further research in this area.

Over the past two decades, recurrent neural networks have been shown to be very powerful computational models for optimization. A number of recurrent neural network models have been designed for solving various constrained optimization problems. Based on the Karush-Kuhn-Tucker optimality conditions and non-smooth functions, recurrent neural networks with discontinuous activation functions are designed for constrained optimization. Chapter 4 presents several recurrent neural networks with discontinuous activation functions for solving linear and quadratic programming problems and a class of non-smooth optimization problems. Moreover, two k-winners-take-all networks are designed based on the presented neural networks.

Chapter 5 presents an application of Wavelet Transform (WT), one of the most frequently used signal processing techniques for the extraction of important information in different frequency sub-bands from the non-stationary Power Quality (PQ) signals. Features derived by WT are used for classification purpose to identify the nature of the PQ disturbances. Panigrahi *et al.*, used the Multi-Layer Feed Forward Neural Network (FNN) for classification. Traditional learning algorithm like back-propagation learning is used for the training of the FNN. Simultaneously an attempt has been made for the design of an evolvable neural network classifier using one of the most promising SI algorithms known as Particle Swarm Optimization (PSO). The basic PSO algorithm was modified to make it adaptive and the proposed algorithm is known as adaptive PSO (APSO). The performance of NNs for the classification is compared for both back-propagation based learning and APSO based learning.

Condition monitoring, fault diagnosis as well as Quality Assurance have become of paramount importance in various industrial sectors such as automobile industries, aeronautics industries, power generating units as well as manufacturing units. Interestingly, all these sectors can use generic solutions approach to achieve a greater autonomy with quality assurance using non-parametric classification and prediction methodologies. Chapter 6 contributed by Yadav *et al.*, provides a generic solution using intelligent methodologies based on statistical techniques, neural networks as well as other machine learning approaches. The methods were justified by using a real life engine data taken from one of the Two Wheeler Industry. The following applications has been investigated at length: Classifying good and bad single cylinder engines based on audio data collected during the running of the engine at various speeds and predicting possible type of the defect in engines. The chapter has been concluded by futuristic approach and wide applicability of the proposed techniques.

Image enhancement techniques are used to improve image quality by maximizing the information content in the input image using intensity transformation functions. For color image enhancement, gray-level enhancement techniques may be applied directly on R, G, B color planes separately, but in this case colors of the enhanced image is distorted as R, G, B planes are enhanced in different scales. So, perception of the original color in the image will be changed, which means hue of the input image is altered. Sometimes enhancement is done in HSI/LHS color space by processing intensity and saturation components keeping the hue unaltered to preserve the original color perception. While performing reverse transformation from these color spaces to RGB space, it usually creates gamut problem. In Chapter 7, a PSO based hue preserving gamut problem free color image enhancement technique is proposed. The algorithm is tested on several color images and the results are compared with two other popular color image enhancement techniques like hue-preserving color image enhancement without gamut problem (HPCIE) and a genetic algorithm based approach to color image enhancement (GACIE). Visual analysis, detail and background variance of the resulted images are reported. It has been found that the proposed method produce better results than the other two methods.

In Chapter 8, Misra *et al.*, presents a procedure for designing classifier model using polynomial neural network (PNN). The PNN is a flexible neural architecture. Number of layers of the PNN is not fixed in advance but it is developed on the fly. This network is also considered as a self-organizing network. The essence of the design procedure of PNN dwells on the group methods of data handling (GMDH) technique. Each node of PNN exhibits a high level of flexibility and realizes a polynomial mapping between input and output variables. This chapter presents how artificial neural network (ANN) is incorporated with PNN to enhance the performance substantially. During the study of PNN, it is observed that a huge amount of time is consumed during training the classifier and the developed model is complex in nature. To alleviate these drawbacks the growth of PNN layers is restricted but the approximation capability is maintained by incorporating ANN. Benchmark databases of different domains have been taken to test the performance of this hybrid model. The simulation results of PNN and its hybrid models are compared, which shows that the performance gain obtained in the hybrid model is much better in comparison to the PNN model.

Chapter 9 by Majhi *et al.*, develops an efficient adaptive prediction model using recently developed Differential Evolution (DE) technique. The forecasting model employs an adaptive linear combiner architecture and DE based learning rule to predict seasonally adjusted (SA) and non seasonally adjusted (NSA) sales data for short and long ranges. The prediction performance of the proposed model is assessed through simulation study and using real life data. For comparison purpose the corresponding results are also obtained using genetic algorithm (GA), bacterial foraging optimization(BFO) and PSO based forecasting models. It is, in general, observed that the new DE forecasting model offers fastest training, best sales prediction and least mean square error after training compared to other three evolutionary computing based models.

Das and Dehuri presents a survey on PSO for single and multi-objective problems in Chapter 10. A considerable number of algorithms have been and are being proposed for single and multi-objective problems based on either tuning or introducing various parameters of PSO. These algorithms have their own merits and demerits. Further, since these algorithms are scattered in various journals, proceedings and book chapters, it is a herculean task for a reader to assemble in one place and learn. Hence, this study can reduce the effort and time of the reader to digest various approaches of PSO. Additionally, the authors identify some application areas, where PSO has given a clear edge over other metaheuristic approaches for solving single and multi-objective optimization problems. This can motivate and lead the readers for extensive application of PSO in new domains of interest.

Microarray technology is a powerful tool for geneticists to monitor interactions among tens of thousands of genes simultaneously. There has been extensive research on coherent subspace clustering of gene expressions measured under consistent experimental settings. However, these methods assume that all experiments are run using the same batch of microarray chips with similar characteristics of noise. Algorithms developed under this assumption may not be applicable for analyzing data collected from heterogeneous settings, where the set of genes being monitored may be different and expression levels may not be directly comparable even for the same gene. In Chapter 11, Mishra *et al.*, propose a biclustering model that imitates the ecosystem taking into account the features of biological data for mining subspace coherent patterns from heterogeneous gene expression data. They have implemented the system using an ACO. The algorithm decides the number of bi-clusters automatically. This processes the input biological data, runs the ACO, assigns biclusters to the genes and conditions simultaneously and displays the output.

Satchidananda Dehuri
Susmita Ghosh
Sung-Bae Cho

Contents

Chapter 1

Swarm Intelligence and Neural Networks

Satchidananda Dehuri[*], Sung-Bae Cho[†] and Susmita Ghosh[‡]

*Department of Information and Communication Technology,
Fakir Mohan University, Vyasa Vihar,
Balasore-756019, ORISSA, India
satchi.lapa@gmail.com*

†*Soft Computing Laboratory,
Department of Computer Science, Yonsei University,
262 Seongsanno, Seodaemun-gu,
Seoul 120-749, South Korea.
sbcho@cs.yonsei.ac.kr*

‡*Department of Computer Science and Engineering,
Jadavpur University, Kolkata 700 032, INDIA.
susmitaghoshju@gmail.com*

This chapter discusses the recent development of swarm intelligence and neural networks. Swarm intelligence deals with the collective behavior of systems having many individuals that interact locally with each other and with their environment, and that use forms of decentralized control and self-organization to achieve their targets. Specifically, this chapter is restricted with particle swarm optimization (PSO), ant colony optimization (ACO) and bee colony optimization (BCO) techniques of swarm intelligence.

Artificial neural networks (ANNs) are known as the universal approximators and computational models with particular characteristics such as the ability to learn or adapt. In particular, it is very difficult to obtain an optimal neural network architecture for solving a complex problem by a human expert. For instance, automatic determination of the optimal number of hidden layers and hidden neurons in each (hidden) layer is the most critical task. Several earlier attempts can provide a few off-the-self solutions but the application of swarm intelligence introduce a new dimension with better solutions in terms of robustness and time. Some of the potential higher order neural networks and their possible integration with swarm intelligence is another attracting items of this chapter.

1.1. Introduction

The laws that govern the collective behavior of social insects like ants, birds, bees, termites, wasps, etc. continue to mesmerize researchers.

A single ant or bee is not smart, but their colonies are.

While individuals are rather unsophisticated, in cooperation they can solve complex tasks. Task-solving results from self-organization, which often evolves from simple means of communication, either directly or indirectly via changing the environment, the latter referred to as stigmergy. Characteristics of the resulting systems include robustness and flexibility. There is no universally accepted definition of swarm intelligence[1,2] but the following one can provide a deeper insight.

Definition.   Swarm intelligence (SI) is a technique based on the study of collective behavior in decentralized, self-organized systems.

The expression "swarm intelligence" was coined by Beni and Wang in 1989,[3] in the context of cellular robotic systems. SI systems are typically made up of a population of simple agents interacting locally with one another and with their environment. Although there is normally no centralized control structure dictating how individual agents should behave, local interactions between such agents often lead to the emergence of global behavior. Example of systems like this can be found in nature, including ant colonies, bird flocking, bees colony, animal herding, bacteria molding and fish schooling.

Swarm intelligence research encompasses both science and engineering disciplines. Science oriented research is focused on trying to understand and model naturally occurring swarm systems. Recently, a group of researchers from engineering discipline have been studying how to create and control swarms of physical or software-based artifacts. This engineering oriented research often, but not always, takes inspiration from scientific studies of natural swarm systems.

Why are we interested in designing swarms of artifacts (be they software or hardware) and trying to solve problems with them? What is it we hope these systems will provide that current technologies cannot? Put simply, creators of artificial swarms are trying to copy the observable success and efficiency of natural swarm systems. Despite the relative simplicity of individual agents, natural swarm systems can often display extremely complex behavior. Natural swarm systems are highly scalable they are sometimes made up of many millions of individuals. In addition, such systems tend to be flexible and robust. They respond well to rapidly changing environments, and continue to function even if many of the individual agents are incapacitated. Studies have shown that in many cases simple behavioral rules at the level of the individual are sufficient to explain complex group behavior. These models also do not require any global communication they rely only on local sensing and communication. Researchers have started to use similar behavioral models in artificially created swarms.

Swarm intelligence thus provides a new framework for the design and implementation of systems made of many agents that are capable of cooperation for the solution of complex problems. The potential advantages of the swarm intelligence approach are manifold: i) collective robustness: the failure of individual components does not significantly hinder performance; ii) individual simplicity: cooperative behavior makes it possible to reduce the complexity of the individuals; and iii) scalability: the control mechanisms used are not dependent on the number of agents in the swarm.

So far, we have only just scratched the surface of swarm intelligence as a practical engineering approach. The majority of research to date has focused on demonstrating the cooperative problem solving capabilities of swarm intelligent systems. Very encouraging results have been obtained, especially in optimization applications. In particular, ant colony optimization,[4,5] bee colony optimization[6] and particle swarm optimization,[7] where swarms of software agents cooperate to search for good solutions to difficult optimization problems e.g., the weight vector and topological structure of neural networks.

Artificial Neural Networks (ANNs) are known as the universal approximators and a kind of robust computational models with particular characteristics such as the ability to learn or adapt, and generalization. It is quite acceptable that designing a (near) optimal network architecture by a human expert requires a tedious trial and error process. Especially automatic determination of the optimal number of hidden layers and hidden nodes in each (hidden) layer is the most difficult and challenging task. For instance, an ANN with no or too few hidden neurons may not approximate the decision boundary among classes (assume problem is classification). In contrast,

if the ANN has too many neurons/layers, it might be affected severely by
the noise in data due to over-parameterization, which eventually leads to
a poor generalization. On such complex networks proper training can also
be highly time-consuming. The optimum number of hidden neurons/layers
might depend on input/output vector sizes, training and test data sizes,
more importantly the characteristics of the problem, e.g., its non-linearity,
dynamic nature, etc. Several researchers have attempted to design ANNs
automatically with respect to a particular problem.

The earlier attempts fall into two broad categories: constructive and
pruning algorithms,[8–10] where many deficiencies and limitations have been
reported.[11] Efforts have been then focused on evolutionary algorithms
(EAs)[12] particularly over Genetic Algorithms (GAs)[13] and Evolutionary
Programming (EP),[14] for both training and evolving ANNs. Most GA
based methods have also been found quite inadequate for evolving ANNs
mainly due to two major problems: permutation problem and noisy fitness
evaluation.[15] Although EP-based methods,[11,16] might address such problems, they usually suffer from their high complexity and strict parameter
dependence.

However, recently swarm intelligence method like PSO has been successfully applied for training feed-forward[17,18] and recurrent ANNs.[19] Some of
the recent attempt of application of ACO and BCO for neural network[20,21]
is also encouraging.

1.2. Swarm Intelligence

1.2.1. *Particle Swarm Optimization*

Kennedy and Eberhart originally proposed the PSO algorithm for single objective optimization.[7] PSO is a population-based search algorithm based
on the simulation of the social behavior of birds within a flock. Originally it was adopted for neural network training and non-linear function
optimization,[22] and soon became a very popular global optimizer, mainly
in problems in which the decision variables are real numbers.[22,23] It is
worth noting that there have been proposals to use alternative encodings
with PSO (e.g., binary[24] and integer[25]). According to Angeline[26] and as
summarized by M. R. Sierra *et al.*,[27] we can make two main distinctions
between a PSO algorithm and an EA:

- EAs rely on three mechanisms in their processing: parent representation, selection of individuals, and the fine tuning of their

parameters. In contrast, PSO relies on only two mechanisms, since it does not adopt an explicit selection function. The absence of a selection mechanism in PSO is compensated by the use of leaders to guide the search. However, there is no notion of offspring generation in PSO as with EAs.

- EAs use a mutation operator that can set an individual in any direction (although the relative probabilities for each direction may be different). PSO uses an operator that sets the velocity of a particle to a particular direction. This can be seen as a directional mutation operator in which the direction is defined by both the particle's personal best (pbest) and the global best (of the swarm) (gbest). If the direction of pbest is similar to the direction of the gbest, the angle of potential directions will be small, whereas a larger angle will provide a larger range of exploration. In fact, the limitation exhibited by the directional mutation of PSO has led to the use of mutation operators similar to those adopted in EAs.

PSO has become so popular because its main algorithm is relatively simple and easier to implement. It is also straightforward and has been found to be very effective in a wide variety of applications with very good results at a very low computational cost.[22,23]

As a basic principle, in PSO, a set of randomly generated particles in the initial swarm are flown (have their parameters adjusted) through the hyper-dimensional search space (problem space) according to their previous flying experience. Changes to the position of the particles within the search space are based on the socio-psychological tendency of individuals to emulate the success of other individuals. Each particle represents a potential solution to the problem being solved. The position of a particle is determined by the solution it currently represents. The position of each particle is changed according to its own experience and that of its neighbors. These particles propagate towards the optimal solution over a number of generations (moves) based on large amount of information about the problem space that is assimilated and shared by all members of the swarm. PSO algorithm finds the global best solution by simply adjusting the trajectory of each individual toward its own best location (pbest) and towards the best particle of the entire swarm (gbest) at each time step (generation). In this algorithm, the trajectory of each individual in the search space is adjusted by dynamically altering the velocity of each particle according to its own flying experience and the flying experience of the other particles in the search space.

The position vector and the velocity vector of the ith particle in the d-dimensional search space can be expressed as $x_i = (x_{i1}, x_{i2}, ..., x_{id})$ and $v_i = (v_{i1}, v_{i2}, ..., v_{id})$ respectively. According to a user defined fitness function, the best position of each particle (which corresponds to the best fitness value obtained by that particle at time t) is $p_i = (p_{i1}, p_{i2}, ..., p_{id})$, denoted as pbest and the fittest particle found so far in the entire swarm at time t is $p_g = (p_{g1}, p_{g2}, ..., p_{gd})$, denoted as gbest. Then the new velocities and the new positions of the particles for the next fitness evaluation are calculated at time t+1 using the following two self-updating equations:

$$v_{id}(t+1) = wv_{id}(t) + c_1 rand_1()(p_{id}(t) - x_{id}(t)) + c_2 rand_2()(p_{gd}(t) - x_{id}(t))$$

$$(1.1)$$

$$x_{id}(t+1) = x_{id}(t) + v_{id}(t+1) \tag{1.2}$$

where $rand_1()$ & $rand_2()$ are two separately generated uniformly distributed random values in the range [0,1], w is inertia weight (or inertia factor) which is employed to control the impact of the previous history of velocities on the current velocity of a given particle, c_1 & c_2 are constants known as acceleration coefficients (or learning factors); c_1 is the cognitive learning factor (or self confidence factor) which represents the attraction that a particle has toward its own success and c_2 is the social learning factor (or swarm confidence factor) which represents the attraction that a particle has toward the success of its neighbors. Hassan et al.[28] have presented an idea about the ranges of c_1 to be from 1.5 to 2, c_2 to be from 2 to 2.5 and w to be from 0.4 to 1.4. From Eq. (1.1), it is observed that it has three components which are incorporated via a summation approach and effect the new search direction. The first component is known as current motion influence component which depends on previous velocity and provides the necessary momentum for particles to roam across the search space. The second component is known as cognitive (or particle own memory influence) component which represents the personal thinking of each particle and encourages the particles to move toward their own best positions found so far. The third component is known as social (or swarm influence) component which represents the collaborative effect of the particles, in finding the global optimal solution. This component always pulls the particles toward the global best particle found so far.

PSO uses an operator that sets the velocity of a particle to a particular direction. This can be seen as a directional mutation operator in which the direction is defined by both the particle's personal best and the global best (of the swarm). If the direction of the personal best is similar to the direction of the global best, the angle of potential directions will be small, whereas a larger angle will provide a larger range of exploration. In contrast, evolutionary algorithms use a mutation operator that can set an individual in any direction (although the relative probabilities for each direction may be different). In fact, the limitations exhibited by the directional mutation of PSO has led to the use of mutation operators (sometimes called turbulence operators) similar to those adopted in evolutionary algorithms.

The pseudocode for a basic PSO algorithm is illustrated as follows:

Algorithm.   Particle Swarm Optimization

(1) BEGIN
(2) Parameter settings and initialization of swarm.
(3) Evaluate fitness and locate the leader (i.e., initialize *pbest* and *gbest*).
(4) I = 0 /* I = Iteration count */
(5) WHILE (the stopping criterion is not met, say, I < Imax)
(6) DO
(7) FOR each particle
(8) Update position & velocity (flight) as per equations (1.1) & (1.2)
(9) Evaluate fitness
(10) Update *pbest*
(11) END FOR
(12) Update leader (i.e., *gbest*)
(13) I++
(14) END WHILE
(15) END

First, the swarm is initialized. This initialization includes both positions and velocities. The corresponding pbest of each particle is initialized and the leader is located (the gbest solution is selected as the leader). Then, for a maximum number of iterations, each particle flies through the search space updating its position (using Eqs. (1.1) and (1.2)) and its pbest and, finally, the leader is updated too.

1.2.2. Ant Colony Optimization

1.2.2.1. Basic Concepts of ACO

In the early 1990s, ant colony optimization (ACO) was introduced by Dorigo and colleagues as a novel nature inspired metaheuristic for the solution of NP-hard problems.[5] ACO belongs to the class of metaheuristics, which are approximate algorithms used to obtain good enough solutions to very hard combinatorial problems in a reasonable amount of computation time. The basic principles of ACO is inspired from the foraging behavior of real ants. When searching for food, ants initially explore the area surrounding their nest in a random manner. As soon as an ant finds a food source, it evaluates the quantity and the quality of the food and carries some of it back to the nest. During the return trip, the ant deposits a chemical pheromone trail on the ground. The quantity of pheromone deposited, which may depend on the quantity and quality of the food, will guide other ants to the food source. The indirect communication between the ants via pheromone trails enables them to find shortest paths between their nest and food sources. This characteristic of real ant colonies is exploited in artificial ant colonies in order to solve NP-hard problems.

ACO is implemented as a group of sophisticated intelligent agents which simulate the ants behavior, walking around the graph representing the problem to solve using mechanisms of cooperation and adaptation. ACO algorithm requires to define the following[4]:

(1) The problem needs to be represented appropriately, which would allow the ants to incrementally update the solutions through the use of a probabilistic transition rules, based on the amount of pheromone in the trail and other problem specific knowledge. It is also important to enforce a strategy to construct only valid solutions corresponding to the problem definition.

(2) A problem-dependent heuristic function η that measures the quality of components that can be added to the current partial solution.

(3) A rule set for pheromone updating, which specifies how to modify the pheromone value τ.

(4) A probabilistic transition rule based on the value of the heuristic function η and the pheromone value τ that is used to iteratively construct a solution.

ACO was first introduced as a method to solve the popular NP-complete Travelling Salesman Problem (TSP). Starting from its start node, an ant iteratively moves from one node to another. When being at a node, an ant chooses to go to a unvisited node at time t with a probability given by

$$p_{i,j}^k(t) = \frac{(\tau_{i,j}(t))^\alpha (\eta_{i,j}(t))^\beta}{\sum_{j \in N_i^k} (\tau_{i,j}(t))^\alpha (\eta_{i,j}(t))^\beta}, j \in N_i^k \qquad (1.3)$$

where N_i^k is the feasible neighborhood of the ant k, that is, the set of cities which ant k has not yet visited; $\tau_{i,j}(t)$ is the pheromone value on the edge (i, j) at the time t, α is the weight of pheromone; $\eta_{i,j}(t)$ is a priori available heuristic information on the edge (i, j) at the time t, β is the weight of heuristic information. Two parameters α and β determine the relative influence of pheromone trail and heuristic information. $\tau_{i,j}(t)$ is determined by

$$\tau_{i,j}(t+1) = \rho.\tau_{i,j}(t) + \sum_{k=1}^n (\Delta\tau_{i,j}^k(t+1)), \forall(i, j) \qquad (1.4)$$

$$\Delta\tau_{i,j}^k(t+1) = \left(\begin{array}{cc} \frac{Q}{L_k(t)} & if\ the\ edge(i, j)\ chosen\ by\ ant\ k \\ 0 & otherwise \end{array} \right) \qquad (1.5)$$

where ρ is the pheromone trail evaporation rate $(0 < \rho < 1)$, n is the number of ants.

The following guidelines can be made while attacking problems by ACO algorithms.

- Represent the problem in the form set of components and transition or by means of weighted graph on which ant builds solutions.
- Define appropriately the meaning of the pheromone trial i.e., the type of decision they bias. A good definition of pheromone trials is not a trivial task and requires insight into problem to be solved.
- Define appropriately the heuristic preference for each decision that an ant has to take while constructing a solution. Heuristic information is crucial for good performances if local search algorithms are not available.
- If possible, implement an efficient local search algorithm with ACO algorithm. Results of many ACO applications to NP-hard problem show the best performance when ACO is coupled with local search algorithm.
- Choose a specific ACO algorithms and apply it to the problem being solved with suitably setting parameters of ACO algorithm.

1.2.3. Bee Colony Optimization

1.2.3.1. Honey Bee in Nature

The bees are the biological entities living in hives in very organized colonies. This highly organized behavior of honey bee enables the colonies to solve problems beyond capability of individual entities by functioning collectively and interacting primitively amongst members of the group. A group of honey bees can extend themselves over a long distance and in multiple directions simultaneously for foraging a large number of flower patches (food sources). A colony prospers by deploying its foragers to optimal fields. Such a colony is characterized by self organization, adaptiveness, and robustness. Seeley[29] proposed a behavioral model of self organization for a colony of honey bees. In principles, flower patches with plentiful amounts of nectar or pollen that can be collected with less effort should be visited by more bees, whereas patches with less nectar or pollen should receive fewer bees.

The foraging process begins in a colony by scout bees being sent to search for promising flower patches (food sources). Scout bees visiting flower patches randomly from one patch to another. During the harvesting season, a colony continues its exploration, keeping a percentage of the swarm as scout bees. When they return to the hive, those scout bees that found a patch, whose profitability is measured and if it is beyond a certain threshold then deposit their nectar or pollen and go to the "dance floor" to perform a dance known as the "waggle dance". The profitability rating is a function of nectar quality, nectar bounty and distance from the hive.

This enlisting signal is essential for colony communication, and contains three pieces of information regarding a flower patch (food source): the direction in which it will be found, its distance from the hive and its quality rating (or fitness). This information helps the colony to send its bees to flower patches precisely, without using references. Each individuals knowledge of the outside environment is gleaned solely from the waggle dance. This distance enables the colony to evaluate the relative merit of different patches according to both the quality of the food they provide and the amount of energy needed to harvest it. After waggle dancing on the dance floor, the dancer (i.e., the scout bee) goes back to the flower patch with flower bees that were waiting inside the hive. More flower bees are sent to more promising patches. This allows the colony to gather food quickly and efficiently.

While harvesting from a patch, the bees monitor its food level. This is necessary to decide upon the next waggle dance when they return to the hive. If the patch is still good enough as a food source, then it will be advertised in the waggle dance and more bees will be recruited to that source.

1.2.3.2. *Basic Algorithm of Honey Bee*

The basic algorithm of this multiple agent approach is given below. The algorithm require a number of parameters to be set, namely: number of scout bees (N_b), number of sites selected (S_s), number of best sites selected (S_b), number of bees recruited for best S_b sites R_b^s, number of bees recruited for the other ($S_s - S_b$) selected sites (R_b^r), initial size of patches (P_s) which includes site and its neighborhood and stopping criterion. The algorithm start with the N_b scout bees being placed randomly in the search space. The fitness of the sites visited by the scout bees are evaluated in step 2.

Algorithm. Bee Colony Optimization

    (1) Initialize population with random scouts;

    (2) Evaluate the fitness of the swarm;

    (3) Repeat

        (a) Select sites for neighborhood search;

        (b) Recruit Bees for selected sites, more Bees for best selected sites;

        (c) Evaluate fitness;

        (d) Select the fittest bee from each patch;

        (e) Assign remaining bees to search randomly and evaluate their fitness;

    (4) Until (termination criteria met)

In step 3(a) bees that have the highest fitness are choosen as selected bees and sites visited by them are choosen for neighborhood search. Then, in steps 3(d) and 3(e), the algorithm conducts searches in the neighborhood of the selected sites, assigning more bees to search near to the best S_b sites. The bees can be choosen directly according to the fitness associated with the sites they are visiting. Alternatively, the fitness values are used to determine the probability of the bees being selected. Searches in the neighborhood of the best S_b sites which represent more promising solutions are made more detailed by recruiting more bees to follow them than the other selected bees. Together with scouting, this differential recruitment is a key operation of the bees algorithm.

However, in step 3(d), for each patch only the bee with the acceptable fitness will be selected to form the next bee population. In nature, there is no such restriction. This restriction is introduced to reduce the number of less important points to be explored. In step 3(e), the remaining bees in the population are assigned randomly around the search space scouting for new potentials solutions. These steps are repeated until a stopping criterion is met. At the end of each iteration, the colony will have two parts to its new population-representatives from each selected patch and other scout bees assigned to conduct random searches.

1.3. Neural Networks

1.3.1. *Evolvable Neural Network*

The advent of EAs has inspired new resources for optimization problem solving, such as the optimal design of artificial neural networks (ANNs). Since EAs are heuristic and stochastic based on populations made up of individuals with a specified behavior similar to biological phenomenon, they are robust and efficient at exploring an entire solution space of optimization problems. EAs have been successfully used to evolve weights, structure, and learning parameters of ANNs in recent years.[30,31] Yao and Liu[15] proposed a prominent evolutionary ANN approach called EPNet by using evolutionary programming (EP) algorithm. Weights and structure are evolved simultaneously by using partial training, mutation of weights and addition or removal of connections or nodes. EPNet encourages smaller networks, as removals are attempted before additions, and behavioral link is maintained between parents and offspring through partial training and node splitting. Castillo *et al.*,[30] proposed a method (G-Prop, genetic back-propagation (BP)) that attempts to search the initial weights and hidden-layer size of multilayer perceptrons (MLPs). The application of the G-Prop algorithm to several real-world and benchmark problems showed that MLPs evolved by G-Prop are smaller and achieve a better generalization than other perceptron training algorithms. Palmes *et al.*[31] proposed a mutation- based genetic ANN (MGNN) algorithm. MGNN can evolve structure and weights of ANNs simultaneously. MGNN implements a stopping criterion where overfitness occurrences are monitored through sliding-windows to avoid premature learning and overlearning. These references indicated that EAs were successful in evolving ANNs.

Recall that the successful applications of PSO to some optimization problems such as function minimization and ANN design[18] have demonstrated its potential. It is considered to be capable to reduce the ill effect of the BP algorithm of feedforward ANNs (e.g., very slow convergence speed in training, easily to get stuck in a local minimum, etc.), because it does not require gradient and differentiable information. Salerno[32] used PSO to evolve parameters (i.e., weights and bias of neurons) of ANNs for solving the XOR problem and parsing natural language. Lu *et al.*[33] adopted PSO to train MLPs to predict pollutant levels of air and their tendency. Their results demonstrated that PSO-based ANN has a better training performance, faster convergence rate, as well as a better predicting ability than BP-based ANN. Juang[34] proposed a hybrid of GA and PSO (HGAPSO) for training recurrent networks. HGAPSO used PSO to enhance the elites generated by GA to generate higher quality individuals. The performance of HGAPSO was compared to both GA and PSO in recurrent networks design problems, demonstrating its superiority. Da and Ge[35] proposed an improved PSO-based ANN with simulated annealing (SA) technique for solving a rock-engineering problem. Their results showed that SAPSO-based ANN has a better training and generalization performance than PSO-based ANN. Comparisons between PSO and GA for evolving recurrent ANNs were done analytically by Settles *et al.*[19] Their comparisons indicated that the GA is more successful on larger networks and the PSO is more successful on smaller networks. However, in comparison with the wide applications of GA in evolutionary ANNs, the applications of PSO for evolving ANNs are relatively sparse.

A hybrid of ACO and BP (ACO-BP) algorithms is proposed by Shi and Li[36] to evolve NNs. The ACO-BP algorithm firstly uses ACO algorithm to search the near optimal solution and then adopts BP algorithm to find the accurate solution. The former attempts to avoid being trapped in the local optima and the later can rapidly find the accurate solution to accelerate its evolving speed. The main idea of neural networks with ant colony algorithm optimization is: the path of ants to food sources is mapped to neural network weights and threshold parameters set, then the optimal path corresponds to optimal parameters of neural networks. Through the selection path evaluation of ant (that is, neural network parameters), the path on the pheromone concentration (expressed as the adjustment of neural network parameters) was updated. When all the ants choose the same path or algorithm to a predetermined number of cycles, then the algorithm is terminated.

Karaboga *et al.*,[6] has introduced another SI techniques for training the ANNs. They proposed an Artificial Bee Colony (ABC) Algorithm which has good exploration and exploitation capabilities in searching optimal weight set for training neural networks.

1.3.2. *Higher Order Neural Network*

ANNs have become one of the most acceptable soft computing tools for approximating the decision boundaries of a classification problem.[37,38] This well-liked behavior stems from a number of reasons like their capability to capture nonlinear relationships between input-output of patterns, their biological plausibility as compared to conventional statistical models,[39,40] their potential for parallel implementations, their celebrated robustness and graceful degradation, etc. In fact, a multilayer perceptron (MLP) with a suitable architecture is capable of approximating virtually any function of interest.[41] This does not mean that finding such a network is easy. On the contrary, problems, such as local minima trapping, saturation, weight interference, initial weight dependence, and overfitting, make neural network training difficult.

An easy way to avoid these problems consists in removing the hidden layers. This may sound a little inconsiderate at first, since it is due to them that nonlinear input output relationships can be captured. Encouragingly enough, the removing procedure can be executed without giving up nonlinearity, provided that the input layer is endowed with additional higher order units.[42,43] This is the idea behind higher order neural networks (HONNs)[44] like functional link neural networks (FLNNs),[45] ridge polynomial neural networks (RPNNs),[46,47] and so on.

In this Section we will discuss few of the recently proposed higher order neural networks quickly as a basic principle of enhancing the understandability of the designed method and consequently for empirical comparative analysis.

1.3.3. *Pi (Π)-Sigma (Σ) Neural Networks*

PSNNs belong to the class of higher order neural networks. The term *pi-sigma* comes from the fact that these networks use product of sums of input components as in higher order processing unit. PSNNs were first introduced by Shin and Ghosh.[48] The primary motivation of PSNNs were to develop a systematic method for maintaining the fast learning property and powerful mapping capability while avoiding the combinatorial increase in the number of weights and processing units required. A PSNN is a feed forward network with a single hidden layer of linear cells, that uses product units in the output layer. PSNNs have a highly regular structure and need a much smaller number of weights as compared to other single layer higher order networks. The presence of only one layer of adaptive weights results in fast training. The PSNNs are categorized into two types: analog pi-sigma neural networks (APSNNs) and the binary pi-sigma neural networks (BPSNNs). The APSNN has been successfully used for function approximation and pattern classification, and a generalization of the APSNN is shown to have a universal approximation capability. The BPSNN is capable of realizing any Boolean function.[49] More details about the architecture and working principle of PSNN can be found in Ref. 48.

1.3.4. *Functional Link Artificial Neural Network*

FLNNs are higher order neural networks without hidden units introduced by Klassen and Pao in 1988. Despite their linear nature, FLNNs can capture non-linear input-output relationships, provided that they are fed with an adequate set of polynomial inputs, or the functions might be a subset of a complete set of orthonormal basis functions spanning an n-dimensional representation space, which are constructed out of the original input attributes.

In contrast to the linear weights of the input patterns produced by the linear links of artificial neural network, the functional link acts on an element of a pattern or on the entire pattern itself by generating a set of linearly independent functions, then evaluating these functions with the pattern as the argument. Thus class separability is possible in the enhanced feature space. Let us consider a two-dimensional input sample $x = [x_1, x_2]$. This sample has been mapped to a higher dimensional space by functional expansion using polynomials with certain degrees. For example, two attributes yield six polynomials up to degree 2 (i.e., $(1, x_1, x_2, x_1^2, x_2^2, x_1.x_2)$). In general for a D-dimensional classification problem there are $\frac{(D+r)!}{D!.r!}$ possible polynomials up to degree r. For most of the real life problems, this is a too big number, even for degree 2, which obviously discourages us to achieving our goal. However, we can still resort to constructive and pruning algorithms in order to address this problem. In fact Sierra *et al.*[50] has proposed a new algorithm for the evolution of functional link networks which makes use of a standard GAs to evolve near minimal linear architectures. Moreover, the complexity of the algorithm still needs to be investigated.

However, the dimensionality of many problems itself is very high and further increasing the dimensionality to a very large extent may not be an appropriate choice. So, it is advisable and also a new research direction to choose a small set of alternative functions, which can map the function to the desired extent with an output of significant improvement. FLNN with trigonometric basis functions, as proposed in Ref. 45 is obviously an example. Chebyshev FLNN is also another improvement in this direction.

1.3.5. *Ridge Polynomial Neural Networks (RPNNs)*

RPNNs are a new class of higher-order feedforward neural networks and were introduced by Shin and Ghosh[46] in 1992. This network can approximate any multivariate continuous function on a compact set in multi-dimensional input space with any degree of accuracy. The RPNN is a special form of ridge polynomial. RPNN is a generalized neural architecture of $\Pi - \Sigma$ Neural Network. It is constructed by embedding different degrees of PSNNs as the basic building blocks. Similar to the PSNNs, RPNNs have only a single layer of adaptive weights and they preserve all the advantages of PSNNs.

In general there may not be much a priori information about the decision functions/boundaries to be approximated and also it is very difficult to choose an appropriate network size. Thus, using a fixed network architecture an unknown decision boundary f in R^D can be approximated by the direct use of the RPNN model of degree up to n based on

$$f(x) \approx s(\Sigma_{i=1}^n h_i(x)), \tag{1.6}$$

where

$$h_i(x) = \Pi_{j=1}^i (xw_j^T + w_{j0}), i = 1, 2, ..., n, \tag{1.7}$$

and $s(.)$ is a suitable linear or non-linear activation function, $w_j \in R^D$ and $w_{j0} \in R$ are determined by learning process. Since each h_i is obtainable as the output of a PSNN of degree i with linear output units, the learning algorithms for the PSNN can be used for the RPNN.

In order to improve the various limitations of traditional neural network, HONNs can be considered. HONNs most often runs faster than feed forward neural networks. Examples in implementation of two and three input XOR functions by using second order neural network proved that it is several time faster than feed forward neural network.[51,52] Various types of HONNs have been applied widely in different research areas including time series data, business, finance, and economics,[52-55] clustering,[56,57] classification,[58,59] pattern recognition and mathematical function approximation.[60-62]

Several works have been done on hybridization to implement, and improve HONNs parameters performance in pattern recognition and classification.[59,63-67] A new approach for determining the number of hidden neurons was also proposed and applied in data mining using adaptive HONN model by Xu and Chen.[68]

1.4. Summary and Discussion

In this chapter we have discussed PSO, ACO and BCO under the umbrella of swarm intelligence for possible integration in artificial neural networks. In addition we discussed the early attempt of evolving neural networks. Some of the recent attempt of integrating swarm intelligence in ANNs has also been discussed. Higher order neural networks is also discussed but there are enough scope to combine both HONN and SI.

References

1. J. Kennedy, R. C. Eberhart, and Y. Shi, *Swarm Intelligence.* (Morgan Kaufmann, 2001).
2. J.-B. Waldner, *Nanocomputers and Swarm Intelligence.* (Wiley, 2008).
3. G. Beni and J. Wang. Swarm intelligence in cellular robotic systems. In *Proc. of NATO Advanced Workshop on Robots and Biological Systems*, Tuscany, Italy, (1989).
4. M. Dorigo and T. Stutzle, *Ant Colony Optimization.* (MIT Press, Cambridge, MA, 2004).
5. M. Dorigo and C. Blum, Ant colony optimization theory: A survey, *Theoretical Computer Science.* **344**(255), 243–278, (2005).
6. D. Karaboga and B. Akay, A survey: Algorithms simulating bee swarm intelligence, *Artificial Intelligence Review.* **31**, 61–85, (2009).
7. J. Kennedy and R. Eberhart. Particle swarm optimization. In *Proc. of IEEE Int. Conference on Neural Networks*, pp. 1942–1948, (1995).
8. N. Burgess, A constructive algorithm that converges for real valued input patterns, *International Journal of Neural Systems.* **5**(1), 59–66, (1994).
9. M. Frean, The upstart algorithm: A method for constructing and training feed forward neural networks, *Neural Computation.* **2**(2), 198–209, (1990).
10. R. Reed, Prunning algorithms-a survey, *IEEE Transactions on Neural Networks.* **4**(5), 740–747, (1993).

11. P. J. Angeline, G. M. Sauders, and J. B. Pollack, An evolutionary algorithm that constructs recurrent neural networks, *IEEE Transactions on Neural Networks*. **5**, 54–65, (1994).

12. T. Back and H.-P. Schwefel, An overview of evolutionary algorithm for parameter optimization, *Evolutionary Computation*. **1**, 1–23, (1993).

13. D. Goldberg, *Genetic Algorithms in Search, Optimization and Machine Learning*. (Addison-Wesley, Reading, MA, 1989).

14. A. E. Eiben and J. E. Smith, *Introduction to Evolutionary Computing*. (Springer, 2003).

15. X. Yao and Y. Liu, A new evolutionary system for evolving artificial neural networks, *IEEE Transactions on Neural Networks*. **8**(3), 694–713, (1997).

16. Y. X., Evolving artificial neural networks, *Proceedings of the IEEE*. **87**(9), 1423–1447, (1999).

17. S. Kiranyaz, T. Ince, A. Yildirim, and M. Gabbouj, Evolutionary artificial neural networks by multi-dimensional particle swarm optimization, *Neural Networks*. **22**, 1448–1462, (2009).

18. J. Yu, S. Wang, and L. Xi, Evolving artificial neural networks using an improved PSO and DPSO, *Neurocomputing*. **71**(71), 1054–1060, (2008).

19. M. Settles, B. Rodebaugh, and T. Soule. Comparision of genetic algorithm and particle swarm optimizer when evolving a recurrent neural network. In *Proc. of Genetic and Evolutionary Computation Conference, LNCS-2723*, pp. 151–152, (2003).

20. D. Karaboga, B. Akay, and C. Ozturk. Artificial bee colony (ABC) optimization algorithm for training feed-forward neural networks. In *Proc. of MDAI 2007, LNAI-4617*, pp. 318–329, (2007).

21. Y. Liu, M.-G. Wu, and J.-X. Qian. Evolving neural networks using the hybrid of ant colony optimization and BP-algorithms. In *Proc. of ISNN Adnavces in Neural Networks LNCS-3971*, pp. 714–722, (2006).

22. J. Kennedy and R. C. Eberhart, *Swarm Intelligence*. (Morgan Kaufmann Publishers, San Francisco, California, 2001).

23. A. P. Engelbrecht, *Computational Intelligence: An Introduction*. (John Wiley & sons, England, 2005).

24. J. Kennedy and R. C. Eberhart. A discrete binary version of the particle swarm algorithm. In *Proceedings of the IEEE Conference on Systems, Man and Cybernetics*, pp. 1764–1771, Canberra, Australia (December, 2003).

25. X. Hu, R. C. Eberhart, and Y. Shi. Swarm intelligence for permutation optimization: A case study on n-queen problem. In *Proceedings of the IEEE Swarm Intelligence Symposium*, pp. 243–246, Indianpolis, Indiana, USA, (2003).

26. P. J. Angeline. Evolutionary optimization versus particle swarm optimization: Philosophy and performance differences. In *Proceedings of the 7th International Conference on Evolutionary Programming*, pp. 601–610, San Diego, California, USA, (1998).

27. M. R. Sierra and C. C. A. C., Multi-objective particle swarm optimizers: A survey of the state-of-the-art, *International Journal of Computational Intelligence Research*. **2**(3), 287–308, (2006).

28. R. Hassan, B. Cohanim, O. de Weck, and G. Venter. A comparision of particle swarm optimization and genetic algorithm. In *46th AIAA/ASME/ ASCE/AHS/ASC Structures, Structural Dynamic and Materials Conference*, pp. 1897–1909, Austin, Texas (April, 2005).

29. T. D. Seeley, *The Wisdom of the Hive: The Social Phisiology of Honey Bee Colonies.* (Harvard University Press, Massachusetts, 1995).

30. P. A. Castillo, J. Carpio, J. J. Merelo, A. Prieto, V. Rivas, and G. Romero, G-prop: Global optimization of multi-layer perceptrons using GAs, *Neuro-computing.* **35**, 149–163, (2000).

31. P. P. Palmes, T. C. Hayasaka, and S. Usui, Mutation based genetic neural network, *IEEE Transaction on Neural Networks.* **16**(3), 587–600, (2005).

32. J. Salerno. Using the particle swarm optimization technique to train a recurrent neural model. In *Proc. of Ninth International Conference on Tools with Artificial Intelligence (ICTAI'97)*, IEEE Press, (1997).

33. W. Z. Lu, H. Y. Fan, and S. M. Lo, Application of evolutionary neural network method in predicting pollutant level in downtown area of hong kong, *Neurocomputing.* **51**, 387–400, (2003).

34. C. F. Juang, A hybrid genetic algorithm and particle swarm optimization for recurrent network design, *IEEE Transactions on Systems Man and Cybernetics.* **32**, 997–1006, (2004).

35. Y. Da and X. R. Ge, An improved PSO based ANN with simulated annealing technique, *Neurocomputing Letter.* **63**, 527–533, (2005).

36. H. Shi and W. Li. Artificial neural networks with ant colony optimization for assessing performance of residential buildings. In *Proc. of International Conference on Future BioMedical Information Engineering*, pp. 379–382, IEEE Press, (2009).

37. S. Haykin, *Neural Networks- A Comprehensive Foundation.* (Prentice Hall, Englewood Cliffs, NJ, 1999).

38. O. L. Mangasarian and E. W. Wild, Non-linear knowledge based classification, *IEEE Transactions on Neural Networks.* **19**(10), 1826–1832, (2008).

39. K. Fukunaga, *Introduction to Statistical Pattern Recognition.* (Academic Press, New York, 1990).

40. S. Theodorodis and K. Koutroumbas, *Pattern Recognition.* (Academic Press, San Diego, USA, 1999).

41. K. Hornik, Approximation capabilities of multilayer feedforward networks, *Neural Networks.* **4**(2), 251–257, (1991).

42. C. L. Giles and T. Maxwell, Learning, invariance and generalization in higher-order neural networks, *Appl. Opt.* **26**(23), 4972–4978, (1987).

43. Y. H. Pao, *Adaptive Pattern Recognition and Neural Network.* (Addison-Wesley, Reading, MA, 1989).

44. E. Antyomov and O. Y. Pecht, Modified higher order neural network for invariant pattern recognition, *Pattern Recognition Letters.* **26**, 843–851, (2005).

45. B. B. Misra and S. Dehuri, Functional link neural network for classification task in data mining, *Journal of Computer Science.* **3**(12), 948–955, (2007).

46. Y. Shin and J. Ghosh, Ridge polynomial networks, *IEEE Transactions Neural Networks.* **6**(2), 610–622, (1995).

47. Y. Shin and J. Ghosh. Approximation of multivariate functions using ridge polynomial networks. In *Proceedings of International Joint Conference on Neural Networks II*, pp. 380–385, IEEE Press, (1992).

48. Y. Shin and J. Ghosh. The pi-sigma networks: an efficient higher order neural network for pattern classification and function approximation. In *Proceedings of International Joint Conference on Neural Networks I*, pp. 13–18, (1991).

49. Y. Shin and J. Ghosh. Realization of boolean functions using binary pi-sigma networks. In *Proceedings of Conference on Artificial Neural Networks in Engineering*, St. Louis, (1991).

50. A. Sierra, J. A. Macias, and F. Corbacho, Evolution of functional link networks, *IEEE Tranasctions on Evolutionary Computation*. **5**(1), 54–65, (2001).

51. M. M. Gupta, N. Homma, Z. Hou, A. M. G. Solo, and T. Goto. Fundamental theory of artificial higher order neural networks. In *M. Zhang (Eds.), Artificial Higher Order Neural Networks for Computer Science and Engineering: Trends for Engineering and Applications*, pp. 368–388, Idea Group Inc. (IGI), (2009).

52. A. Hussain and P. Liatsis. A novel recurrent polynomial neural network for financial time series prediction. In *M. Zhang (Eds.), Artificial Higher Order Neural Networks for Computer Science and Engineering: Trends for Engineering and Applications*, pp. 190–211, Idea Group Inc. (IGI), (2009).

53. J. Fulcher, M. Zhang, and S. Xu. Application of higher order neural networks to financial time series prediction. In *J. B. Kamruzzaman (Eds.) Artificial Neural Networks in Finance and Manufacturing*, pp. 80–108, Idea Group Inc. (IGI), (2006).

54. R. Ghazali, A. J. Hussain, D. Al-jumeily, and M. Merabti. Dynamic ridge polynomial neural networks in exchange rates time series forecasting. In *B. Beliczynski, et al. (Eds.), Adaptive and Natural Computing Algorithms*, pp. 123–132, Berlin/Heidelberg, Springer, (2007).

55. S. Xu. Adaptive higher order neural network models and their applications in business. In *M. Zhang (Eds.), Artificial Higher Order Neural Networks for Computer Science and Engineering: Trends for Engineering and Applications*, pp. 314–329, Idea Group Inc. (IGI), (2009).

56. K. Ramanathan and S. U. Guan, Multi-order neurons for evolutioanry higher order clustering growth, *Neural Computation*. **12**, 3369–3391, (2007).

57. H. Lipson and H. T. Siegelmann, Clustering irregular shapes using higher order neural networks, *Neural Computation*. **12**, 2331–2353, (2000).

58. S. Dehuri and S.-B. Cho, Evolutionarily optimized features in functional link neural network for classification, *Expert Systems with Applications*. **37(6)**, 4379–4391, (2010).

59. M. Fallahrezhad, M. H. Moradi, and S. Zaferanlouei, A hybrid higher order neural classifier for handling classification problems, *Expert Systems with Applications*. **38**, 386–393, (2011).

60. E. Artyomov and O. Yadid-Pech, Modified higher order neural network for invariant pattern recognition, *Pattern Recognition Letters*. **26(6)**, 843–841, (2005).

61. G. L. Foresti and T. Dolso, An adaptive higher order neural tree for pattern recognition, *IEEE Transactions on Systems, Man and Cybernetics-Part B: Cybernetics.* **34(2)**, 988–996, (2004).

62. G. A. Rovithakis, M. Mariadakis, and M. Zervakis, Higher order neural network structure selection for function approximation applications using genetic algorithms, *IEEE Transactions on Systems, Man, and Cybernatics Part B: Cybernatics.* **34(1)**, 150–158, (2004).

63. H. C. Tsai, Hybrid higher order neural networks, *Applied Soft Computing.* **9(3)**, 874–881, (2009).

64. A. Abdelbar, Achieving superior generalization with a higher order neural network, *Neural Computing and Applications.* **7(2)**, 141–146, (1998).

65. Z. He and M. Y. Siyal, Improvement on higher order neural netwroks for invariant object recognition, *Neural Processing Letters.* **10(1)**, 49–55, (1999).

66. J. Jakubowski and et al., Higher order statistics and neural networks for tremor recognition, *IEEE Transactions on Biomedical Engineering.* **49(2)**, 152–159, (2002).

67. T. Kaita, S. Tomita, and J. Yamanaka, On a higher order neural network for distortion invariant pattern recognition, *Pattern Recognition Letters.* **23(8)**, 977–984, (2002).

68. S. Xu and L. Chen. Application of new adaptive higher order neural networks in data mining. In *Proceedings of International Conference on Computer Science and Software Engineering, Vol.1*, pp. 115–118, China, (2008).

Chapter 2

Neural Network and Swarm Intelligence for Data Mining

S. N. Omkar* and J. Senthilnath

Department of Aerospace Engineering,
Indian Institute of Science, Bangalore-560012, Karnataka, India
*omkar@aero.iisc.ernet.in

This chapter describes the classification task of data mining using nature inspired techniques such as neural network and swarm intelligence algorithm. Neural network and swarm intelligence techniques have an advantage over conventional statistical techniques like maximum likelihood classifier, nearest neighbor classifier etc, because they are distribution free, i.e., no knowledge is required about the distribution of data. The extraction of knowledge from the dataset in the form of weights and rules using various neural network and swarm intelligence methods is increasing over the last few years. In this chapter, we present the knowledge discovery process using three types of testbeds such as the iris data, the pima diabetes data and the shuttle data. The performance analysis of the classification task of data mining is done based on per class accuracy as well as overall accuracy. Finally, we evaluate the computational complexity of neural network and swarm intelligence techniques.

2.1. Introduction

Data mining is the process of automatic extraction of knowledge from data. Knowledge and mining information from large data-base has been recognized by many researchers as a key research topic in machine learning, statistics and database system with an emphasis on the efficiency, robustness and scalability.

The main goal of data mining is to extract knowledge that is not only accurate, but also comprehensible for the user.[1] Comprehensibility is important whenever extracted knowledge is to be used for supporting a decision made by a user. Here the challenges involved are handling of different

types of data, efficiency and scalability of data mining algorithms, usefulness and certainty of data mining results, expression of various kinds of
data mining results, interactive mining knowledge at multiple abstraction
levels and mining information from different sources of data.

There are several data mining tasks, including clustering, classification,
regression, dependence modeling, etc.[2] Each of these tasks can be regarded
as a kind of problem to be solved by a data mining algorithm. Therefore,
the first step in designing a data mining algorithm is to define which task
the algorithm will address.

The three important steps for classification task of data mining are as
follows — the kinds of knowledge to be mined, the kinds of databases to
work on and the kinds of techniques to be utilized.[3]

The 'kinds of knowledge to be mined' presents a clear picture on different data mining requirements and techniques. Methods for mining different
kinds of knowledge are their ability to learn the underlying data relationships and express them into useful and understandable information (i.e., in
the form of mathematical expression, simple rules, weight etc). This information is not typically retrievable by standard techniques but is covered
through the use of nature inspired techniques, including neural network and
swarm intelligence.

In this chapter, we describe neural network[4] such as multilayer perceptron and radial basis function and swarm intelligence techniques[5] such
as ant miner, artificial bee colony and particle swarm optimization for the
classification task of data mining. In this task, the goal is to assign each
object (record or instance) to a specific "class" i.e., one of many predetermined groups (classes). This is of fundamental importance in a number of
engineering applications.

The data classification mainly depends on two data sets: i) training set:
It consists of an input vector and an answer vector, and is used together
with different learning methods to extract knowledge from the datasets; ii)
testing set: In any classification technique it is common to train the system
using known samples. The set of samples used for validation is commonly
called the test set. The knowledge is extracted (weights, rules, etc.) from
the training dataset using classification techniques, the performance of testing samples are evaluated. A program's fitness is often based solely upon
its performance on the test set. A common performance measure is the
number of samples in each group it gets right. Consider a training set,
consisting of sample patterns which are representative of all classes, along
with class membership information for each pattern. Using the training

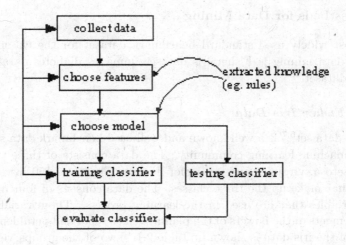

Fig. 2.1.: Classification design cycle.

set, we extract knowledge for membership in each class and create a classifier, which can then be used to assign other patterns to their respective classes.

Figure 2.1 shows the design cycle of a classification system, which usually entails the repetition of a number of different activities as follows: i)Data Collection: This depends on some experimentally collected adequate set of attributes, instances and class type; ii) Feature Choice: Depends on the characteristics of each class group, the entire dataset is segregated for training and testing the system; iii) Model Choice: The class of model includes various classification techniques; here we use multilayer perceptron, radial basis function network, ant miner, artificial bee colony, particle swarm classification, etc.; iv) Training: the dataset are trained using classification techniques to learn the system; by extracting knowledge. This knowledge forms self-regulatory, efficient and stable for a wide range of attributes and instances; v) Evaluation: the samples are validated using the extracted knowledge and the performance is evaluated based on per class accuracy as well as overall accuracy.

In this chapter, we discuss the testbeds for data mining in Sec. 2.2, neural network for data mining in Sec. 2.3, followed by swarm intelligence for data mining in Sec. 2.4 and comparative study in Sec. 2.5. Conclusion is given in Sec. 2.6.

2.2. Testbeds for Data Mining

The most widely used standard benchmark dataset for the classification task of data mining is Fisher iris dataset, pima — diabetes dataset and shuttle dataset.[6,7]

2.2.1. *Fisher Iris Data*

The iris data set[6,7] is a well known and well used benchmark data set used by the machine learning community. The data consists of three varieties of iris, setosa, virginica and versicolor flowers. There are 150 instances of plants that make up the three classes. The data consists of four independent variables that are used in the learning process. These variables are measurements of the flowers of the plants such as sepal and petal lengths. A subset of the iris data is shown in Table 2.1. Two of the groups, versicolor and virginica are known not to be linearly separable.

We have considered $\frac{2}{3}$rd of the data available for training. Using this knowledge, we have classified the complete iris data. After training, we obtain the extracted knowledge in the form of rules, weights, etc., that can be used for classifying iris data.

This data is illustrated in Fig. 2.2, sepal length, sepal width and petal length are taken as x, y and z-axis respectively. It is confirmed that setosa class is linearly separable from other two.

2.2.2. *Pima — Diabetes Data*

The pima — diabetes data set[6,7] has 768 instances of 8 attributes and two classes which are to determine if the detection of diabetes is positive (class A) or negative (class B). The 8 attributes are namely, Number of

Table 2.1.: Subset of iris dataset.

Sepal length	Sepal width	Petal length	Petal width	Iris variety
5.1	3.5	1.4	0.2	Setosa
4.9	3.0	1.4	0.2	Setosa
6.4	2.9	4.3	1.3	Versicolor
6.6	3.0	4.4	1.4	Versicolor
7.2	3.0	5.8	1.6	Virginia
7.4	2.8	6.1	1.9	Virginia

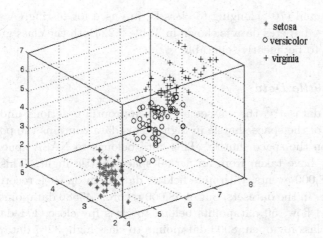

Fig. 2.2.: Iris data with sepal length, sepal width and petal length as the three axes.

times pregnant, Plasma glucose concentration a 2 hours in an oral glucose tolerance test, Diastolic blood pressure (mm Hg), Triceps skin fold thickness (mm), 2-Hour serum insulin (mu U/ml), Body mass index (weight in kg/(height in m)^2), Diabetes pedigree function and Age (years). In our study, 768 data points are used in which 500 (samples) data points belong to class A and 268 (samples) data points belong to class B. Further the complete data is segregated as training data and testing data. In our experiment, 378 data points belonging to class A and 198 data points belonging to class B are used for training. The rest of the data points (122 belonging

Table 2.2.: Subset of pima — diabetes dataset.

Number of times pregnant	Glucose tolerance test	Diastolic blood pressure	Triceps skin fold thickness	serum insulin	Body mass index	Diabetes pedigree function	Age	Detection diabetes
1	85	66	29	0	26.6	0.351	31	Class A
1	89	66	23	94	28.1	0.167	21	Class A
5	116	74	0	0	25.6	0.201	30	Class A
6	148	72	35	0	33.6	0.627	50	Class B
8	183	64	0	0	23.3	0.672	32	Class B
0	137	40	35	168	43.1	2.288	33	Class B

to class A) and (70 belonging to class B) are used for testing. A subset of
the pima — diabetes data is shown in Table 2.2. Both the class groups are
known not to be linearly separable.

2.2.3. *Shuttle Data*

In shuttle dataset[6,7] the classes are the appropriate actions under given
conditions of the space shuttle in flight. This data contains the position of
radiators on the space shuttle. This data belongs to NASA space center,
which they have taken from an actual space shuttle flight. This dataset
contains 58,000 events, with nine field of data item on each record. There
are 7 classes in the datasets, out of 58,000 records, 45586 datapoints belong
to class rad flow, 50 datapoints belong to class fpv close, 171 datapoints
belong to class fpv open, 8903 datapoints to class high, 3267 datapoints to
class bypass, 10 datapoints to class bpv close and 13 datapoints to the class
bpv open. The training dataset contains 43500 dataset and testing dataset
contains 14500. In the data approximately 80% of the data belongs to class
1. Therefore the default accuracy is about 80%. A subset of the shuttle
data is shown in Table 2.3.

2.2.4. *Classification Efficiency*

To classify and evaluate the performance (i.e., individual class efficiency
and overall classification accuracy) for the above dataset we use neural
network and swarm intelligence. Initially, the dataset is used to arrive at the
classification matrix which is of size $n * n$, where n is the number of classes.
A typical entry q_{ij} in the classification matrix shows how many samples
belonging to class i have been classified into class j. For a perfect classifier,

Table 2.3.: Subset of Shuttle dataset.

A_1	A_2	A_3	A_4	A_5	A_6	A_7	A_8	A_9	Class
50	-1	89	-7	50	0	39	40	2	1
42	23	77	0	28	0	34	48	14	2
77	-44	107	0	62	-4	30	45	16	3
55	0	81	0	-6	11	25	88	64	4
79	0	84	0	-36	-196	4	120	116	5
64	1532	106	-2	34	-35	42	72	30	6
37	-1063	106	-1	34	-2	69	72	4	7

the classification matrix is diagonal. However due to misclassification we get off-diagonal elements. The individual efficiency of class i is defined as

$$\frac{q_{ii}}{\sum q_{ji}} \tag{2.1}$$

for all j. The overall efficiency is defined as

$$\frac{\left(\sum q_{ii}\right)}{N} \tag{2.2}$$

where N is the total number of elements in the dataset.

In the next section, we solve these above classification dataset using multi-layer perceptron, radial basis function, ant miner, particle swarm optimization and artificial bee colony method. We will also present a comparison of the result and computational complexity.

2.3. Neural Network for Data Mining

Neural network systems are computational models and systems based on the principles of the human brain; learning and generalization are the main features of neural network.

Traditionally, the term neural network has been used to refer to a network or circuit of biological neurons. The modern usage of the term often refers to artificial neural networks which are composed of artificial neurons or nodes. An Artificial Neural Network (ANN), also called a Simulated Neural Network (SNN) is an interconnected group of artificial neurons that uses a mathematical or computational model for information processing based on a connectionistic approach to computation. In most cases an ANN is an adaptive system that changes its structure based on external or internal information that flows through the network. In more practical terms, neural networks are non-linear statistical data modeling or decision making tools. They can be used to model complex relationships between inputs and outputs or to find patterns in data. The most widely used neural network methods for classification are Multi-layer Perceptron[8] and Radial Basis Function.[9]

2.3.1. *Multi-Layer Perceptron (MLP)*

Multilayer Perceptron (MLP)[4] are a very popular class of neural networks. A multilayer feed-forward network has an input layer, one or more hidden layers, and an output layer. A MLP with one input layer, n number of hidden layers and one output layer is shown in Fig. 2.3.

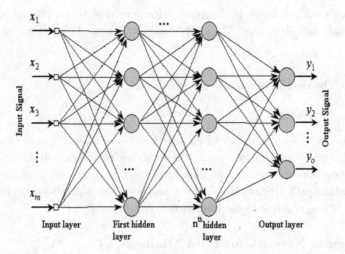

Fig. 2.3.: Multilayer perceptron neural network.

Computations take place in the hidden and output layers only. The input signal propagates through the network in a forward direction, layer-by-layer. They have been successfully applied to many difficult and diverse problems.

A multilayer perceptron has three distinctive features: 1) Neurons in the hidden and output layer are modeled with a smooth nonlinear activation function. It may be noted that the activation functions in the output layer could be linear; 2) One or more hidden layer(s) — this enables learning complicated mappings and tasks, and gives the ability to classify nonlinear separable problems; 3) High degree of connectivity.

These properties give the multilayer perceptron its computational power. Each neuron in the network includes a smooth nonlinear activation function (continuous-differentiable everywhere). These nonlinearities are important; otherwise the network could be reduced to a linear single-layer perceptron. Generally, a sigmoidal nonlinearity defined by the logistic function:

$$y_j = \frac{1}{1 + exp(-v_j)} \tag{2.3}$$

where, v_j is local field (weighted sum of inputs + bias) is used. Use of logistic function is also biologically motivated — accounts for the refractory phase of real neurons.

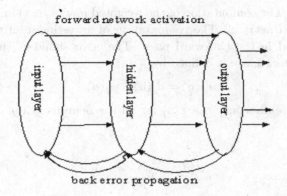

Fig. 2.4.: Propagation of signals in MLP network.

On one hand, hidden layer/s with nonlinear neurons are imperative for MLP, while on the other hand, distributed nonlinearities make the theoretical analysis of a MLP network difficult. Also, back-propagation learning[4,8] is more difficult and in its basic form slow because of the hidden layers.

Multilayer perceptron are typically trained using so-called error back-propagation algorithm. This is a supervised error-correction learning algorithm and is a generalization of the Least Mean Square (LMS) algorithm. Learning consists of two passes through the different layers of a MLP network — forward and backward. In the forward pass, the output (response) of the network to an input vector is computed. During this, all the synaptic weights are kept fixed. During the backward pass, the error signal is propagated backward through the network and the weights are adjusted using an error-correction rule. After the weight adjustment, the output of the network should have moved closer to the desired response in a statistical sense.

2.3.1.1. *Back-Propagation Algorithm*

Two kinds of signals appear in the MLP network as shown in Fig. 2.4: 1) Function signals: Input signals propagating forward through the network, producing in the last phase, output signals; 2) Error signals: Originate at output neurons, and propagate layer by layer backward through the network.

Input layer nodes simply act as post office — they just forward the inputs to hidden layer. Each hidden or output neuron performs two computations: 1) The computation of the weighted inputs (net) and functional value of net (f(net)); 2) The computation of an estimate of the gradient vector, needed in the backward pass. The error signal at the output of neuron j at iteration n is defined by,

$$e_j(n) = d_j(n) - y_j(n). \tag{2.4}$$

The total instantaneous error $\varepsilon(n)$ for all the neurons in the output layer is therefore,

$$J(n) = \frac{1}{2} \sum_{j \in C} e_j^2(n), \tag{2.5}$$

where the set C contains all the neurons in the output layer.

The objective is to derive a learning algorithm for minimizing error with respect to the free parameters.[4,8] In the first hidden layer, $m = L$ is the number of input signals $x_i(n)$, $i=1,...,L$. In the output layer, $m = N$ is the number of outputs. The outputs (components of the output vector) are denoted by $y_j(n)$. These outputs are then compared with the respective desired responses $d_j(n)$, yielding the error signals $e_j(n)$.

In the forward pass, computation starts from the first hidden layer and terminates at the output layer. In the backward pass, computation starts at the output layer, and ends at the first hidden layer. The local gradient is computed for each neuron by passing the error signal through the network layer by layer. The delta rule is used for updating the synaptic weights. The weight updates are computed recursively layer by layer. Both the phases of computation are performed for each training (input) vector. The input vector is fixed through each round-trip (forward pass followed by a backward pass). After this, the next training (input) vector is presented to the network and this process is iteratively performed for the complete training set. From the training dataset the knowledge in the form of optimum weight matrix is extracted. The trained network should also generalize well. A network generalizes well when its input-output mapping is (almost) correct also for test data. The test data is not used in training the network. But it is assumed that the test data comes from the same population (distribution) as the training data.

2.3.1.2. *Fisher Iris Data Classification Using MLP*

Multilayer Perceptron Neural Network trained using Back Propagation algorithm (MLPNN-BP) with a single hidden layer, 4 inputs, 12 neurons in the hidden layer and 1 neuron in the output layer has been employed for iris data classification. The MLPNN-BP parameters are varied until they produce most favorable classification result. The optimum values for the parameters of the most favorable results are as follows:

Number of input nodes (with bias) = 5
Number of hidden nodes = 12
Hidden layer activation = Sigmoidal function
Output layer activation = Tanh function
Number of output nodes = 1
Training stopping criteria = Mean Square Error (0.001)

From the training dataset the knowledge in the form of optimum weight matrix is extracted. For this network configuration the testing datasets are applied and the efficiencies are evaluated. As we can notice from Table 2.4, the classification matrix for the testing samples belonging to setosa is getting classified without any misclassifications and hence has an individual efficiency of 100%. Two samples belonging to versicolor source is misclassified as virginia and hence has an individual efficiency of 96%. A sample belonging to virginia is misclassified as versicolor and hence has an individual efficiency 98%. An overall efficiency of iris data classification using MLPNN-BP is 98%.

Table 2.4.: Classification matrix for iris data.

	Setosa	Versicolor	Virginia	Efficiency
Setosa	50	0	0	100%
Versicolor	0	48	2	96%
Virginia	0	1	49	98%

Overall efficiency = 98%

2.3.1.3. *Pima — Diabetes Data Classification Using MLP*

Multilayer Perceptron Neural Network trained using Back Propagation algorithm (MLPNN-BP) with a single hidden layer, 9 inputs, 12 neurons in the hidden layer and 1 neuron in the output layer has been employed for pima — diabetes data classification. The MLPNN-BP parameters are varied until they produce most favorable classification result. The optimum values for the parameters for the most favorable results are as follows:

Number of input nodes (with bias) = 9
Number of hidden nodes = 12
Hidden layer activation = Sigmoidal function
Output layer activation = Tanh function
Number of output nodes = 1
Training stopping criteria = Mean Square Error (0.0017)

From the training dataset the knowledge in the form of optimum weight matrix is extracted. For this network configuration the testing datasets are applied and the efficiencies are evaluated. As we can notice from Table 2.5, the classification matrix for the testing samples, where 7 samples belonging to class A are getting misclassified to class B and hence have an individual efficiency of 94.3% and 32 samples belonging to class B source is misclassified as class A hence has an individual efficiency of 54.3%. An overall efficiency of pima — diabetes data classification using MLPNN-BP is 79.7%.

Table 2.5.: Classification matrix for pima diabetes data.

	Class A	Class B	Efficiency
Class A	115	7	94.3%
Class B	32	38	54.3%

Overall efficiency = 79.7%

2.3.1.4. *Shuttle Data Classification Using MLP*

Like previous two data classification here also MLPNN-BP has a single hidden layer. There are 9 inputs, 12 neurons in the hidden layer and 1 neuron in the output layer which has been employed to train shuttle dataset. The MLPNN-BP parameters are varied until they produce most favorable classification result. The optimum values for the parameters for the most favorable results are as follows:

Number of input nodes (with bias) = 10
Number of hidden nodes = 12
Hidden layer activation = Sigmoidal function
Output layer activation = Tanh function
Number of output nodes = 1
Training stopping criteria = Mean Square Error (0.0001)

From the training dataset the knowledge in the form of optimum weight matrix is extracted. For this network configuration the testing datasets are applied and the efficiencies are evaluated. As we can notice from Table 2.6, the testing samples belonging to Class C, E & F is getting classified without any misclassifications and hence has an individual efficiency of 100%. A sample belonging to Class B and Class G is misclassified as Class D & A and hence has an individual efficiency 92.3% and 50% respectively. The first row in Table 2.6 indicates that 149 samples belonging to Class A is misclassified as Class D hence has an individual efficiency of 98.7%. 27 samples belonging to Class D has been misclassified as Class B with an individual efficiency 98.7%. An overall efficiency of shuttle data classification using MLPNN-BP is 98.8%.

2.3.2. *Radial Basis Function Network*

The Radial Basis Function Network (RBFN)[9] model consists of three layers; the input, hidden and output layers. The nodes within each layer are fully connected to the previous layer, as shown in Fig. 2.5. The input attributes are each assigned to a node in the input layer and pass directly to the hidden layer without weights. The hidden nodes or units contain the radial basis functions (RBF), also called transfer functions, and are analogous to the sigmoid functions commonly used in the back-propagation network models. They are represented by the bell-shaped curve in the hidden nodes shown in Fig. 2.5.

The RBF is similar to the Gaussian density function which is defined by a 'width' parameter and a 'centre' position. The Gaussian function gives the highest output when the incoming variables are closest to the

Table 2.6.: Classification matrix for shuttle data.

	Class A	Class B	Class C	Class D	Class E	Class F	Class G	Efficiency
Class A	11329	0	0	149	0	0	0	98.7%
Class B	0	12	0	1	0	0	0	92.3%
Class C	0	0	39	0	0	0	0	100%
Class D	0	27	0	2128	0	0	0	98.7%
Class E	0	0	0	0	809	0	0	100%
Class F	0	0	0	0	0	4	0	100%
Class G	1	0	0	0	0	0	1	50%

Overall efficiency = 98.8%

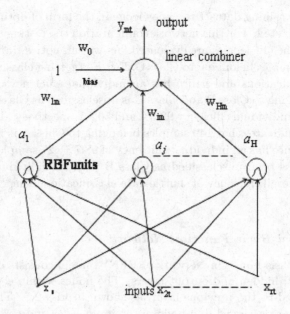

Fig. 2.5.: Radial basis function network model.

centre position and decreases monotonically as the distance from the centre increases. The width of the RBF unit controls the rate of decrease;
for example, a small width gives a rapidly decreasing function and a large
value gives a slowly decreasing function.

Consider an observation used to train the model to have r inputs
variables such as forecasted weather readings, previous load values, etc.,
because all inputs are connected to each hidden node, each hidden
node has an r-dimensional centre, but only one width value is used to
scale all r-dimensions. Let X_t be the incoming vector with components,
$x_{1t}, x_{2t}, ..., x_{rt}$. The output of the ith unit, $a_i(X_t)$, in the hidden layer for
the above input pattern is equal to

$$a_i(X_t) = exp(-\sum_{j=1}^{r} \frac{[x_{jt} - \hat{x}_{ji}]^2}{\sigma_i^2}) \qquad (2.6)$$

where

$\hat{x}_{ji}$ = centre of ith RBF unit for input variable j
σ_i = width of ith RBF unit
x_{jt} = jth variable of input pattern t

The hidden neurons compute essentially a Gaussian function which has maximum value when the input exactly coincides with the weight vector. As the difference between the weight vector and the input increases, the neuron's output approaches zero. The rate of this decrease in output is governed by σ_i.

The connection between the hidden units and the output units are weighted sums as shown in Fig. 2.5. The output value y_{mt} of the mth output node is equal to the summation of the weighted outputs of the hidden units and the biasing term of the output node, and is described by Eq. (2.7).

$$y_{mt} = \sum w_{im} a_i(X_t) + w_0 \qquad (2.7)$$

Where

y_{mt} = output value of mth node in output layer for tth incoming pattern
w_{im} = weight between ith RBF unit and mth output node
w_0 = biasing term at mth output node

The parameters of the RBF units are determined in three steps of the training activity. First, the unit centres are determined by k-means clustering algorithm. Then, the widths are determined by a nearest-neighbor method. Finally weights connecting the RBF units and the output units are calculated using multiple linear regression techniques.[9,10]

2.3.2.1. *Fisher Iris Data Classification Using RBFN*

Radial Basis Function Network (RBFN) with a single hidden layer, 4 inputs, 12 neurons in the hidden layer and 1 neuron in the output layer has been employed for iris data classification. The RBFN parameters are varied until they produce most favorable classification result. The optimum values for the parameters for the most favorable results are as follows:

Number of input nodes = 4
Number of hidden nodes = 15
Spread of Gaussian function=10

From the training dataset the knowledge in the form of optimum network structure (i.e., based on the number of RBF units in the hidden layer and the value for the nearest neighbour) is extracted. For this network configuration the testing datasets are applied and the efficiencies are evaluated. As we can observe from Table 2.7, the classification matrix for the testing samples belonging to setosa is getting classified without any misclassifications and hence has an individual efficiency of 100%. A sample belonging

Table 2.7.: Classification matrix for iris data.

	Setosa	Versicolor	Virginia	Efficiency
Setosa	50	0	0	100%
Versicolor	0	49	1	98%
Virginia	0	3	47	94%

Overall efficiency = 97.3%

to versicolor source is misclassified as virginia and hence has an individual efficiency of 98%. Three samples belonging to virginia is misclassified as versicolor and hence has an individual efficiency 94%. An overall efficiency of iris data classification using RBFN is 97.3%.

2.3.2.2. *Pima — Diabetes Data Classification Using RBFN*

Radial Basis Function Network (RBFN) with a single hidden layer, 8 inputs, 12 neurons in the hidden layer and 1 neuron in the output layer has been employed for pima — diabetes data classification. The RBFN parameters are varied until they produce most favorable classification result. The optimum values for the parameters for the most favorable results are as follows:

Number of input nodes (bias) = 8
Number of hidden nodes = 15
Spread of Gaussian function = 5

From the training dataset the knowledge in the form of optimum network structure (i.e., based on the number of RBF units in the hidden layer and the value for the nearest neighbour) is extracted. For this network configuration the testing datasets are applied and the efficiencies are evaluated. As we can observe from Table 2.8, the classification matrix for the testing samples, where 16 classes belonging to class A are getting misclassified as class B and hence have an individual efficiency of 86.8% and 30 samples belonging to class B source is misclassified as class A hence has an individual efficiency of 57.1%. An overall efficiency of pima — diabetes data classification using RBFN is 76%.

2.3.2.3. *Shuttle Data Classification Using RBFN*

Radial Basis Function Network (RBFN) with a single hidden layer, 9 inputs, 12 neurons in the hidden layer and 1 neuron in the output layer has

Table 2.8.: Classification matrix for pima diabetes data.

	Class A	Class B	Efficiency
Class A	106	16	86.8%
Class B	30	40	57.1%

Overall efficiency = 76%

been employed for shuttle data classification. The RBFN parameters are varied until they produce most favorable classification result. The optimum values for the parameters for the most favorable results are as follows:

Number of input nodes = 9
Number of hidden nodes = 15
Spread of Gaussian function=15

From the training dataset the knowledge in the form of optimum network structure (i.e., based on the number of RBF units in the hidden layer and the value for the nearest neighbour) is extracted. For this network configuration the testing datasets are applied and the efficiencies are evaluated. As we can observe from Table 2.9, the classification matrix for the testing samples belonging to Class E & F is getting classified without any misclassifications and hence has an individual efficiency of 100%. A sample belonging to Class G is misclassified as Class F and hence has an individual efficiency 50%. The first row in Table 2.9 indicates that 156 & 93 samples belonging to Class A is misclassified as Class B & C respectively hence has an individual efficiency of 97.8%. 136 samples belonging to Class D has been misclassified as Class B with an individual efficiency 93.7%. An overall efficiency of shuttle data classification using RBFN is 97.3%.

Table 2.9.: Classification matrix for Shuttle data.

	Class A	Class B	Class C	Class D	Class E	Class F	Class G	Efficiency
Class A	11229	156	93	0	0	0	0	97.8%
Class B	1	11	0	1	0	0	0	84.6%
Class C	0	0	30	9	0	0	0	76.9%
Class D	0	136	0	2019	0	0	0	93.7%
Class E	0	0	0	0	809	0	0	100%
Class F	0	0	0	0	0	4	0	100%
Class G	0	0	0	0	0	1	1	50%

Overall efficiency = 97.3%

2.4. Swarm Intelligence for Data Mining

In social insect colonies, each individual seems to have its own agenda; and yet the group as a whole appears to be highly organized. Apparently, algorithms based on swarm intelligence and social insects begin to show their effectiveness and efficiency to solve difficult problems. A swarm is a group of multi-agent system such as ants, birds, bees etc., in which simple agents coordinate their activities to solve the complex problem of the allocation of labor to multiple forage sites in dynamic environments. This collective behavior that emerges from a group of social insects has been dubbed Swarm Intelligence.[11,12] In case of Swarm Intelligence; Ant Miner,[13,14] Artificial Bee Colony(ABC)[15,16] and Particle Swarm Optimization (PSO)[17] are widely used to solve pattern classification problem.

2.4.1. Ant Miner

A novel algorithm called Ant-Miner, has been proposed[13,14] which can be used for data mining. Ant Miner is a rule based data classification technique which is based on one of the early studies of swarm intelligence; investigating the foraging behavior of ants. An important and interesting behavior of ant colonies is their foraging behavior, and, in particular, how ants can find the shortest paths between food sources and their nest. While walking from food sources to the nest and vice versa, ants deposit on the ground a substance called pheromone, forming in this way a pheromone trail. Ants can smell pheromone, and when choosing their way, they tend to choose, in probability, paths marked by strong pheromone concentrations. The pheromone trail allows the ants to find their way back to the food source (or to the nest). Also, it can be used by other ants to find the location of the food sources found by their nest mates.

It has been shown experimentally that this pheromone trail following behavior can give rise, once employed by a colony of ants, to the emergence of shortest paths. That is, when more paths are available from the nest to a food source, a colony of ants may be able to exploit the pheromone trails left by the individual ants to discover the shortest path from the nest to the food source and back.

These techniques are used to solve a multitude of problems ranging from the classic Traveling Salesman problem to the difficult data classification problem. In this work, the concept of Ant Miner algorithm[13,14] has been adopted to obtain the rules that can be used for classifying the standard benchmark dataset. A brief description of Ant Miner algorithm is given below.

In the context of the classification task of data mining, discovered knowledge is often expressed in the form of IF-THEN rules, as follows:

IF <conditions> THEN <class>

The rule antecedent (IF part) contains a set of conditions, usually connected by a logical conjunction operator (AND). Each rule is evaluated based on "term", so that the rule antecedent is a logical conjunction of terms in the form:

ALGORITHM I: A High-Level Description of Ant-Miner
Training_Set = {all training cases};
/*Rule list is initialized with an empty list*/
Discovered_Rule_List = [ ];
WHILE (Training Set > Max uncovered cases)
   t = 1; /* ant index */
   j = 1; /* convergence test index */
   /*Initialize all trails with the same amount of pheromone*/
REPEAT
   → Ant_t starts with an empty rule and incrementally constructs a classification rule R_t by adding one term at a time to the current rule;
   → Determine quality;
   → Prune rule R_t;
   → Update the pheromone of all trails by increasing pheromone in the trail followed by Ant (proportional to the quality of R_t) and decreasing pheromone in the other trails (simulating pheromone evaporation);
   IF (R_t is equal to R_{t-1}) /* update convergence test */
      THEN j = j + 1;
      ELSE j = 1;
   END IF
   → t = t + 1; /* increment index of the ant */
UNTIL (t ≥ No_of_ants) OR (j ≥ No_rules_converg)
Choose the best rule R_{best} among all rules R_t constructed by all the ants;

> Add rule R_{best} to Discovered_Rule_List;
> Training_Set=Training_Set -{set of cases correctly covered by R_{best}};
> END WHILE

IF term1 AND term2 AND...

Each term is a triple <attribute, operator, value>, such as <Attribute1 = 24>.

The rule consequent (THEN part) specifies the class predicted for cases whose predictor attributes satisfy all the terms specified in the rule antecedent. From a data mining viewpoint, this kind of knowledge representation has the advantage of being intuitively comprehensible for the user, as long as the number of discovered rules and the number of terms in rule antecedents are not large.

Ant-Miner[13,14] follows a sequential covering approach to discover a list of classification rules covering all, or almost all, the training cases. At first, the list of discovered rules is empty and the training set consists of all the training cases. Each iteration of the WHILE loop of Algorithm, corresponding to a number of executions of the REPEAT-UNTIL loop, discovers one classification rule. The rule with the best quality is added to the list of discovered rules, and the training cases that are correctly covered by this rule (i.e., cases satisfying the rule antecedent and having the class predicted by the rule consequent) are removed from the training set. This process is iteratively performed while the number of uncovered training cases is greater than a user-specified threshold, called Max_uncovered_cases.

Each iteration of the REPEAT-UNTIL loop of Algorithm consists of three steps, comprising rule construction, rule pruning, and pheromone updating, detailed as follows.

2.4.1.1. Rule Construction

The rule construction has two parts namely, antecedent and consequent. Each rule is evaluated based on a quality term.

First, Ant t starts with an empty rule, that is, a rule with no term in its antecedent, and adds one term at a time to its current partial rule or the tentative rule. The current partial rule constructed by an ant corresponds to the current partial path followed by that ant. Similarly, the choice of a term to be added to the current partial rule corresponds to the choice of the direction in which the current path will be extended. The choice of the term to be added to the current partial rule depends on both a problem-dependent heuristic function (h)[13,14] and on the amount of pheromone (t) associated with each term, as will be discussed in detail in the next sections. Ant t keeps adding one-term-at-a time to its current partial rule until one of the following two stopping criteria is met: 1) Any term to be added to the rule would make the rule cover a number of cases lesser than a user-specified threshold, called Min_cases_per_rule (minimum number of cases covered per rule); 2) All attributes have already been used by the ant, so that there are no more attributes to be added to the rule antecedent. Note that each attribute can occur only once in each rule, to avoid invalid rules such as "IF (Sex = male) AND (Sex = female)...".

Once the rule antecedent is completed, the system chooses the rule consequent (i.e., the predicted class) that is most appropriate for the rule among all the classes. The class having the maximum number of cases satisfying the rule is assigned as the rule consequent.

The quality of the rule is calculated based on the equation:

$$Quality(Q) = \frac{TP}{TP+FN} * \frac{TN}{FP+TN}, \tag{2.8}$$

where,

- TP (true positives) is the number of cases covered by the rule that have the class predicted by the rule.
- FP (false positives) is the number of cases covered by the rule that have a class different from the class predicted by the rule.
- FN (false negatives) is the number of cases that are not covered by the rule but that have the class predicted by the rule.
- TN (true negatives) is the number of cases that are not covered by the rule and that do not have the class predicted by the rule.

2.4.1.2. *Rule Pruning*

The main goal of rule pruning is to remove irrelevant terms that might have been unduly included in the rule due to stochastic variations in the term selection procedure. Rule pruning potentially increases the predictive power of the rule, by avoiding over-fitting of the training data.[13,14] Another motivation for rule pruning is that it improves the simplicity of the rule, since a shorter rule is easier to be understood by the user. As soon as the current ant completes the construction of its rule, the rule pruning subroutine is called. The basic idea is to iteratively remove superfluous terms one at a time, from the rule without decreasing the quality of the rule.

More precisely, to start with, we consider the rule in its entirety. Then this rule pruning module tentatively tries to remove each of the terms of the rule-each one in turn and the quality of the resulting rule is computed using the rule-quality function It should be noted that this step might involve replacing the class in the rule consequent, since the majority class in the cases covered by the pruned rule can be different from the majority class in the cases covered by the original rule. The term whose removal most improves the quality of the rule is identified and removed. This completes the first iteration. In the next iteration, the above mentioned procedure is repeated on the improved rule. This process is repeated until the rule has just one term or until there is no term whose removal will improve the quality of the rule. The simplified rule thus obtained is called the pruned rule.

2.4.1.3. *Pheromone Updation*

Whenever an ant constructs its rule and that rule is pruned, the amount of pheromone in all segments of all paths must be updated. This pheromone updating is supported by two basic ideas, namely: 1) The amount of pheromone associated with each term occurring in the rule found by the ant (after pruning) is increased in proportion to the quality of that rule; 2) The amount of pheromone associated with each term that does not occur in the rule is decreased, simulating pheromone evaporation in real ant colonies. These two ideas can be implemented so that the following ant starts to construct its rule, using the new amounts of pheromone to guide its search.

This process is repeated until one of the following two conditions is met: 1) The number of constructed rules is equal to or greater than the user-specified threshold Number_of_ants, which will be set initially; 2) The current Ant t has constructed a rule that is exactly the same as the rule constructed by the previous Number_rules_converg — 1 ants. Here Number_rules_converg stands for the number of rules used to test convergence of the ants and can be set by the user. If Number_rules_converg ants construct the same rule, it can be conveniently assumed that the remaining ants will also form the same rules and the iteration can be terminated.

Once the REPEAT-UNTIL loop is completed, the best rule among the rules constructed by all ants is added to the list of discovered rules, as mentioned earlier, and the system starts a new iteration of the WHILE loop, by reinitializing all trails with the same amount of pheromone.

2.4.1.4. *Ant-Miner's Parameter Setting*

Ant-Miner has the following four user-defined parameters: 1) Number of ants (Number_of_ants): This is also the maximum number of complete candidate rules constructed and pruned during an iteration of the WHILE loop of the algorithm, since each ant is associated with a single rule. In each iteration, the best candidate rule found is considered a discovered rule. The larger Number_of_ants, the more candidate rules are evaluated per iteration, but the slower the system is; 2) Minimum number of cases per rule (Min_cases_per_rule): Each rule must cover at least Min_cases_per_rule cases to enforce at least a certain degree of generality in the discovered rules. This helps to avoid an over- fitting of the training data; 3) Maximum number of uncovered cases in the training set (Max_uncovered_cases): The process of rule discovery is iteratively performed until the number of training cases that are not covered by any discovered rule is smaller than this threshold; 4) Number of rules used to test convergence of the ants (Number_rules_converg): If the current ant has constructed a rule that is exactly the same as the rule constructed by the previous (Number_rules_converg-1) ants, then the system concludes that the ants have converged to a single rule (path). The current iteration of the WHILE loop of Algorithm I is therefore stopped and another iteration is started.

Ant-Miner is parameter sensitive and it can be observed that these parameters have an influence on the quality of the result and the computational time. The selection of parameter itself is quite a research problem and the importance of these parameters and a method to obtain the optimal values for them has been discussed in Ref. 18. In this chapter, we have made no serious attempt to optimize the setting of these parameters. Such an optimization could be tried in future research. We have selected the parameters based on intuition, earlier experience and trail and error to get good results. It is interesting to note that even the above non optimized parameters' setting has produced quite good results, as will be shown later.

2.4.1.5. *Fisher Iris Data Classification Using Ant Miner*

The parameters -Number_of_ants,Min_cases_per_rule,Max_uncovered_cases, Number_rules_to_converge — are used to extract rules and the overall classification efficiencies hence obtained are recorded. Among these, the

Table 2.10.: The rules obtained from Ant-Miner for iris data.

Rule Antecedent	Rule consequent	Quality of the rule
Petal Ln < 24 && Petal Wd < 18	1	1
Petal Ln > 20 && 8 < Petal Ln ≤ 17	2	0.9118
Sepal Width > 24 && Petal Wd ≥ 17	3	0.8862

parameters corresponding to the best classification accuracy are chosen. The optimum values chosen for these parameters are as follows:

Number of ants = 50
Min no. of cases per rule = 5
Max uncovered cases = 4
No of rules for convergence = 8

The Ant-Miner is trained with the above-mentioned training dataset to generate rules. After the training process, Ant-Miner generates simple and easily understandable rules as shown in Table 2.10. The classification matrices obtained after applying the derived rules for the testing data are shown in Table 2.11. Each rule extracted has a quality[13,14] associated, which is a measure of how best the rule can classify the input samples correctly. For setosa simple rules are obtained with the best quality. This implies that setosa is linearly separable. The quality of the rule, for versicolor and virginia, is less in comparison with that of setosa. This is indicative of the fact that versicolor and virginia are not linearly separable. The rules extracted are applied to the testing datasets and the efficiencies are evaluated. As we can notice from Table 2.11, the classification matrix generated for the testing data, there is no misclassification in setosa, whereas there are few misclassifications between versicolor and virginia with individual efficiency 98% and 90% respectively. An overall efficiency of iris data classification using Ant Miner is 96%.

2.4.1.6. *Pima — Diabetes Data Classification Using Ant Miner*

The parameters -Number_of_ants,Min_cases_per_rule,Max_uncovered_cases, Number_rules_to_converge — are used to extract rules and the overall classification efficiencies hence obtained are recorded. Among these, the

Table 2.11.: Classification matrix for iris data.

	Setosa	Versicolor	Virginia	Efficiency
Setosa	50	0	0	100%
Versicolor	0	49	1	98%
Virginia	0	5	45	90%

Overall efficiency = 96%

Table 2.12.: The rules obtained from Ant-Miner for pima diabetes data.

Rule Antecedent	Rule consequent	Quality of the rule
106 ≥ Tolerance < 137 && BP < 109 && 24 > Skin FT < 31 && Diabetes PF ≥ 0.078 && Age < 72	1	55.5%
N Preg < 5 && 28 ≥ Tolerance < 121 && insulin ≤ 73 && Body MI ≤ 59 && Age < 32	2	53.2%

parameters corresponding to the best classification accuracy are chosen. The optimum values chosen for these parameters are as follows:

Number of ants = 150
Min no. of cases per rule = 10
Max uncovered cases = 10
No of rules for convergence = 5

The Ant-Miner is trained with the above-mentioned training dataset to generate rules. After the training process, Ant-Miner generates simple and easily understandable rules as shown in Table 2.12. The classification matrices obtained after applying the derived rules for the testing data are shown in Table 2.13. Each rule extracted has a quality[13,14] associated, which is a measure of how best the rule can classify the input samples correctly. The quality of the rule, for Class A and Class B, is less. This is indicative of the fact that both the class are not linearly separable. The rules extracted are applied to the testing datasets and the efficiencies are evaluated. As we can notice from Table 2.13, the classification matrix generated for the testing data, there is misclassification between both the classes with individual efficiency 77.05% and 71.43% respectively. An overall efficiency of pima-diabetes data classification using Ant Miner is 75%.

Table 2.13.: Classification matrix for pima diabetes data.

	Class A	Class B	Efficiency
Class A	94	28	77.1%
Class B	20	50	71.4%

Overall efficiency = 75%

2.4.1.7. *Shuttle Data Classification Using Ant Miner*

The parameters -Number_of_ants,Min_cases_per_rule,Max_uncovered_cases, Number_rules_to_converge — are used to extract rules and the overall classification efficiencies hence obtained are recorded. Among these, the parameters corresponding to the best classification accuracy are chosen. The optimum values chosen for these parameters are as follows:

Number of ants = 500
Min no. of cases per rule = 40
Max uncovered cases = 40
No of rules for convergence = 25

The Ant-Miner is trained with the above-mentioned training dataset, the knowledge extracted is in the form of rules. After the training process, Ant-Miner generates simple and easily understandable rules as shown in Table 2.14. The classification matrices obtained after applying the derived rules for the testing data are shown in Table 2.15. Each rule extracted has a

Table 2.14.: The rules obtained from Ant-Miner for shuttle data.

Rule antecedent	Rule consequent	Quality of the rule
$A_1 \leq 53$ && $A_2 > -4$ && $A_4 > -3$ && $A_7 \geq 0$ && $A_8 \leq 128$ && $A_9 < 6$	1	89%
$A_1 \leq 61$ && $23 > A_3 < 79$ && $A_4 \geq 0$ && $A_7 > 0$	2	81%
$-712 > A_2 < -20$ && $A_5 > -24$ && $A_8 < 101$	3	89.1%
$53 \geq A_1 < 58$ && $A_2 \leq 0$ && $A_7 < 55$	4	90.7%
$A_1 \geq 45$ && $A_2 > -11$ && $A_4 > -9$ && $A_7 < 11$	5	91.2%
$A_7 < 65$ && $A_9 \geq 29$	6	97.6%
$A_1 < 47$ && $A_5 \leq 34$ && $A_9 < 23$	7	99.2%

Table 2.15.: Classification matrix for shuttle data.

	Class A	Class B	Class C	Class D	Class E	Class F	Class G	Efficiency
Class A	11389	42	47	0	0	0	0	99.2%
Class B	0	12	0	1	0	0	0	92.3%
Class C	0	0	39	0	0	0	0	100%
Class D	38	0	0	2117	0	0	0	98.2%
Class E	2	0	0	0	807	0	0	99.8%
Class F	0	0	0	0	0	4	0	75%
Class G	1	0	0	0	0	0	1	100%

Overall efficiency = 99.1%

quality[13,14] associated, which is a measure of how best the rule can classify the input samples correctly. For all the class the simple rules are obtained with the best quality. This implies that the data is less complex. The rules extracted are applied to the testing datasets and the efficiencies are evaluated. As we can observe from Table 2.15, the classification matrix for the testing samples belonging to Class C & G is getting classified without any misclassifications and hence has an individual efficiency of 100%. A sample belonging to Class B is misclassified as Class D and hence has an individual efficiency 92.3%. Similarly a sample belonging to Class F is misclassified as Class A and hence has an individual efficiency 75%. Two samples belonging to Class E has been misclassified as Class A with an individual efficiency 99.8%. The first row in Table 2.15 indicates that 42 & 47 samples belonging to Class A is misclassified as Class C & D respectively hence has an individual efficiency of 99.2%. 38 samples belonging to Class D has been misclassified as Class A with an individual efficiency 98.2%. An overall efficiency of shuttle data classification using Ant-Miner is 99.1%.

2.4.2. *Artificial Bee Colony*

An important and interesting behavior of bee colonies is their foraging behavior, and in particular, how bees find a food source based on the amount of nectar and successfully bring back to the hive. In a real bee colony the bees are grouped as scout bees, employed bees and onlookers. Initially, the foraging process begins in a colony by scout bees (unemployed bees) which explore food sources by moving randomly. At the entrance of the hive is an area called the dance-floor, where dancing takes place. Upon their return to the hive from a foraging trip, it communicates by performing the so-called waggle dance[19] so as to recruit other bees to go to the food source. A bee waiting on the dance area for making decision to choose a food source is called an onlooker, which seems to learn information from the dance regarding the food source: its nectar amount, the direction in which it will be found and its distance.[19,20] If the scouts discover rich food source then the scout bees are selected and classified as the forager bee (employed bee). After waggle dancing the forager bee leaves the hive to get nectar with their fellow bees that were waiting inside the hive. The number of follower bees assigned to nectar depends on the overall quality of the nectar. Upon arrival, the bees take a load of nectar and return to the hive relinquishing the nectar to a food-storer (onlooker) bee. In this way a good food source is exploited, and the number of foragers at this site is reinforced.

In a robust search process, exploration and exploitations process must be carried out together. In the Artificial Bee Colony (ABC) algorithm,[15,16] the scout bees control the exploration process, while the employed bees and onlookers' carryout the exploitation process in the search space. The number of employed bees and the onlookers is equal to the number of solution in the population. The employed bee whose food source has been exhausted becomes a scout bee. The position of an enhanced nectar amount of a food represents a possible solution to the optimization problem.

At the first step, create a population of n artificial (scout) bees placed randomly in the search space representing the food source position, where n denotes the size of population. After initialization, the population of the positions (solutions) is subjected to repeated iteration of the search processes of the employed bees, the onlooker bees and scout bees. This search process can be divided into two phases:

i) Exploration phase: For each solution x_{ij}, where $i = 1,2...n$ and j is dimensional vector. The scout bees explore a new food source with x_i. This operation can be defined as in Eq. (2.9).

$$x_i^j = x_j^{min} + (x_j^{max} - x_j^{min}) * rand(0,1). \qquad (2.9)$$

Here the value of each component in every x_i vector should be clamped to the range $[x_{min}, x_{max}]$ to reduce the likelihood of scout bees leaving the search space (S). The population spread is restricted within the search space S i.e., $x_{ij} \in S$ and in Eq. (2.9), x_{min} and x_{max} is the lower and upper limit respectively of the search scope on each dimension.

ii) Exploitation phase: In this phase, assuming the scout bees which has explored food source are selected as an employed bees which produces a modification on the position (solution) in her memory depending on the local information (visual information) and tests the nectar amount (fitness value) of the new source (new solution). If the nectar amount of the new one is higher than that of the previous one, the bee memorizes the new position and forgets the old one. Otherwise it memorizes the position of the previous one. After all employed bees complete the search process; they communicate the nectar information of the food sources and their position information with the onlooker bees on the dance area. An onlooker bee evaluates the nectar information taken from all employed bees and chooses a food source with better nectar amount. As in the case of the employed bee, onlooker bee also produces a modification on the position in her memory and checks the nectar amount of the candidate source. Providing that its nectar is higher than that of the previous one, the bee memorizes the new position and forgets the old one.

An artificial onlooker bee chooses a food source depending on the new positions, using Eq. (2.10).

$$P_i = \begin{cases} v_i, if(f(x_i) \geq f(v_i)) \\ x_i, if(f(x_i) \leq f(v_i)) \end{cases} \tag{2.10}$$

In order to select the better nectar position found by an onlooker, O_b is defined as

$$O_b = argmin_{P_i} f(p_i) \tag{2.11}$$

where P_i is the best fitness value of the solution i which is proportional to the nectar amount of the food source in the position i and n is the number of food sources which is equal to the number of employed bees.

In order to produce a candidate food position from the old one in memory, the ABC uses the following Eq. (2.12):

$$v_{ij} = x_{ij} + \alpha(x_{ij} - x_{kj}), \tag{2.12}$$

where $k=1,2,...,n$ and $j=1,2,...,D$ are randomly chosen indexes. Although k is determined randomly, it has to be different from i. α is a adaptively generated random number. It controls the production of neighbour food sources around x_{ij} and represents the comparison of two food positions visually by a bee. As can be seen from Eq. (2.12), as the difference between the parameters of the x_{ij} and x_{kj} decreases, the perturbation on the position x_{ij} gets decrease, too. Thus, as the search approaches to the optimum solution in the search space, the step length is adaptively reduced. The food source of which the nectar is abandoned by the bees is replaced with a new food source by the scout bees.

2.4.2.1. *ABC for Multi-Layer Perceptron Neural Network*

MLP are a class of feed forward neural networks trained with the standard back-propagation algorithm or other techniques like ABC. In traditional training algorithm, the MLPNN are usually trained using back propagation (MLPNN-BP) algorithm which is a gradient based optimization algorithm. The back propagation algorithm has difficulties in handling local optima and cannot yield optimal adjustable weights for MLP networks.[4] For this study, we consider only a single hidden layer perceptron network based classifier, with eight neurons in the hidden layer.

Training basically involves presenting the training samples as input vectors through a neural network, calculating the error of the output layer, and then adjusting the weights of the network to minimize the error. Each "training epoch" involves one exposure of the network to a training sample from the training set, and adjustment of each of the weights of the network each layer by layer. The "outside" samples make up the "validation" set.

In this model, single hidden layer perceptron network based architecture is used for the data classification. This is coupled with a ABC based learning algorithm to train the network. ABC has been extensively used for training artificial neural network and proved to be more efficient than many other gradient based training algorithms.[21,22] This can be mainly attributed to stochastic nature of the algorithm which makes it very robust and flexible.

In ABC every bee explores a possible solution — in the current case the optimum weight matrix for the given network configuration. The training error *err(x)* is used as the fitness value; this indicates the extent of conformance of the network output with actual output. Minimising the error (fitness value) will lead to the best set of weights for the given network configuration and the network is said to be trained. For any classifier, its performance is dependent on the chosen loss function. Selection of proper loss function *err(x)* for a given problem is often difficult.[23] Different classification techniques in machine learning employ different loss functions to get better classification accuracy. Commonly neural network classifiers employ mean square error minimization[24] or cross entropy.[25] In our study, we use the loss function such as root mean sum of squared residuals (error) in the training data as the fitness values of the ABC. This serves as a qualitative performance measure of the network learning and is given in Eq. (2.13).

$$rmse = \sqrt{\frac{1}{N} \sum_{N} (\bar{y}(k) - \hat{y}(k))} \qquad (2.13)$$

where, $\bar{y}(k)$ is the time value of the output, $\hat{y}(k)$ is the estimated output of the neural network and N is the number of data points used in the training set.

Our objective is to minimize this fitness value. At each time step the randomness amplitude and speed of convergence of each bees is changed towards its food source. The random factor prevents the swarm getting stuck in the wrong place and speed of convergence is used to identify the rate at which bees converge to a solution.

2.4.2.2. *Fisher Iris Data Classification Using ABC*

The ABC trained MLPNN classifier has been employed for iris data classification. The ABC parameters which includes number of bees (n), randomness amplitude of bees (α), enhanced nectar amount (β), speed of convergence (γ) and maximum iteration (it) are varied until they produce most

```
Pseudo-code: A High-Level Description of MLPNN-ABC
/* Initialize number of bees (n) with randomness amplitude (α) and
speed of convergence (γ) for enhanced nectar amount (β)*/
Initialize n, α, β and γ
For each bees
    b = 1;
    /*Evaluate the objective for each pattern*/
    For each patterns
            /*Calculate and store the root mean sum of squared error
            (RMSE)*/
            err(b) = rmse(error)
    End for
    /*If the fitness value err is better for the current solution
    say e then assign the current solution as the randomness
    amplitude value*/
    If (err(b) < e(x))
            e(x) = err(b);
            update α;
    End if
    /*The whole process is repeated by updating speed of
    convergence*/
```

```
    b=b+1;
    Minerror = min(e);
    Update β, γ;
    /*IF the RMSE goes below the specified threshold or maximum
    iteration is reached*/
End
```

favorable classification result. The optimum values for the above parameters for the most favorable results are as follows:

Number of bees, $n = 100$
Randomness amplitude of bees, $\alpha = 0.5$
Enhanced nectar amount, $\beta = 0.8$
Speed of convergence is adaptively generated for each iteration, $\gamma=[0.5,..,1]$

The optimum weight matrix for the given network configuration are applied to the testing data sets and the efficiencies are evaluated. As we can notice from Table 2.16, the classification matrix for the testing, samples belonging to setosa are getting classified without any misclassifications and hence have an individual efficiency of 100%. A sample belonging to versicolor is misclassified as virginia and hence has an individual efficiency of 98%. Two samples belonging to virginia have misclassified as versicolor and hence have an individual efficiency 96%. An overall efficiency of iris data classification using ABC is 98%.

2.4.2.3. *Pima — Diabetes Data Classification Using ABC*

The ABC trained MLPNN classifier has been employed for pima-diabetes data classification. The ABC parameters which includes number of bees (n), randomness amplitude of bees (α), enhanced nectar amount (β), speed

Table 2.16.: Classification matrix for iris data.

	Setosa	Versicolor	Virginia	Efficiency
Setosa	50	0	0	100%
Versicolor	0	49	1	98%
Virginia	0	2	48	96%

Overall efficiency = 98%

Table 2.17.: Classification matrix for pima diabetes data.

	Class A	Class B	Efficiency
Class A	106	16	86.9%
Class B	18	52	74.3%

Overall efficiency = 82.3%

of convergence (γ) and maximum iteration (it) are varied until they produce most favorable classification result. The optimum values for the above parameters for the most favorable results are as follows:

Number of bees, $n = 200$
Randomness amplitude of bees, $\alpha = 1.65$
Enhanced nectar amount, $\beta = 1.65$
Speed of convergence is adaptively generated for each iteration, $\gamma=[0.5,..,1]$

The optimum weight matrix for the given network configuration are applied to the testing data sets and the efficiencies are evaluated. As we can notice from Table 2.17, the classification matrix for the testing, samples belonging to both the classes are getting misclassified and hence have an individual efficiency of 86.9% and 74.3% respectively. An overall efficiency of pima-diabetes data classification using ABC is 82.3%.

2.4.2.4. *Shuttle Data Classification Using ABC*

The ABC trained MLPNN classifier has been employed for shuttle data classification. The ABC parameters which includes number of bees (n), randomness amplitude of bees (α), enhanced nectar amount (β), speed of convergence (γ) and maximum iteration (it) are varied until they produce most favorable classification result. The optimum values for the above parameters for the most favorable results are as follows:

Number of bees, $n = 1000$
Randomness amplitude of bees, $\alpha = 1.85$
Enhanced nectar amount, $\beta = 1.85$
Speed of convergence is adaptively generated for each iteration, $\gamma=[0.9,..,1]$

The optimum weight matrix for the given network configuration are applied to the testing data sets and the efficiencies are evaluated. As we can notice from Table 2.18, the classification matrix for the testing samples belonging to Class C, E, F & G is getting classified without any misclassifi-

Table 2.18.: Classification matrix for shuttle data.

	Class A	Class B	Class C	Class D	Class E	Class F	Class G	Efficiency
Class A	11402	20	56	0	0	0	0	99.3%
Class B	1	12	0	0	0	0	0	92.3%
Class C	0	0	39	0	0	0	0	100%
Class D	0	0	3	2152	0	0	0	99.8%
Class E	0	0	0	0	809	0	0	100%
Class F	0	0	0	0	0	4	0	100%
Class G	0	0	0	0	0	0	2	100%

Overall efficiency = 99.5%

cations and hence has an individual efficiency of 100%. A sample belonging to Class B is misclassified as Class A and hence has an individual efficiency 92.3%. Three samples belonging to Class D has been misclassified as Class C with an individual efficiency 99.8%. The first row in Table 2.18 indicates that 20 & 56 samples belonging to Class A is misclassified as Class B & C respectively hence has an individual efficiency of 99.3%. An overall efficiency of shuttle data classification using ABC is 99.5%.

2.4.3. *Particle Swarm Optimization*

Particle Swarm Optimization (PSO)[17] is an evolutionary optimization algorithm proposed by Kennedy and Eberhart in the mid 1990s while attempting to simulate the choreographed, graceful motion of swarms of birds as part of a socio-cognitive study investigating the notion of 'collective intelligence' in biological populations.

The back-propagation algorithm uses gradient information and hence is plagued by the problem of local minima.[4] To avoid this problem, a host of other algorithms have been explored for MLPNN training. Evolutionary algorithms inspired from nature are one such alternative.[26,27] These are a new range of computational algorithms that have emerged from the behaviour of social insects. Social insects are usually characterized by their self-organization and the absence of central control. Still, complex group behaviour emerges from the interactions of individuals who exhibit simple behaviours by themselves. In social insects, every individual is self-autonomous. They can only obtain local information, and interact with their geographical neighbours. All these features characterize swarm intelligence. Examples of systems like this can be found in nature, including ant colonies, bird flocking, animal herding, fish schooling, and etc. Particle swarm optimization (PSO) is one of the evolutionary algorithms which have been used for training artificial neural network.[28]

In this model, single hidden layer perceptron network based architecture is used for the standard benchmark data classification. This is coupled with a PSO based learning algorithm to train the network.

In PSO every swarm particle explores a possible solution. It adjusts its flight according to its own and its companion's flying experience to zero for the optimum solution. The best position in the course of flight of each swarm is the best solution that is found by the particle. The best position of the whole flock is the global best solution. The former is called pBest, and the latter gBest. Every particle continuously updates itself through the above mentioned best solutions. Thus a new generation of community comes into being, which has moved closer towards a better solution, ultimately converging to the optimal solution. In the practical operation, the fitness function, which is determined by the optimization problem, assesses the extent to which the particle is good or bad. If the scale of swarm is N, then the position of the ith, $(i=1,2,3,..,N)$ particle is expressed as X_i. The best position discovered by the particle is expressed as $pBest_i$. The index of the position of the particle of the swarm, with best solution is expressed as $gBest_i$. Therefore, a swarm particle — i will update its own speed and position according to the following equations:

$$V_{(i+1)} = w * V_i + C_p * r_1 * (pbest_i - X_i) + C_g * r_2 * (gbest_i - X_i) \quad (2.14)$$

$$X_{(i+1)} = X_i + V_{(i+1)} \quad (2.15)$$

where C_p is the cognitive learning rate and C_g is the social learning rate. The factors r_1 and r_2 are randomly generated within the range (0,1) and w is the inertia factor. As indicated earlier each swarm particle represents a possible solution, and in the current case when applied to feed forward neural network (FNN) training, each particle represents a possible FNN configuration, i.e., its weights. Therefore, each vector has a dimension equal to the number of weights in the FNN. The mean square error — training error is used as the fitness value, and the weights are modified so that the training error is minimized. Hence, finally arriving at a set of weights, that result in the least error for a given network configuration.

2.4.3.1. *Fisher Iris Data Classification Using PSO*

PSO trained MLPNN classifier is extremely sensitive to control parameter choices. The parameter values used for training and testing data set

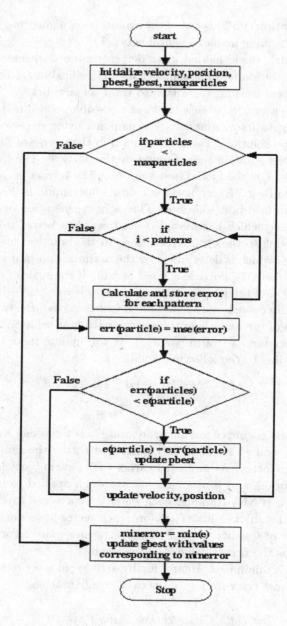

Fig. 2.6.: Flowchart of PSO trained MLPNN implementation.

Table 2.19.: Classification matrix for iris data.

	Setosa	Versicolor	Virginia	Efficiency
Setosa	50	0	0	100%
Versicolor	0	47	3	94%
Virginia	0	2	48	96%

Overall efficiency = 96.6%

are as follows:

Individualistic factor = 1.85
Socialistic factor = 1.85
Inertial factor = [1,...,0.8] (adaptively generated for each iteration)
No of Swarm particles = 24

The optimum weight matrix for the given network configuration are applied to the testing data sets and the efficiencies are evaluated. As we can notice from Table 2.19, the classification matrix for the testing, samples belonging to class A are getting classified without any misclassifications and hence have an individual efficiency of 100%. Three samples belonging to class B is misclassified as class C and hence has an individual efficiency of 94%. Two samples belonging to class C have misclassified as class B and hence have an individual efficiency 96%. An overall efficiency of iris data classification using PSO is 96.6%.

2.4.3.2. *Pima — Diabetes Data Classification Using PSO*

PSO trained MLPNN classifier is extremely sensitive to control parameter choices. The parameter values used for training and testing data set are as follows:

Individualistic factor = 1.85
Socialistic factor = 1.85
Inertial factor = [1,...,0.8] (adaptively generated for each iteration)
No of Swarm particles = 24

The optimum weight matrix for the given network configuration are applied to the testing data sets and the efficiencies are evaluated. As we can notice from Table 2.20, the classification matrix for the testing samples belonging to both the classes are getting misclassified and hence have an individual efficiency of 89.34% and 71.42% respectively. An overall efficiency of pima-diabetes data classification using PSO is 82.8%.

Table 2.20.: Classification matrix for pima diabetes data.

	Class A	Class B	Efficiency
Class A	109	13	89.3%
Class B	20	50	71.4%

Overall efficiency = 82.8%

2.4.3.3. *Shuttle Data Classification Using PSO*

PSO trained MLPNN classifier is extremely sensitive to control parameter choices. The parameter values used for training and testing data set are as follows:

Individualistic factor = 2.55
Socialistic factor = 2.55
Inertial factor = [1,...,0.6] (adaptively generated for each iteration)
No of Swarm particles = 150

The optimum weight matrix for the given network configuration are applied to the testing data sets and the efficiencies are evaluated. As we can notice from Table 2.21, the classification matrix for the testing samples belonging to Class E, F & G is getting classified without any misclassifications and hence has an individual efficiency of 100%. A sample belonging to Class C is misclassified as Class A and hence has an individual efficiency 92.3%. Two samples belonging to Class B has been misclassified as Class A & D with an individual efficiency 84.6%. Also six samples belonging to Class D has been misclassified as Class E with an individual efficiency 99.7%. The first row in Table 2.21 indicates that 39 & 47 samples belonging to Class A is misclassified as Class C & D respectively hence has an individual efficiency of 99.3%. An overall efficiency of shuttle data classification using PSO is 99.3%.

2.5. Comparative Study

Table 2.22 shows the comparison of overall efficiency using nature inspired techniques for most widely used standard benchmark dataset for the classification task of data mining.[6,7]

The overall accuracy for testing samples of iris, pima diabetes and shuttle data using all the nature inspired method has performed better and also matched with the Watkins *et al.*[7] The pima-diabetes dataset is complex in

Table 2.21.: Classification matrix for shuttle data.

	Class A	Class B	Class C	Class D	Class E	Class F	Class G	Efficiency
Class A	11392	0	39	47	0	0	0	99.3%
Class B	1	11	0	1	0	0	0	84.6%
Class C	1	0	38	0	0	0	0	97.4%
Class D	0	0	0	2149	6	0	0	99.7%
Class E	0	0	0	0	809	0	0	100%
Class F	0	0	0	0	0	4	0	100%
Class G	0	0	0	0	0	0	2	100%

Overall efficiency = 99.3%

Table 2.22.: Classification accuracy using nature inspired techniques.

Nature inspired techniques	Algorithm	Iris	Pima diabetes	Shuttle
Neural network	MLP	98%	79.7%	98.8%
	RBF	97.3%	76%	97.3%
Swarm Intelligence	Ant-Miner	96%	75%	99.1%
	ABC	98%	82.3%	99.5%
	PSO	96.7%	82.8%	99.3%

comparison with that of iris and shuttle data.[7] Here the overall accuracy performed better incase of ABC and PSO.

Let us analyze the time complexity for worst case needed to extract the knowledge from the data classification using neural network and swarm intelligence methods.

(1) Multi-layer perceptron (MLP): The MLP algorithm is performed at a $O(N^2)$ complexity, where N is the dimensionality of the data matrix used. Compared to other classification methods, MLP have a much lower computational complexity.[29]

(2) Radial Basis Function (RBF): In this, the computational complexity is $O(P(Nn+nm))$ where P is the number of training patterns, N the number of the input layer nodes, and m the number of the output layer nodes. $O(Nn)$ is the time to calculate for the middle layer (hidden layer) and $O(nm)$ the time to calculate the output layer. Further, it takes $O(n^3)$ to calculate the inverse of the covariance

matrix and $O(mn^3)$ to multiply matrices to obtain the weights. In total the RBF method needs the time of $O(P(Nn + nm) + mn^3)$ to calculate the optimum weights.[10]

(3) Ant Miner: the computational complexity for worst case to extract knowledge in the form of rules using Ant miner[13] is $O(r * z * a^3 * n)$ where, r is the number of rules, z is the number of ants, a is the number of attributes in the data set and n is the number of cases.

(4) Artificial Bee Colony (ABC): The ABC employed to train a multi-layer perceptron neural network (MLPNN-ABC) is computationally intensive. Here we need $O(P(Nn+nm))$ where P is the number of training patterns, N is the number of the input layer nodes, and m is the number of the output layer nodes. $O(Nn)$ is the time to calculate for the middle layer (hidden layer) and $O(nm)$ the time to calculate the output layer.[28] In the proposed approach for classification, ABC is employed to train a multilayer perceptron neural network. The bee movement is based on randomness amplitude (distance) and speed of convergence (direction) towards its food source (nectar). The random factor prevents the swarm getting stuck in the wrong place and speed of convergence is used to identify the rate at which bees converge to a solution i.e., for enhanced nectar amount. The computation time required to retrieve better nectar of all the bees is $O(m^2)$. It is important to notice that m may vary along the learning iterations, such that at each generation the algorithm has a different computational cost. The computational complexity of ABC is $O(nm^2i)$, where n is number of instance, m is scale of dataset and i is the number of iterations. In total the MLPNN-ABC method needs the time of $O(P(Nn + nm) + nm^2i)$ to calculate the global optimum weights.

(5) Particle Swarm Optimization (PSO): all members of the swarm update towards global best, has a complexity of $O(kn^2)$, where n is the number of swarm and k is the number of objectives. In this way, the complexity of the updating process for the complete run of $O(kmn^2)$, where M is the total number of iterations.[30] In the proposed approach for classification, PSO is employed to train a multilayer perceptron neural network. The algorithm performed at an $O(n^2)$ complexity, where N is the dimensionality of the data matrix used. In total the MLPNN-PSO method needs the time of $O(n^2 + kmn^2)$ to calculate the global optimum weights.

2.6. Conclusions and Outlook

This chapter is mainly focused on the classification task of data mining using nature inspired techniques such as neural network and swarm intelligence. The knowledge extraction using dataset is the main goal of data mining. Here we extract knowledge in the form of rules and weights that is not only accurate, but also comprehensible for the user.

For this, well-established algorithms such as MultiLayer Perceptron, Radial Basis Function Network, Ant Miner, Artificial Bee Colony and Particle Swarm Optimization are used on three types of testbeds — the iris data, the diabetes data and the shuttle data. For each of these methods, the performance is evaluated based on individual class efficiency, overall classification accuracy and computational complexity.

Acknowledgments

We would like to thank for the support we received from Aeronautical Research and Development Board, Defence Research and Development Organisation, New Delhi, India.

We also like to thank T. R. Raghavendra, Rahul Khandelwal, Dheevatsa Mudigere and M. Manoj Kumar for valuable comments to prepare this chapter.

References

1. A. A. Freitas and S. H. Lavington, Mining Very Large Databases with Parallel Processing, Norwell, MA: Kluwer, (1998).
2. U. M. Fayyad, G. Piatetsky-Shapiro, and P. Smyth, From data mining to knowledge discovery: An overview, in Advances in Knowledge Discovery & Data Mining, (Eds. Cambridge, MA: MIT Press, 1996), p. 1-34.
3. M. S. Chen, J. Han, and P. S. Yu, Data mining: An overview from a database perspective, IEEE Transactions on Knowledge and Data Engineering, vol. 8(6), (1996), p. 866-883.
4. S. Haykin, Neural Networks - A Comprehensive Foundation, (Second Edition. Macmillan College, New York, 1994).
5. Eric Bonabeau, Macro Dorigo, and Guy Theraulaz, Swarm intelligence-from natural to artificial system, (Oxford University Press, New York, 1999).
6. C. L. Blake and C. J. Merz, UCI repository of machine learning databases, http://www.ics.uci.edu/~mlearn/MLRepository.html, (1998).
7. A. Watkins and L. Boggess, A New Classifier Based on Resource Limited Artificial Immune Systems, Proceedings of the Congress on Evolutionary Computation, (2002), p. 1546-1551.

8. Marchant, J. A. and Onyango, C. M, Comparison of a Bayesian classifier with a multilayer feed-forward neural network using the example of plant/weed/soil discrimination, Computers and Electronics in Agriculture, vol. 39, (2003), p. 3-22.
9. Ranaweera DK, Hubele NF and Papalexopoulos AD, Application of radial basis function for short term load forecasting, IEE Proc Gener Trans, vol. 142(1), (1995), p. 45-50.
10. Y. S. Hwang and S. Y. Bang, An Efficient Method to Construct a Radial Basis Function Neural Network Classifier, Vol. 10(8), (1997), p. 1495-1503.
11. E. Bonabeau and G. Theraulaz, (Swarm smarts. Scientific American 2000), p. 72-79.
12. S. H. Zahiri and S. A. Seyedin, Swarm intelligence based classifiers, Journal of the Franklin Institute, vol. 344, (2007), p. 362-376.
13. R. S. Parpinelli, H. S. Lopes and A. A. Freitas, Data mining with an ant colony optimization algorithm, IEEE Transactions on Evolutionary Computation vol. 6, (2002), p. 321-332.
14. S. N. Omkar and T. R. Raghavendra, Rule extraction for classification of acoustic emission signals using Ant Colony Optimization, Engineering Applications of Artificial Intelligence, Vol. 21(8), (2008), p. 1381-1388.
15. D. Karaboga and B. Basturk, On the performance of artificial bee colony (ABC) algorithm, Applied Soft Computing, Vol. 8 (1), (2008), p. 687-697.
16. S. N. Omkar, J. Senthilnath, Rahul Khandelwal, G. Narayana Naik and S. Gopalakrishnan, Artificial Bee Colony (ABC) for multi-objective design optimization of composite structures, Applied Soft Computing, In Press.
17. J. Kennedy and R. C. Eberhart, Particle swarm optimization, in Proceedings of IEEE International Conference on Neutral Networks, (1995), p. 1942-1948.
18. Z. F. Hao, R. C. Cai and H. Huang, An adaptive parameter control strategy for ACO. In: Proceedings of the Fifth International Conference on Machine Learning and Cybernetics, Dalian, (2006), p. 13-16.
19. K. V. Frisch, Bees: Their Vision, Chemical Senses and Language, Revised Edition (Ithaca, N.Y.: Cornell University Press, 1976).
20. Von Frisch K, The Dance Language and Orientation of HoneyBees, (Harvard University Press, 1967).
21. D.T. Pham, E. Koc, A. Ghanbarzadeh, and S. Otri, Optimisation of the Weights of Multi-Layered Perceptrons Using the Bees Algorithm, Proceedings of 5th International Symposium on Intelligent Manufacturing Systems, (2006), p. 38-46.
22. S. N. Omkar and J. Senthilnath, Artificial Bee Colony for Classification of Acoustic Emission Signal Sources, International Journal of Aerospace Innovations , Vol. 1 (3), (2009), p. 129-143.
23. N. Cristianini and J. S. Taylor, An Introduction to Support Vector Machines (Cambridge University Press, Cambridge, UK, 2000).
24. M. D. Richard and R. P. Lippmann, Neural network classifiers estimate Bayesian a posteriori probabilities, Neural Comput., 3, (1991), p. 461-483.
25. A. Cichocki and R. Unbenhauen, Neural Networks for Optimization and Control (Baffins Lane, U.K.: Wiley, 1993).

26. Eric Bonabeau, Macro Dorigo, and Guy Theraulaz, Swarm intelligence-from natural to artificial system, (Oxford University Press, New York, 1999).

27. J. Kennedy, R. C. Eberhart and Y. Shi, Swarm intelligence, (Morgan Kaufmann Publishers, San Francisco, 2001).

28. S. N. Omkar, J. Senthilnath, Dheevatsa Mudigere, Manoj Kumar M, Crop Classification using Biologically Inspired Techniques with High Resolution Satellite Image, Journal of the Indian Society of Remote Sensing, Springer India, Vol. 36,(2008), p. 172-182.

29. Balakrishnan, D. and Puthusserypady, S. (2005). Multilayer perceptrons for the classification of brain computer interface data, Proceedings of the IEEE Bioengineering Conference, (2005), p. 118 - 119.

30. M Reyes-Sierra and C. C. Carlos, Multi-objective particle swarm optimizers: A survey of the state-of-the-art, International Journal of Computational Intelligence Research, vol. 2(3), (2006), p. 287-308.

Chapter 3

Multi-Objective Ant Colony Optimization:
A Taxonomy and Review of Approaches

Guillermo Leguizamón* and Carlos A. Coello Coello[†,‡]

*LIDIC - Universidad Nacional de San Luis, San Luis, Argentina
UMI LAFMIA 3175 CNRS at CINVESTAV-IPN,
Departamento de Computación,
Av. IPN No. 2508. Col. San Pedro Zacatenco,
México D.F. 07300, México
legui@unsl.edu.ar

†CINVESTAV-IPN (Evolutionary Computation Group),
Departamento de Computación,
Av. IPN No. 2508. Col. San Pedro Zacatenco,
México D.F. 07300, México
ccoello@cs.cinvestav.mx

Ant Colony Optimization (ACO) is one of the most representative meta-heuristics derived from the broad concept known as Swarm Intelligence (SI) where the behavior of social insects is the main source of inspiration. Being a particular SI approach, the ACO metaheuristic is mainly characterized by its distributiveness, flexibility, capacity of interaction among simple agents, and its robustness. The ACO metaheuristic has been successfully applied to an important number of discrete and continuous single-objective optimization problems. However, this metaheuristic has shown a great potential to also cope with multi-objective optimization problems as evidenced by the several proposals currently available in that regard. This chapter is aimed at describing the most relevant and recent developments on the use of the ACO metaheuristic for solving multi-objective optimization problems. Additionally, we also derive a refined taxonomy of the types of ACO variants that have been used for multi-objective optimization. Such a taxonomy intends to serve as a quick guide for those interested in using an ACO variant for solving multi-objective optimization problems. In the last part of the chapter, we provide some potential paths for further research in this area.

---

‡He is also affiliated at the UMI LAFMIA 3175 CNRS at CINVESTAV-IPN.

3.1. Introduction

The field of Computational Intelligence (CI), and particularly the algorithms based on the concept of Swarm Intelligence (SI) has been intensively studied and successfully applied to optimization problems. Among these problems are those that include multiple objectives, which usually are very common in many application areas. SI-based algorithms involve several characteristics that make them particularly suitable for solving multi-objective optimization problems (MOOPs), e.g., inherently decentralized, the members of the swarm can be in charge of different objectives, different levels and types of interactions can be defined in order to share individual search experience with the rest of the swarm, etc. The most representative and developed SI algorithms include Particle Swarm Optimization (PSO) (see Ref. 1 for more details) and the Ant Colony Optimization (ACO) metaheuristic.[2,3]

In the case of ACO algorithms, an important number of proposals have shown, with different levels of success, the applicability of these algorithms to multi-objective optimization problems. As an example, two interesting reviews on this topic can be found in García-Martinez *et al.*,[4] and in Angus and Woodward.[5] Similarly, a section in Coello Coello *et al.*'s book is devoted to the ACO metaheuristic[6] as an example of an alternative metaheuristic[a] for solving MOOPs.

The remainder of this chapter is organized as follows. In the next section, we present a general overview of the ACO metaheuristic for discrete and continuous problems. Section 3.3 gives a general introduction to multi-objective optimization in which the most relevant concepts are described. In Sec. 3.4, we present an up-to-date review of the ACO metaheuristic for multi-objective optimization problems in which the specific components of the proposals dealing with the multi-objective aspects are highlighted. A refined taxonomy of multi-objective ACO approaches is presented in Sec. 3.5. Some of the promising research areas within this topic are briefly described in Sec. 3.6. Finally, the conclusions of this chapter are provided in Sec. 3.7.

3.2. Ant Colony Optimization

The Ant Colony Optimization (ACO) metaheuristic[3] embodies a broad class of algorithms whose design is mainly based on the foraging behavior of real ants. ACO algorithms were originally designed and have a long tradi-

---

[a]By "alternative" the author means, with respect to evolutionary algorithms.

tion in solving a specific type of combinatorial optimization problems (i.e., problems for which the solution construction process can be implemented by simulating a walk through a construction graph). The seminal works of the use the ACO metaheuristic were devoted to the Traveling Salesperson Problem (TSP), a classical NP-complete problem whose main characteristics can be easily exploited to show the applicability of this metaheuristic. For example, in Dorigo[7] there is a description of the first ACO algorithm designed to solve the TSP, the so-called Ant System (AS). After that, several improvements to the AS for solving TSP were proposed: *elitist*-AS, an AS with an *elitist strategy* for updating the pheromone trail levels, AS_{rank} (a rank-based version of Ant System), $\mathcal{M}ax$-$\mathcal{M}in$ Ant System ($\mathcal{M}\mathcal{M}AS$), and the Ant Colony System (ACS).[3]

In the following, we present the main aspects to be considered when applying the ACO metaheuristic to a particular discrete problem. First, it is important to define an appropriate problem representation, i.e., the construction graph and the way this represents the different problem components and connections among them as well as the definition (if any) of the problem information to be exploited. Second, the behavior of the artificial ants should be defined in order to show how each ant will walk through the construction graph to build the corresponding solutions.

Algorithm 3.1   Outline of the ACO metaheuristic

1: Initialize();
2: **while** termination-condition is NOT TRUE **do**
3:     BuildSolutions();
4:     PheromoneUpdate();
5:     DaemonActions(); // Optional
6: **end while**

A general outline of the ACO metaheuristic is displayed in Algorithm 3.1 in which four main activities are considered. The way in which those activities are implemented defines the possible algorithms that can be obtained, i.e., AS, elitist-AS, AS-rank, $\mathcal{M}\mathcal{M}AS$, ACS, or any other. The main activities could vary from one algorithm to another, however, they can be described in a general way as follows:

- **Initialize():** As in any typical metaheuristic algorithm, some basic tasks need to be done before starting the exploration of the search space. In this case, the initialization of pheromone trail structure, the heuristic values (when available and used), and any other structure necessary to complete the problem representation.

- **BuildSolutions():** This activity involves the release of an independent colony of artificial ants in charge of incrementally building a solution to the problem. Each ant, at each step of the construction process, makes a local stochastic decision about the next component to be included in the solution under construction.

- **PheromoneUpdate():** The acquired experience achieved by the colony at each iteration is considered in this activity. High quality solutions will positively affect the amount of pheromone trail, i.e., those edges that are part of solutions found will receive an increased amount of pheromone trail according to the goodness of these solutions. This is known as the *global pheromone update*. As in nature, a process of pheromone evaporation takes place (usual implementations of this metaheuristic decrease the amount of pheromone trail for all edges in the construction graph). Thus, the amount of pheromone corresponding to those edges that are not part of any solution at the current iteration will show a gradually diminishing pheromone intensity. It should be noticed that some ACO algorithms, such as ACS, apply a local pheromone update rule which does not depend on the solution quality. Instead, a fixed amount is deposited as soon as an edge in the construction graph is selected to make the move (the next component added to the solution under construction).

- **DaemonActions():** As single ants can not carry out some centralized actions, many ACO algorithms include some specific activities called *daemon actions*. Examples of these activities are: activation of a local search procedure or a collection of global information (e.g., use of a set of the best ranked solutions) that could be used to reinforce some entrances in the pheromone trail structure.

Let us assume that the problem under consideration is the TSP. Figure 3.1 shows a possible construction graph for an instance of size $n = 5$.

Each vertex represents the problem components (i.e., the cities) and each edge represents the connections (routes) between the cities where the distance ($d_{ij} = d_{ji}$) or other value (e.g., cost) is associated to each edge. From the perspective of the ACO metaheuristic, two additional values are associated to each connection, the desirability of choosing edge (i, j) repre-

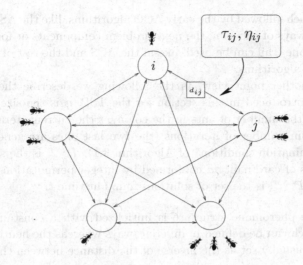

Fig. 3.1.: Construction graph for an instance of the symmetric TSP of size $n = 5$.

sented by τ_{ij}, and the problem's information, a heuristic value represented by η_{ij}. In the case of the TSP, the usual value is $\eta_{ij} = 1/d_{ij}$. These two values are then used in the construction solution process to estimate the probability p_{ij} of choosing city j from city i by ant k:

$$p_{ij}^k = \begin{cases} \dfrac{\tau_{ij}^{\alpha} \eta_{ij}^{\beta}}{\sum_{h \in \mathcal{N}^k(i)} \tau_{ih}^{\alpha} \eta_{ih}^{\beta}} & \text{if } j \in \mathcal{N}^k(i) \\ 0 & \text{otherwise,} \end{cases} \tag{3.1}$$

where α and β are the parameters that respectively represent the importance of the pheromone trail and the heuristic information, and $\mathcal{N}^k(i)$ represents the set of cities that can be visited by ant k, i.e., the feasible cities. On the side of the pheromone update process, the following equation (usually called global updating) can be applied:

$$\tau_{ij}(t+1) = (1 - \rho) \cdot \tau_{ij}(t) + \Delta\tau_{ij}, \tag{3.2}$$

where ρ is the evaporation rate and $\Delta\tau_{ij}$ represents the amount of pheromone trail deposited in edge (i, j) according to the quality of the solutions found by the whole colony that include that edge as part of the solution found. It should be noticed that Eqs. (3.1) and (3.2) represent the basic approach followed by the early ACO algorithms, like the AS. However, alternative ways of choosing the next problem components or for updating the pheromone trail can be used, like in the ACS and the rest of the family of the ACO algorithms.

From another point view, in the following we describe the activities previously introduced in this section for the TSP. n is the size of the instance, a is the number of ants in the colony, t the current iteration, t_{max} the maximum number of iterations (the two last ones are generally used in the "termination condition" of Algorithm 3.1). $P^{a \times n}$ is the space of all possible sets of cardinality a conformed by integer permutations of size n, and $S(t) \subset P^{a \times n}$, is the set of solutions found at time t.

- The pheromone structure is initialized with a constant value τ_0 which can be defined in different ways whereas the heuristic values are usually set as the inverse of the distance between the cities.

$$\tau(0) = \tau_0 \times \mathbf{1}^{n \times n} \text{ and } \eta_{ij} = 1/d_{ij} \text{ for } i, j \in \{1, \ldots, n\}.$$

- The solution construction process, for each ant, takes into account the current amount of the pheromone trails and the heuristic information to obtain the new sample of solutions. The whole process can be expressed as:

$$S(t) = \mathcal{BS}(\tau(t), \eta) \text{ for } t \in \{0, \ldots, t_{max}\},$$

where $\mathcal{BS} : \mathbb{R}^{n \times n} \times \mathbb{R}^{n \times n} \to P^{a \times n}$ manages the a ants to incrementally build the solutions based on Eq. (3.1).

- After the whole colony has built a solutions (this applies for global updating) the new pheromone trail values are calculated as follows:

$$\tau(t + 1) = \mathcal{PU}(\tau(t), S(t)), \text{ for } t \in \{0, \ldots, t_{max} - 1\}$$

where $\mathcal{PU} : \mathbb{R}^{n \times n} \times P^{a \times n} \to \mathbb{R}^{n \times n}$ performs the pheromone updating process based on Eq. (3.2).

For continuous problems, several proposals have been considered and studied from the perspective of the ACO metaheuristic. The first ACO extension designed to operate on continuous search spaces was introduced by Bilchev *et al.*[8] Since then, several other proposals have been introduced (see Refs. 9–16). Particularly, one relevant proposal is that introduced by Socha[17]) and further extended by Socha & Dorigo.[18] Although the different proposed versions of the ACO metaheuristic for continuous problems have been applied with different levels of success, to the best of authors' knowledge there is no unifying approach for continuous problems as in the case of discrete problems. This has motivated more research in this area, aiming to have a more standard and widely recognized platform for the ACO metaheuristic in continuous domains. Indeed, it is expected that in the near future, more powerful and competitive versions of continuous ACO algorithms will be available as has happened in the case of other metaheuristics such as Evolutionary Algorithms and Particle Swarm Optimization.

To conclude this section, it is important to notice that ACO algorithms have been traditionally applied to single-objective problems. However, many researchers have reported encouraging results regarding the application of such algorithms to multi-objective problems. The next section is devoted to present some basic concepts about multi-objective optimization. After that, we describe the most relevant and recent advances in the design and application of MOACO algorithms.

3.3. Basic Concepts of Multi-Objective Optimization

A multi-objective optimization problem (MOOP) can be formulated as[b]:

$$\text{minimize } \vec{f}(\vec{x}) := [f_1(\vec{x}), f_2(\vec{x}), \ldots, f_k(\vec{x})] \tag{3.3}$$

subject to:

$$g_r(\vec{x}) \leq 0 \quad r = 1, 2, \ldots, m \tag{3.4}$$

$$h_j(\vec{x}) = 0 \quad j = 1, 2, \ldots, p \tag{3.5}$$

where $\vec{x} = [x_1, x_2, \ldots, x_n]^T$ is the vector of decision variables, $f_i : \mathbb{R}^n \to \mathbb{R}$, $i = 1, \ldots, k$ are the objective functions and $g_r, h_j : \mathbb{R}^n \to \mathbb{R}$, $r = 1, \ldots, m$, $j = 1, \ldots, p$ are the constraint functions of the problem.

To describe the concept of optimality in which we are interested, we will introduce next a few definitions.

Definition 1. Given two vectors $\vec{x}, \vec{y} \in \mathbb{R}^k$, we say that $\vec{x} \leq \vec{y}$ if $x_i \leq y_i$ for $i = 1, \ldots, k$, and that $\vec{x}$ **dominates** $\vec{y}$ (denoted by $\vec{x} \prec \vec{y}$) if $\vec{x} \leq \vec{y}$ and $\vec{x} \neq \vec{y}$.

---

[b]Without loss of generality, we will assume only minimization problems.

Definition 2. We say that a vector of decision variables $\vec{x} \in \mathcal{X} \subset \mathbb{R}^n$ is **non-dominated** with respect to $\mathcal{X}$, if there does not exist another $\vec{x}' \in \mathcal{X}$ such that $\vec{f}(\vec{x}') \prec \vec{f}(\vec{x})$.

Definition 3. We say that a vector of decision variables $\vec{x}^* \in \mathcal{F} \subset \mathbb{R}^n$ ($\mathcal{F}$ is the feasible region) is **Pareto-optimal** if it is non-dominated with respect to $\mathcal{F}$.

Definition 4. The **Pareto Optimal Set** $\mathcal{P}^*$ is defined by:

$$\mathcal{P}^* = \{\vec{x} \in \mathcal{F} | \vec{x} \text{ is Pareto-optimal}\}$$

Definition 5. The **Pareto Front** $\mathcal{PF}^*$ is defined by:

$$\mathcal{PF}^* = \{\vec{f}(\vec{x}) \in \mathbb{R}^k | \vec{x} \in \mathcal{P}^*\}$$

We thus wish to determine the Pareto optimal set from the set $\mathcal{F}$ of all the decision variable vectors that satisfy (3.4) and (3.5). Note however that in practice, not all the Pareto optimal set is normally desirable (e.g., it may not be desirable to have different solutions that map to the same values in objective function space) or achievable.

3.4. The ACO Metaheuristic for MOOPs in the Literature

An important number of proposals of ACO algorithms have shown their applicability to multi-objective optimization problems with different degrees of success. Many of those proposals were reviewed by García-Martinez et al.,[4] by Angus and Woodward,[5] and by Coello Coello et al.[6] In García-Martinez et al.[4] a taxonomy of the ACO metaheuristic for MOOPs is proposed based on two criteria (number of structures to store the pheromone trail and number of heuristic functions) and considering a number of existing ACO algorithms for MOOPs (MOACOs). In addition, the authors include a comparative study of some of the reviewed algorithms and two multi-objective evolutionary algorithms, an improved Strength Pareto Evolutionary Algorithm (SPEA2)[19] and a Fast Elitist Non-Dominated Sorting Genetic Algorithm for Multi-Objective Optimization (NSGA-II).[20] For the experimental study, it was considered the bi-criteria TSP. In a more recent report, Angus and Woodward[5] present an alternative and extended taxonomy of the ACO metaheuristic for MOOPs which includes five attributes to classify a particular algorithm as seen in Table 3.1. Based on

Table 3.1.: The taxonomy proposed by Angus and Woodward.[5]

Attribute	Values
Pheromone matrix	Multiple, Single
Solution construction	Targeted, Dynamic, Fixed
Evaluation	Pareto, Non-Pareto
Pheromone Update	Individual, Global
Pareto archive	Offline, Online, Elitist, None

that taxonomy, the authors describe a set of relevant ACO algorithms for MOOPs as well as their main characteristics.

In the remainder of this section, we extend the literature review presented in García-Martinez et al.,[4] Angus and Woodward,[5] and Coello Coello et al..[6] by considering some recent and relevant proposals of the ACO metaheuristic for MOOPs. In the next section, we discuss a refined taxonomy of ACO algorithms for MOOPs.

It is worth noticing that there exist few applications of ACO algorithms for multi-objective problems in continuous domains. For example, Angus[21] proposed a Population-based ACO algorithm for Multi-Objetive Function Optimization (PACO-MOFO). PACO-MOFO is based on the Crowding Population-based ACO algorithm (CPACO)[22] (which was designed for discrete domains) and $ACO_{\mathbb{R}}$.[18] PACO-MOFO uses an a posteriori preference articulation method and implements two niching approaches: crowding (for the population replacement) and fitness sharing (for the selection mechanism). The authors compared the proposed algorithm with NSGA-II[20] on a set of four widely known benchmark problems: MOP1, MOP2, MOP3, and MOP6 (see Refs. 6 and 23). Based on the summary attainment analysis performance assessment measure,[24] the results show a similar performance of PACO-MOFO and NSGA-II on problems MOP2, MOP3, and MOP6, but not problem MOP1 where NSGA-II outperforms PACO-MOFO. An additional experiment was conducted to study the algorithm's scalability for which problem MOP2 was considered. In this case, PACO-MOFO showed some interesting results regarding the quality of the solutions found in the center of the Pareto front in comparison with those found by NSGA-II and in view of the capability of PACO-MOFO to find solutions when considering a higher dimensionality problem. Another proposal of MOACOs in continuous domains is the work of Garcia-Nareja and Bullinaria[25] in which they present an extended version of $ACO_{\mathbb{R}}$.[18] To deal with several objectives they maintain an archive of selected solutions (the approximation of the Pareto optimal set). The concept of *dominance depth* is used to compare the quality of the solutions. The criteria chosen to determine which solution should be eliminated from the archive is the *crowding distance*. The proposed algorithm is applied to a well-known set of multi-objetive problems and compared to NSGA-II[20] and the Multi-Objective Particle Swarm Optimizer (MOPSO).[26] The solutions found by the extended $ACO_{\mathbb{R}}$ were comparable to those found by NSGA-II and MOPSO.

As one would expect, the use of ACO in discrete MOOPs is, by far, the most commonly reported in the specialized literature. Next, we present some of the most recent proposals in this regard.

Liu *et al.*,[27] presented an ant algorithm, called MO-ant, to generate Pareto fronts for multi-objective siting of Emergency Service Facilities (EFSs). They used the concept of Pareto ranking[28] to determine which were the most qualified solutions to the problem of their interest. The geographic area under study was represented as a grid map from which the ants had to find the best sites to allocate the EFSs. MO-ant does not use any heuristic information and it manages one pheromone matrix, in which each position in that matrix represents the desirability of allocating an EFS at the corresponding position in the real geographic area. In addition, this algorithm applies at each iteration a two-phase local search procedure involving a Pareto ranking of the solutions generated. During the first phase of the local search, it applies the so-called *Neighborhood Random Search* (NRS) by which the ants randomly move from one cell to another one within a certain distance. After that, all the solutions are re-evaluated and a Pareto ranking is obtained. The second phase of the local search consists in the application of the so-called *Adaptive Enumeration Neighborhood Search* (AENS) to the first solution in the set of previously ranked solutions. AENS aims at finding a better position than the current one for every ant in the colony by considering all the cells within a certain distance. If the total sum of the objective values has improved, the ant moves to the cell that produced such a global improvement. This procedure is applied until no more improvements can be observed. The new set of solutions found by AENS are then Pareto ranked. The pheromone matrix is updated by taking into account the rank of the solutions found. As can be seen, the search of the Pareto front is mainly guided by the local search procedure as well as by the pheromone values. MO-ant was applied in a real-world scenario: the multi-objective siting of fire stations in Singapore for which three objectives were considered. The results obtained were compared with respect to a set of Pareto ranked solutions obtained out of 10, 000 randomly generated solutions.

Bui et al.[29] investigated the effect of elitism in a multi-objective ACO algorithm under local,[c] global, and mixed non-dominated solutions. The MOACO algorithm used one pheromone matrix for each of the objectives. The authors presented a systematic study on five instances of the multi-objective TSP (mTSP) considering the following elitist alternatives: a) local-best, b) global-best, and c) local-set (all solutions, dominated and non-dominated, generated in one iteration). These elitist alternatives are then used in the pheromone updating process in one of the following ways: i) local-set, ii) local-best, iii) local-best + local-set, iv) global-best, and v) global-best+local-set. In addition, the authors proposed an adaptive mechanism (aging) to control the use of an external archive for the global-best elitist alternative. The results indicate the importance of including elitism in MOACOs, for which global elitism was the mechanism that provided more effective information. Also, the adaptive strategy of aging showed an extra improvement on the quality of the solutions found.

Benlian and Zhiquan[30] proposed a MOACO-based data association method for bearings-only multi-target tracking. The algorithm uses multiples matrices, one for each of the two objectives considered: distance and slope difference. The heuristic information is shared by both pheromone trail matrices and the corresponding values are calculated as an aggregating function that combines the heuristic values of each objective. The conducted experiments showed an improved performance of the proposed MOACO with respect to the joint Maximum Likelihood method.

Mora et al.[31] reported a comparison of six different ACO algorithms for the Bi-criteria Military Path-Finding Problem in which they defined two objectives: minimize resources while maximizing safety on a map corresponding to a simulated battlefield. Four of the compared algorithms are variants of an enhanced version of the Compañía de Hormigas Acorazadas (CHAC) or Armoured Ant Company,[32] called hexa-CHAC (hCHAC).[33] The original CHAC is basically an ACS including one pheromone matrix for each objective as well as the corresponding heuristic information. In CHAC, two different state transition rules were tested, one based on an aggregating function that combines heuristic and pheromone information regarding the two objectives (CSTR) and the other one, based on the dominance

---

[c]The term local is used by the authors to refer to the current iteration.

among solutions (DSTR). An additional parameter ($0 \leq \lambda \leq 1$) was used in CSTR and DSTR to control the relative importance of the two objectives of the problem under study. For the tested scenarios, CHAC with CSTR showed a better performance. However, such a performance was less robust than that of CHAC with DSTR. On the other hand, hCHAC represented an algorithm capable of dealing with more realistic problem conditions and constraints and the main difference with CHAC was that the scenarios were modeled as a grid of hexagons which implies a change in the construction graph, and the possibility of working with real-world images by defining an underlying information layer. The other two algorithms compared in Mora et al.,[31] were the so-called mono-hCHAC (a version the uses an aggregating function combining the two objectives into a scalar value) and the Multi-Objective Ant Colony System (MOACS).[34] From this comparison, hCHAC-CSTR was considered to be the best overall performer for this particular problem, i.e., the enhanced CHAC algorithm applying a combined state transition rule during the solution construction phase.

McMullen and Tarasewich[35] presented an application of the ACO metaphor to solve a multi-objective assembly line balancing problem. The approach adopted by the authors consisted of using an *ad hoc* aggregating function that combined the four objectives considered: required crew size, system utilization, probability of jobs being assembled within a certain time frame, and cost of the system design. This function (called *metric*) is involved in the computation of pheromone levels associated to a particular task i when considering its assignment to the work center j. In addition, the pheromone levels depend on a value observed regarding the historical precedence of task i with respect to the remaining task in work center j. The authors used the concept of "efficient frontier"[d] to measure the solutions quality by considering two entities: required crew size and a value obtained by a combination of system utilization, probability of jobs being assembled within a certain time frame, and cost of the system design.

Xing et al.[36] proposed a fuzzy multi-objective ACO algorithm with linguistically quantified decision functions for flexible job shop scheduling with an interactive decision maker (DM). The main contribution of this work is the interaction of the ACO algorithm with a DM at each iteration in order to bias the search. Particularly, the DM takes into account the distance of the best solution found at each iteration with respect to the aspiration level considering a fuzzy metric. In this case, a classical single-

---

[d]Efficient frontier is the term used in Operations Research to denote the Pareto front of a problem.

objective ACO algorithm is used as the search engine within a more complex process that includes a close interaction with the DM to solve a multi-objective problem.

Angus[22] extended a Population-based ACO algorithm[37] (PACO) with a Crowding population replacement scheme for the multi-objective the TSP (CPACO). The proposed CPACO algorithm uses only one pheromone matrix and individual heuristic values for each objective. The basic idea of the crowding scheme is taken from Evolutionary Computation (EC). In the CPACO algorithm, the pheromone matrix at each iteration is updated in the following way: i) the set of solutions in the population are ranked (the same approach used in NSGA-II[20]), ii) all the elements of the pheromone matrices are re-initialized to a value τ_{init}, and iii) the ranked solutions in the population produce an updating bias in the re-initialized pheromone matrices using the inverse of their ranks. Thus, the better the rank obtained for a solution, the more the amount of pheromone laid on the corresponding matrix entrances determined by that solution. The way in which CPACO combines one pheromone matrix and several heuristic values (one for each objective) follows the proposal of Barán and Schaerer.[34] CPACO was studied on a set of bi-objective test instances of TSP as well as a 4-objective version of TSP. The comparative study of CPACO with respect to the original PACO showed an improved performance of CPACO on the bi-objective instances of TSP. Although the results were not as good as for the first set of instances, CPACO showed an acceptable performance on the 4-objective TSP. The authors indicated that the pheromone update process and the use historical information are mechanisms that require further research.

In Alaya *et al.*,[38] a generic ACO algorithm for multi-objective problems was presented. The proposed ant algorithm (called m-ACO) follows the design of a $\mathcal{MMAS}$ and it is parameterized with the number of colonies and the number of pheromone matrices. As the values of these parameters can vary, the authors instantiated them on four different ways: m-$\text{ACO}_1(m+1, m)$, m-$\text{ACO}_2(m+1, m)$, m-$\text{ACO}_3(1,1)$, and m-$\text{ACO}_4(1, m)$, where m represents the number of objectives. The two first variants are similar in the sense that they use $m+1$ colonies and m pheromone matrices. The difference is that the first algorithm produces solutions in colony $m+1$ by using a randomly selected pheromone matrix (i.e., a random objective) to build the solutions whereas in the second variant, the pheromone values associated to colony $m+1$ correspond to the sum of the all the pheromone values associated to each objective. In the third variant, m-$\text{ACO}_3(1,1)$, only one colony and one pheromone matrix are involved. This is, in fact, very similar to the solution of a single-objective problem. However, the heuristic values are obtained in this case as the sum of the heuristic values associated to each objective. In addition, the pheromone update process only takes into account the non-dominated solutions. Thus, the components of each of the non-dominated solutions indicated which entrances of the pheromone matrix would be rewarded. The amount of pheromone added to each component was the same for all the non-dominated solutions. Finally, the last variant (m-ACO$_4(1, m)$) used one colony and m pheromone matrices. First, each ant of the (only) colony randomly selects one objective and then builds the solution using the corresponding pheromone matrix. The heuristic values are always calculated as in the third variant. At each iteration, i.e., once the colony has built the solutions, the pheromone matrices are updated by considering in turn the best m solutions with respect to each of the m objective function values. The experimental study reported the results of the four variants described above when applied to different instances of the multi-objective multidimensional knapsack problem. The variant m-ACO$_4(1, m)$ globally achieved the best performance on the tested instances. This variant was compared with several non-elitist multi-objective evolutionary algorithms[e] (MOGA,[39] NSGA,[40] SPEA,[41] HLGA,[42] and VEGA[43]) based on the coverage of two sets measure[44] for which m-ACO$_4(1, m)$ found the best results except for some instances when compared with SPEA.[41] Although the presentation of m-ACO seems to be a generalization for multi-objective subset problems, it could also be generalized for any discrete combinatorial optimization problem.

Afshar et al.[45] presented a bi-colony ACO algorithm with a Non-dominated Archive (NA-ACO) which combines a novel interaction of the two involved colonies to evolve the solutions. The NA-ACO algorithm maintains one pheromone matrix and an archive of non-dominated solutions. Each colony is in charge of optimizing one of the two objectives. NA-ACO works as follows: the first colony produces a set of solutions that are moved and evaluated according to the objective function assigned to the

---

[e]Elitism, in the context of multi-objective metaheuristics, normally consists of using an external archive (usually called a "secondary population") that can (or cannot) interact in different ways with the main (or "primary") population of the multi-objective metaheuristic. The main purpose of this archive is to store all the non-dominated solutions generated throughout the search process, while removing those that become dominated later in the search (called local non-dominated solutions). The approximation of the Pareto optimal set produced by an algorithm is thus the final contents of this archive. Practically all modern multi-objective evolutionary algorithms (i.e., those designed after 1999) are elitist.[6]

second colony. From this set of solutions, the second colony selects the best one and uses this solution to update the pheromone matrix. After that, a new set of solution is generated and moved to the first colony for evaluation according to the corresponding objective function. The new best ant is then considered for pheromone updating. The above process is repeated for N cycles after which the values of the objective functions are computed, in order to find the non-dominated solutions to be stored in the archive. The pheromone matrix is re-initialized and then updated according to the new solutions in the archive. The whole process is repeated for M iterations. NA-ACO algorithms was first studied on problems ZDT1 and ZDT2,[46] and compared with NSGA-II,[20] SPEA,[41] and the Pareto Archived Evolution Strategy (PAES).[47] In addition, NA-ACO was tested on two bi-objective water-resource problems for which encouraging results were found when compared with a weighted-sum method.

In an extended abstract, Eppe[48] presented a mechanism that integrated the decision maker's preferences into multi-objective Ant Colony Optimization. The main contribution of this work was the use of a preference function (based on the PROMETHEE methodology[49]) to define a normalized and aggregated preference index. This preference index is applied to a solution at the component level for the TSP. Although no actual results were reported by the author, the proposed mechanism seems to be capable to deal with multi-objective problems.

Chica *et al.*,[50] incorporated preferences to a multi-objective ACO algorithm for a variant of the time and space assembly line balancing problem (TSALBP-1/3) which was previously studied by the authors in Ref. 51. The preferences incorporated were represented by *a priori* information of the problem provided by the plant experts and was used to guide the search. The authors selected a Multiple Ant Colony System (MACS) as their search engine (see Ref. 34). The main goal of the new proposed algorithm was to reduce the size of the Pareto optimal set and increase the quality of the Pareto front. For the authors a high quality Pareto front is one which presents a focused and reduced set of solutions for the decision maker. The generation of such a high quality Pareto front required the redefinition of the concept of Pareto dominance in order to include some specific preferences given by the experts and applied when there were some solutions with the same objective function values. In the case of TSALBP-1/3, the information considered was the adoption of the same value for the area and number of stations for a fixed cycle time in the assembly line. The new dominance definition takes into account the workload of the plant and the required space for the workers' instruments. Accordingly, two measures were formulated and the corresponding preference-based dominance definitions were provided. MACS was tested in a real world problem: the assembly process of the Nissan Pathfinder engine. The results were compared to those of the Multi-Objective Randomised Greedy Algorithm (MORGA)[51] and to a modified version of the ACS reported in Ref. 51. The results showed an improved performance of MACS with respect to the other two algorithms with respect to which it was compared, regarding each of the metrics considered in the experimental study. Following this same line of research, Chica et al.[52] extended this proposal by incorporating the elicitation of preferences regarding the economic factors related to the location of the plant. These preferences were included in the objective space in order to achieve a more focused Pareto front and then make easier the task of the decision maker. They used six scenarios around the world and the preferences were based on the Evolutionary Multi-Objective (EMO) preferences given in Refs. 53 and 54.

Häckel et al.[55] developed an ACO algorithm for solving the multi-objective shortest path problem. The authors considered in their experimental study a problem with three criteria. However, their proposal can be extended to a higher number of objectives. Interestingly, they used only one pheromone matrix. However, their corresponding trail values were weighted differently according to each ant and criteria. This weighting policy was also applied to the heuristic values which were calculated by considering four alternatives: i) standard (inverted edge weights), ii) LAH/bc, iii) LAH/wc, and iv) LAH/ac; where LAH stands for Look-Ahead Heuristic considering the best (bc), worst (wc), and average (ac) cases. The corresponding LAH values were separately calculated using dynamic programming for each of the criteria involved. Two ACO algorithms were tested: the Ant System (AS) and the Ant Colony System (ACS). The ACS with LAH/wc showed the best overall performance regarding the set of ad hoc metrics adopted by the authors.

Chaharsooghi and Kermani[56] proposed a multi-objective ACO algorithm for solving the Multi-Objective Resource Allocation Problem (MORAP) in which the objectives considered were: maximization of workers' efficiency and minimization of the resources cost. The ACO algorithm used only one pheromone matrix and only one structure to store the heuristic values that represented a combined calculation obtained as the product of the heuristic values associated to the two separate objectives: efficiency and cost. The pheromone matrix was updated considering a modified updating rule that took into account the concept of non-dominance between solutions, current iteration, and an *ad hoc* parameter to control the influence of the current generation on the amount of pheromone trail to be deposited. Since the approach adopted for pheromone updating can generate negative values when calculating the probability used during the construction solution phase, a mechanism was incorporated to deal with this problem. The proposal was compared and showed an improved performance with respect to a hybrid GA (hGA) using only one particular instance of MORAP. In a related work of Chaharsooghi and Kermani[57] a similar approach was adopted for solving the multi-objective Multidimensional Knapsack Problem. The main difference on the ACO algorithm proposed here is that this algorithm used multiple colonies, each one updating its own pheromone matrix associated to the two objectives considered here. The probability of selection used in the phase of solution construction was based exclusively on a combination of the pheromone values for each objective and no heuristic information was considered. This version of the ACO algorithm was applied on a set of instances of the knapsack problem and compared with NSGA-II.[20] Based on two different metrics, the authors claimed that the ACO algorithm outperformed NSGA-II when considering the solutions on the extreme portions of the Pareto front. However, by ruling out these extreme solutions, NSGA-II actually outperformed the proposed ACO algorithm.

Vieira *et al.*[58] dealt with the feature selection problem as a multi-criteria problem with a single objective function. Two criteria were considered: the size of the subset of features and the features that are to be selected to build a fuzzy classifier. The ant algorithm used two colonies, two pheromone matrices and two different heuristic values, one for each criterion. The objective function to be minimized was an aggregation of the two criteria which were combined in order to measure the classification error rate and the number of features. The ant algorithm worked as follows: the first colony was in charge of selecting the number of features (one for each ant) and then, this selected number was used for the corresponding ant in the second colony in order to find the features used to build the classifier. The ant algorithm was tested on data sets taken from the UCI repository[59] and compared with some previous works. The achieved performance of the proposed ant algorithm was similar or better to that of the compared algorithms.

In Yang *et al.*,[60] a multi-objective task scheduling approach for Grid over Optical Burst Switching (GOBS) networks was proposed, considering three objectives: 1) completion time, 2) cost for using the resources, and 3) load balancing. The Multi-Objective ACO algorithm (called MOACO in this work) incorporated the idea of Pareto dominance as well as an *ad hoc* operator that combined pheromone exchanges and a sharing niching method. MOACO uses one pheromone and one heuristic information (lineal) structure associated to each of the resources in the network. The pheromone structure was initialized according to the computational capacity of each resource, i.e., the higher the computational capacity, the higher the initial amount of the pheromone trail. Similarly, the heuristic values were calculated combining the corresponding initial amount of pheromone trails (i.e., computational resources) and network resources. Interestingly, the probability of selection of a particular resource included a threshold value below which the corresponding resource could not choose (i.e., it avoided the overuse of computational resources). With respect to the pheromone updating, the authors proposed, as in the Ant Colony System, two pheromone updating rules: local and global. The local one aims at influencing the pheromone values according to the time necessary to execute a job assigned to that particular computational resource considering a recently obtained solution, whereas the global one is similar to the local rule but the solution involved is the one obtained by searching the optimal solution found by one of the following: either pheromone exchanging or the sharing niching method. The experimental study was conducted on one particular instance of GOBS and the results were compared with respect to those of NSGA-II.[20] The results showed that MOACO outperformed NSGA-II. The authors claim that these results are due to the fact that in MOACO the pheromone values are initialized such that they incorporate information about the problem, whereas NSGA-II performs a blinder search. Additional experiments were conducted to compare separately the results of MOACO with respect to those found by a single-objective ACO algorithm considering each objective in turn.

3.5. ACO Variants for MOOP: A Refined Taxonomy

When considering the development of the Evolutionary Multiobjective Optimization (EMO) field, it is clear that ACO has a lot of room left for expansion, particularly if we consider continuous search spaces. The literature review presented by García-Martinez *et al.*,[4] Angus and Woodward,[5] Coello Coello *et al.*,[6] and in this chapter, show important achievements of muti-objective ACO in several application domains within the last few years. Here, we complement this information with a refinement of the taxonomy proposed by Angus and Woodward.[5] More precisely, we reconsider

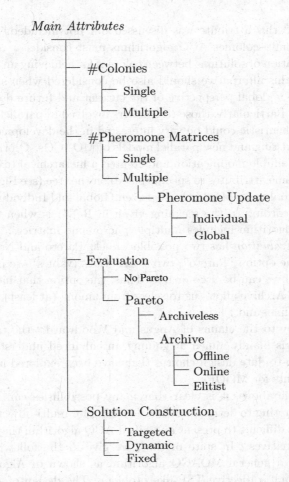

Main Attributes

- #Colonies
  - Single
  - Multiple
- #Pheromone Matrices
  - Single
  - Multiple
    - Pheromone Update
      - Individual
      - Global
- Evaluation
  - No Pareto
  - Pareto
    - Archiveless
    - Archive
      - Offline
      - Online
      - Elitist
- Solution Construction
  - Targeted
  - Dynamic
  - Fixed

Fig. 3.2.: A refined and extended taxonomy of MOACO algorithms.

the different attributes presented in Ref. 5 and some dependencies observed among them. Based on the previous discussion, we propose a taxonomy with hierarchical elements that could better help to visualize the different components of the existing MOACOs. Our taxonomy also adds an additional attribute that was discussed in Ref. 5 but was not included in their proposed taxonomy. The names and meaning of each of the attributes considered remains the same in our taxonomy as those described in Ref. 5.

Before describing our extended taxonomy, let us discuss the inclusion of the new attribute #Colonies. As indicated before, in Angus and

Woodward[5] this attribute was discussed but not included. Although the design of multi-colonies ACO algorithms must consider a certain level of communication of solutions between colonies for achieving an improved performance, this alternative should also be considered when showing to the researchers a global perspective of the current and future developments of MOACOs. Particularly, those researchers involved in parallel models of the ACO metaheuristic could be very interested in the development and application of existing and new parallel models of MOACOs. On the other hand, we made a simple modification and created a hierarchical taxonomy by associating some attributes to specific places on the tree (see Fig. 3.2). For example, the only way of selecting between Global and Individual pheromone update (according to the meaning given in Ref.5) is when the particular MOACO algorithm includes multiple pheromone matrices. Similarly, the attribute *Evaluation* has two possible values: Pareto and Non-Pareto. By choosing the option "Pareto", two alternatives (values) are possible, i.e., a Pareto Archive can be used or not. Thus, the option that includes the use of a Pareto Archive allows us to select from among (at least) three options: Offline, Online, and Elitist.

Similarly to the claims of Angus and Woodward,[5] the taxonomy proposed here is mainly aimed at gaining an enhanced understanding of the different up-to-date design choices that have been explored in the field the ACO variants for MOOPs.

From the above, it is clear that many possibilities can be considered when attempting to design an ACO algorithm for multi-objective problems. This makes difficult to present an unified ACO algorithm that embodies all those alternatives. In spite of that, we give in the following a possible example of a general MOACO algorithm as shown in Algorithm 3.2 for solving a multi-objective TSP-like problem. The design of Algorithm 3.2 includes a single colony of ants, k pheromone matrices and k structures that maintain respectively the amount of pheromone trails and heuristic information associated to each one of the k problem objectives. This algorithm corresponds to the class of MOACO algorithms that use Pareto evaluation for the solutions and maintain an archive, in this case, a set of ranked non-dominated solutions. In the following we describe in detail each one of the main components of the MOACO presented in Algorithm 3.2. Firstly, it must be noticed that t represents the current generation, t_{max} is the maximum number of iterations, N_a is the number of ants in the colony, and $\mathcal{A}$ is the archive that maintains the set of non-dominated solutions and it is updated at each iteration.

Algorithm 3.2   Outline of a possible MOACO for k objectives

1: Init_Pheromone_Trails(τ_{rj}^i); // for $i = 1, \ldots, k$; $r, j = 1, \ldots, n$
2: $\mathcal{A} = \emptyset$; // Archive of ranked non-dominated solutions
3: $t = 0$;
4: **while** ($t \leq T_{max}$) **do**
5:   **for** $h = 1, \ldots, N_a$ **do**
6:     BuildSolution$_h$(S(t), $\tau^1, \ldots, \tau^k, \eta^1, \ldots, \eta^k$);
7:   **end for**
8:   ArchiveUpdate($S(t), \mathcal{A}$);
9:   PheromoneUpdate($\tau^1, \ldots, \tau^k, \mathcal{A}$);
10:   $t = t + 1$
11: **end while**
12: return $\mathcal{A}$; // the Pareto front achieved

Algorithm 3.2 starts by initializing the k pheromone matrices (Init_Pheromone_Trails()). These initial values could depend on the particular objective. The archive of non-dominated solutions is initially empty (line 2). Function BuildSolution$_h$() is executed once for each of the N_a ants. This function is in charge of building a particular solution based on information of the k pheromone matrices and the respective heuristic information for each objective. As a general algorithm design, we do not show the way in which those values are used. Nevertheless, many possibilities can be considered, e.g., building an aggregating function used to calculate the probability values involved at each step on the solution construction process. The N_a solutions obtained at iteration t are stored in $S(t)$. After that, the archive of non-dominated solutions is updated in a such way that the new created solutions in $S(t)$ compete with the solutions stored in $\mathcal{A}$ to obtain a renewed set of ranked non-dominated solutions (ArchiveUpdate()). According to the new set $\mathcal{A}$ of non-dominated solutions, all the pheromone matrices are updated taking into account the respective objective values. Finally, after T_{max} iterations, the achieved set of non-dominated solutions is returned.

3.6. Promising Research Areas

There are several topics within this area that we believe that have a high potential for future research. The main ones are the following:

- **Use in continuous optimization problems:** In contrast with
  the several ACO variants that have been recently proposed for continuous single-objective search spaces (see for example Refs. 15,
  16, 18), such proposals are very scarce within multi-objective optimization. Clearly, the development of multi-objective extensions
  of continuous variants of ACO could bring a variety of novel applications. This could potentially place ACO next to other metaheuristics which are very popular for continuous multi-objective
  optimization (e.g., particle swarm optimization).
- **Parallelization:** ACO algorithms are, by definition, highly distributed algorithms in which a set of ants are in charge of independently building a solution. Thus, their parallelization is relatively
  straightforward, but such a feature has not been yet properly exploited in a multi-objective optimization context. Furthermore,
  ACO algorithms are flexible enough to allow the addition of many
  components that can be combined in many different ways in order
  to obtain improved ACO-based multi-objective optimizers (e.g.,
  several pheromone matrices, multiple colonies, heuristic information (when available), Pareto archives, etc). Again, the experience
  acquired so far regarding parallel MOEAs[6,61,62] could be very valuable here.
- **Hybridization:** In the last few years, the hybridization of different
  types of metaheuristics has become a relatively popular scheme
  within multi-objective optimization. For example, some MOEAs
  have been hybridized with powerful local search engines, giving rise
  to the so-called multi-objective memetic algorithms.[63] MOACOs
  also have a great hybridization potential, not only with local search
  engines, but also with other metaheuristics such as particle swarm
  optimization and genetic algorithms. Such hybrids are expected to
  become more common during the next few years.

3.7. Conclusions

In this chapter we have presented an overview of the use of the Ant Colony
Optimization metaheuristic for solving multi-objective optimization problems. The chapter has provided some introductory concepts about the ACO
metaheuristic and about multi-objective optimization. The most relevant
features of some recent proposals within the area were briefly reviewed,
too. Based on the existing MOACOs and a particular taxonomy proposed

in the literature, we presented a refined taxonomy of MOACO algorithms that could help practitioners to better identify and exploit the potential of the ACO metaheuristic as a multi-objective optimizer. Nevertheless, we hope that this chapter can be found useful not only for practitioners, but also for students and researchers interested in multi-objective optimization using ACO algorithms, for such has been its main purpose.

Acknowledgments

The first author acknowledges the support from the UMI-LAFMIA 3175 CNRS at CINVESTAV-IPN and from the Universidad Nacional de San Luis, Argentina. The second author gratefully acknowledges support from CONACyT project no. 103570.

References

1. R. C. Eberhart, Y. Shi, and J. Kennedy, *Swarm Intelligence*. The Morgan Kaufmann Series in Artificial Intelligence, (Morgan Kaufmann, April 2001), first edition.
2. E. Bonabeau, M. Dorigo, and G. Theraulaz, *Swarm Intelligence - From Natural to Artificial Systems*. Santa Fe Institute Studies in the Sciences of Complexity, (Oxford University Press, 1999).
3. M. Dorigo and T. Stützle, *Ant Colony Optimization*. (Mit-Press, 2004).
4. C. García-Martínez, O. Cordón, and F. Herrera, A taxonomy and an empirical analysis of multiple objective ant colony optimization algorithms for the bi-criteria TSP, *European Journal of Operational Research*. **180**, 116–148, (2007).
5. D. Angus and C. Woodward, Multiple objective ant colony optimisation, *Swarm Intelligence*. **3**(1), 69–85, (2009).
6. C. A. Coello Coello, D. A. Van Veldhuizen, and G. B. Lamont, *Evolutionary Algorithms for Solving Multi-Objective Problems*. (Springer, September 2002), second edition. ISBN 0-387-33254-3.
7. M. Dorigo, V. Maniezzo, and A. Colorni, Ant system: Optimization by a colony of cooperating agents, *IEEE Trans. on Systems, Man, and Cybernetics–Part B*. **26**(1), 29–41, (1996).
8. G. Bilchev and I. Parmee. The Ant Colony Metaphor for Searching Continuous Design Spaces. In ed. T. C. Fogarty, *Evolutionary Computing. AISB Workshop*, pp. 25–39. Springer, Sheffield, UK (April, 1995).
9. N. Monmarché, G. Venturini, and M. Slimane, On how pachycondyla apicalis ants suggest a new search algoritm, *Future Generation Computer Systems*. **16**, 937–946, (2000).
10. J. Li and N. Satofuka, Optimization design of a compressor cascade airfoil using a Navier-stokes solver and genetic algorithms, *Proceedings of the In-*

stitution of Mechanical Engineering Part A—Journal of Power and Energy. **216**(A2), 195–202, (2002).

11. J. Dréo and P. Siarry. A New Ant Colony Algorithm Using the Heterarchical Concept Aimed at Optimization of Multiminima Continuous Functions. In eds. M. Dorigo, G. Di Caro, and M. Sampels, *Proceedings of the Third international Workshop on Ant Algorithms - ANTS 2002*, pp. 216–221. Springer-Verlag. Lecture Notes in Computer Science Vol. 2463, Brussels, Belgium (September, 2002).

12. J. Dréo and P. Siarry, Continuous interacting ant colony algorithm based on dense heterarchy, *Future Generation Comp. Syst.* **20**(5), 841–856, (2004).

13. L. Q. Ling Chen, J. Shen and H. Chen, An improved ant colony algorithm in continuous optimization, *Journal of Systems Science and Systems Engineering.* **12**(2), 224–235, (2003).

14. S. Pourtakdoust and H. Nobahari. An Extension of Ant Colony Systems to Continuos Optimization Problems. In eds. M. Dorigo, M. Birattari, C. Blum, L. M. Gambardella, F. Mondada, and T. Stützle, *Proceedings of Ant Colony Optimization and Swarm Intelligence, 4th International Workshop, ANTS Workshop 2004*, pp. 294–301. Lecture Notes in Computer Science Vol. 3172, Brussels, Belgium, (2004). Springer-Verlag.

15. M. Kong and P. Tian. A direct application of ant colony optimization to function optimization problem in continuous domain. In *ANTS Workshop*, pp. 324–331, (2006).

16. X. Hu, J. Zhang, and Y. Li, Orthogonal methods based ant colony search for solving continuous optimization problems, *J. Comput. Sci. Technol.* **23**(1), 2–18, (2008).

17. K. Socha. ACO for continuos and mixed-variable optimization. In eds. M. Dorigo, M. Birattari, C. Blum, L. M. Gambardella, F. Mondada, and T. Stützle, *Proceedings of Ant Colony Optimization and Swarm Intelligence, 4th International Workshop, ANTS Workshop 2004*, pp. 25–36, Brussels, Belgium, (2004). Springer-Verlag. Lecture Notes in Computer Science Vol. 3172.

18. K. Socha and M. Dorigo, Ant colony optimization for continuous domains, *European Journal of Operational Research.* **185**(3), 1155–1173, (2008).

19. E. Zitzler, M. Laumanns, and L. Thiele. SPEA2: Improving the Strength Pareto Evolutionary Algorithm. In eds. K. Giannakoglou, D. Tsahalis, J. Periaux, P. Papailou, and T. Fogarty, *EUROGEN 2001. Evolutionary Methods for Design, Optimization and Control with Applications to Industrial Problems*, pp. 95–100, Athens, Greece, (2001).

20. K. Deb, A. Pratap, S. Agarwal, and T. Meyarivan, A Fast and Elitist Multi-objective Genetic Algorithm: NSGA–II, *IEEE Transactions on Evolutionary Computation.* **6**(2), 182–197 (April, 2002).

21. D. Angus. Population-based ant colony optimisation for multi-objective function optimisation. In eds. M. Randall, H. A. Abbass, and J. Wiles, *ACAL*, vol. 4828, *Lecture Notes in Computer Science*, pp. 232–244. Springer, (2007). ISBN 978-3-540-76930-9.

22. D. Angus. Crowding Population-based Ant Colony Optimization for the Multi-objective Travelling Salesman Problem. In *Proceedings of the 2007 IEEE Symposium on Computational Intelligence in Multicriteria Decision Making (MCDM'2007)*, pp. 333–340, Honolulu, Hawaii, USA (April, 2007). IEEE Press.

23. D. A. Van Veldhuizen. *Multiobjective Evolutionary Algorithms: Classifications, Analyses, and New Innovations.* PhD thesis, Department of Electrical and Computer Engineering. Graduate School of Engineering. Air Force Institute of Technology, Wright-Patterson AFB, Ohio (May, 1999).

24. J. Knowles. A summary-attainment-surface plotting method for visualizing the performance of stochastic multiobjective optimizers. In *Fifth International Conference on Intelligent Systems Design and Applications (ISDA'2005)*, pp. 552–557. IEEE, (2005).

25. A. Garcia-Najera and J. A. Bullinaria. Extending ACO_R to Solve Multi-Objective Problems. In ed. G. M. Coghill, *Proceedings of the UK Workshop on Computational Intelligence (UKCI 2007)*, London, UK, (2007). Imperial College United Kingdom.

26. C. A. Coello Coello, G. Toscano Pulido, and M. Salazar Lechuga, Handling Multiple Objectives With Particle Swarm Optimization, *IEEE Transactions on Evolutionary Computation.* **8**(3), 256–279 (June, 2004).

27. N. Liu, B. Huang, and X. H. Pan, Using the Ant Algorithm to Derive Pareto Fronts for Multiobjective Siting of Emergency Service Facilities, *Transportation Research Record: Journal of the Transportation Research Board.* **1935**, 120–129, (2005).

28. D. E. Goldberg, *Genetic Algorithms in Search, Optimization and Machine Learning.* (Addison-Wesley Publishing Company, Reading, Massachusetts, 1989).

29. L. T. Bui, J. M. Whitacre, and H. A. Abbass. Performance Analysis of Elitism in Multi-Objective Ant Colony Optimization Algorithms. In *2008 Congress on Evolutionary Computation (CEC'2008)*, pp. 1633–1640, Hong Kong (June, 2008). IEEE Service Center.

30. X. Benlian and W. Zhiquan, A multi-objective-ACO-based data association method for bearings-only multi-target tracking, *Communications in Nonlinear Science and Numerical Simulation.* **12**(8), 1360–1369, (2007).

31. A. M. Mora, J. J. M. Guervós, C. Millán, J. Torrecillas, J. L. J. Laredo, and P. A. C. Valdivieso. Comparing ACO Algorithms for Solving the Bi-criteria Military Path-Finding Problem. In eds. F. A. e Costa, L. M. Rocha, E. Costa, I. Harvey, and A. Coutinho, *Advances in Artificial Life. 9th European Conference (ECAL'2007)*, pp. 665–674. Springer, Lecture Notes in Computer Science, Vol. 4648, Lisbon, Portugal (September 10-14, 2007). ISBN 978-3-540-74912-7.

32. A. Mora, J. Merelo, C. Millan, J. Torrecillas, and J. Laredo. CHAC. A MOACO Algorithm for Computation of Bi-Criteria Military Unit Path in the Battlefield. In eds. D. Pelta and N. Krasnogor, *Proceedings of the First Workshop in Nature Inspired Cooperative Strategies for Optimization (NICSO'06)*, pp. 85–96, Granada, Spain (June, 2006).

33. A. M. Mora, J. J. M. Guervós, C. Millán, J. Torrecillas, J. L. J. Laredo, and P. A. C. Valdivieso. Enhancing a moaco for solving the bi-criteria pathfinding problem for a military unit in a realistic battlefield. In eds. M. Giacobini, A. Brabazon, S. Cagnoni, G. D. Caro, R. Drechsler, M. Farooq, A. Fink, E. Lutton, P. Machado, S. Minner, M. O'Neill, J. Romero, F. Rothlauf, G. Squillero, H. Takagi, S. Uyar, and S. Yang, *EvoWorkshops*, vol. 4448, *Lecture Notes in Computer Science*, pp. 712–721. Springer, (2007). ISBN 978-3-540-71804-8.

34. B. Barán and M. Schaerer. A Multiobjective Ant Colony System for Vehicle Routing Problem with Time Windows. In *Proceedings of the 21st IASTED International Conference on Applied Informatics*, pp. 97–102, Innsbruck, Austria (February, 2003). IASTED.

35. P. R. McMullen and P. Tarasewich, Multi-objective assembly line balancing via a modified ant colony optimization technique, *International Journal of Production Research*. **44**, 27–42, (2006).

36. L.-N. Xing, Y.-W. Chen, and K.-W. Yang. Interactive fuzzy multi-objective ant colony optimization with linguistically quantified decision functions for flexible job shop scheduling problems. In *FBIT '07: Proceedings of the 2007 Frontiers in the Convergence of Bioscience and Information Technologies*, pp. 801–806, Washington, DC, USA, (2007). IEEE Computer Society. ISBN 978-0-7695-2999-8. doi: http://dx.doi.org/10.1109/FBIT.2007.18.

37. M. Guntsch and M. Middendorf. A Population Based Approach for ACO. In *Applications of Evolutionary Computing. EvoWorkshops 2002: EvoCOP, EvoIASP, EvoSTIM/EvoPLAN*, pp. 72–81, Kinsale, Ireland (April, 2002). Springer. Lecture Notes in Computer Science Vol. 2279.

38. I. Alaya, C. Solnon, and K. Ghédira. Ant Colony Optimization for Multi-objective Optimization Problems. In *Proceedings of the 19th IEEE International Conference on Tools with Artificial Intelligence (ICTAI 2007)*, vol. 1, pp. 450–457. IEEE Computer Society Press (October, 2007).

39. C. M. Fonseca and P. J. Fleming. Genetic Algorithms for Multiobjective Optimization: Formulation, Discussion and Generalization. In ed. S. Forrest, *Proceedings of the Fifth International Conference on Genetic Algorithms*, pp. 416–423, San Mateo, California, (1993). University of Illinois at Urbana-Champaign, Morgan Kaufmann Publishers.

40. N. Srinivas and K. Deb, Multiobjective Optimization Using Nondominated Sorting in Genetic Algorithms, *Evolutionary Computation*. **2**(3), 221–248 (fall, 1994).

41. E. Zitzler and L. Thiele, Multiobjective Evolutionary Algorithms: A Comparative Case Study and the Strength Pareto Approach, *IEEE Transactions on Evolutionary Computation*. **3**(4), 257–271 (November, 1999).

42. P. Hajela and C. Y. Lin, Genetic search strategies in multicriterion optimal design, *Structural Optimization*. **4**, 99–107, (1992).

43. J. D. Schaffer. Multiple Objective Optimization with Vector Evaluated Genetic Algorithms. In *Genetic Algorithms and their Applications: Proceedings of the First International Conference on Genetic Algorithms*, pp. 93–100, Hillsdale, New Jersey, (1985). Lawrence Erlbaum.

44. E. Zitzler. *Evolutionary Algorithms for Multiobjective Optimization: Methods and Applications.* PhD thesis, Swiss Federal Institute of Technology (ETH), Zurich, Switzerland (November, 1999).

45. A. Afshar, F. Sharifi, and M. Jalali, Non-dominated archiving multi-colony ant algorithm for multi-objective optimization: Application to multi-purpose reservoir operation, *Engineering Optimization.* **41**(4), 313–325 (April, 2009).

46. E. Zitzler, K. Deb, and L. Thiele, Comparison of Multiobjective Evolutionary Algorithms: Empirical Results, *Evolutionary Computation.* **8**(2), 173–195 (Summer, 2000).

47. J. D. Knowles and D. W. Corne, Approximating the Nondominated Front Using the Pareto Archived Evolution Strategy, *Evolutionary Computation.* **8** (2), 149–172, (2000).

48. S. Eppe. Integrating the decision maker's preferences into multi objective ant colony optimization. In eds. F. Hutter and M. M. de Oca, *2nd Doctoral Symposium on Engineering Stochastic Local Search Algorithms, SLS 2009*, pp. 56–60 (August, 2009).

49. J.-P. Brans and B. Mareschal. PROMETHEE methods. In eds. J. Figueira, S. Greco, and M. Ehrgott, *Multiple Criteria Decision Analysis. State of the Art Surveys*, pp. 163–195. Springer, New York, USA, (2005).

50. M. Chica, Óscar Cordón, S. Damas, J. Pereira, and J. Bautista. Incorporating Preferences to a Multi-objective Ant Colony Algorithm for Time and Space Assembly Line Balancing. In eds. M. Dorigo, M. Birattari, C. Blum, M. Clerc, T. Stützle, and A. F. Winfield, *Ant Colony Optimization and Swarm Intelligence. 6th International Conference, ANTS 2008. Proceedings*, pp. 331–338. Springer, Brussels, Belgium (September, 2008).

51. M. Chica, O. Cordón, S. Damas, J. Bautista, and J. Pereira. Multi-objective, constructive heuristics for the 1/3 variant of the time and space assembly line balancing problem: Aco and randomised greedy. Technical Report AFE-08-01, European Centre for Soft Computing, Asturias (Spain), (2008).

52. M. Chica, Óscar Cordón, S. Damas, and J. Bautista. Integration of an EMO-based Preference Elicitation Scheme into a Multi-objective ACO Algorithm for Time and Space Assembly Line Balancing. In *2009 IEEE Symposium on Computational Intelligence in Multi-Criteria Decision-Making (MCDM'2009)*, pp. 157–162, Nashville, TN, USA (March 30 - April 2, 2009). IEEE Press. ISBN 978-1-4244-2764-2.

53. K. Deb. Solving Goal Programming Problems Using Multi-Objective Genetic Algorithms. In *1999 Congress on Evolutionary Computation*, pp. 77–84, Washington, D.C. (July, 1999). IEEE Service Center.

54. J. Branke and K. Deb. Integrating User Preferences into Evolutionary Multi-Objective Optimization. In ed. Y. Jin, *Knowledge Incorporation in Evolutionary Computation*, pp. 461–477. Springer, Berlin Heidelberg, (2005). ISBN 3-540-22902-7.

55. S. Häckel, M. Fischer, D. Zechel, and T. Teich. A Multi-Objective Ant Colony Approach for Pareto-Optimization Using Dynamic Programming. In *2008 Genetic and Evolutionary Computation Conference (GECCO'2008)*, pp. 33–40, Atlanta, USA (July, 2008). ACM Press. ISBN 978-1-60558-131-6.

56. S. K. Chaharsooghi and A. H. M. Kermani, An effective ant colony optimization algorithm (ACO) for multi-objective resource allocation problem (MORAP), *Applied Mathematics and Computation.* **200**(1), 167–177 (June 15, 2008).

57. S. K. Chaharsooghi and A. H. M. Kermani. An Intelligent Multi-Colony Multi-Objective Ant Colony Optimization (ACO) for the 0-1 Knapsack Problem. In *2008 Congress on Evolutionary Computation (CEC'2008)*, pp. 1195–1202, Hong Kong (June, 2008). IEEE Service Center.

58. S. M. Vieira, J. ao M. C. Sousa, and T. A. Runkler. Multi-Criteria Ant Feature Selection Using Fuzzy Classifiers. In eds. C. A. Coello Coello, S. Dehuri, and S. Ghosh, *Swarm Intelligence for Multi-objective Problems in Data Mining*, chapter 2, pp. 19–36. Springer. Studies in Computational Intelligence. Vol. 242, Berlin, (2009).

59. A. Asuncion and D. Newman. UCI machine learning repository, (2007). URL http://www.ics.uci.edu/$\sim$mlearn/{MLR}epository.html.

60. Y. Yang, G. Wu, J. Chen, and W. Dai, Multi-objective optimization based on ant colony optimization in grid over optical burst switching networks, *Expert Systems with Applications.* **37**(2), 1769–1775 (March, 2010).

61. D. A. Van Veldhuizen, J. B. Zydallis, and G. B. Lamont, Considerations in Engineering Parallel Multiobjective Evolutionary Algorithms, *IEEE Transactions on Evolutionary Computation.* **7**(2), 144–173 (April, 2003).

62. A. Nebro, F. Luna, E.-G. Talbi, and E. Alba. Parallel Multiobjective Optimization. In ed. E. Alba, *Parallel Metaheuristics*, pp. 371–394. Wiley-Interscience, New Jersey, USA, (2005). ISBN 13-978-0-471-67806-9.

63. C.-K. Goh, Y.-S. Ong, and K. C. Tan, Eds., *Multi-Objective Memetic Algorithms.* (Springer, Berlin, Germany, 2009). ISBN 978-3-540-88050-9.

Chapter 4

Recurrent Neural Networks with Discontinuous Activation Functions for Convex Optimization

Qingshan Liu and Jun Wang

School of Automation, Southeast University,
Nanjing 210096, China,
qsliu@seu.edu.cn

Department of Mechanical and Automation Engineering,
The Chinese University of Hong Kong,
Shatin, New Territories, Hong Kong,
jwang@mae.cuhk.edu.hk

Over the past two decades, recurrent neural networks have been shown to be very powerful computational models for optimization. A number of recurrent neural network models have been designed for solving various constrained optimization problems. Based on the Karush-Kuhn-Tucker optimality conditions and non-smooth functions, recurrent neural networks with discontinuous activation functions are designed for constrained optimization. The present chapter presents several recurrent neural networks with discontinuous activation functions for solving linear and quadratic programming problems and a class of non-smooth optimization problems. Moreover, two k-winners-take-all networks are designed based on the presented neural networks.

4.1. Introduction

Optimization problems arise in numerous of science and engineering applications, such as robot control, manufacturing system design, signal and image processing, and pattern recognition.[1,2] Constrained optimization is concerned with optimizing an objective function subject to a set of constraints, which defined the feasible set. Usually, if the feasible set is convex and the objective function is convex on the feasible set, we call the optimization problem is a convex one.

Over the years, a variety of numerical algorithms have been developed for solving linear and nonlinear programming problems, such as the simplex methods for linear programming, active set methods, and interior point methods for nonlinear optimization.[2] Recently, the swarm intelligence techniques have been developed for the global optimization,[3] such as the genetic algorithm (GA), ant colony optimization (ACO), particle swarm optimization (PSO), etc. However, reported results of numerous investigations have shown many advantages over the traditional optimization algorithms, especially in real-time applications. For such real-time applications, recurrent neural networks based on hardware implementation are more competent.[4,5]

In the past two decades, recurrent neural networks for optimization and their engineering applications have been widely investigated. In 1986, Tank and Hopfield[4] applied the Hopfield network for solving linear programming problems, which motivated the development of neural networks for optimization.[5–17] Kennedy and Chua[5] presented a neural network with a finite penalty parameter for nonlinear programming which can converge to approximate optimal solutions. Zhang and Constantinides[18] proposed the Lagrangian network based on the Lagrangian method which had a two-layers structure and this neural network was globally convergent to an optimal solution if only the objective function was strictly convex. Wang[9,10] developed the deterministic annealing network for linear and nonlinear convex programming. Xia[19] proposed some primal neural networks for solving convex quadratic programming problems. The primal-dual neural networks[20,21] with two-layers structure were proposed for solving linear and nonlinear programming problems. The dual neural networks[22,23] as simplified forms of the primal-dual neural networks were presented to solve convex quadratic programming problems utilizing only the dual variables. In order to simplify the architecture of the dual neural network, a simplified dual neural network was proposed for solving quadratic programming problems.[24] Based on the projection method,[25] the projection neural network was proposed for solving general convex programming problems[26,27] which was globally convergent to exact optimal solutions. Recently, Forti, Nistri and Quincampoix[15] proposed a generalized neural network for solving non-smooth nonlinear programming problems based on the gradient and penalty parameter methods. The delayed neural networks[28–30] were proposed for solving convex quadratic programming problems. More recently, we developed some one-layer recurrent neural networks[16,17] for solving linear and quadratic programming problems. The one-layer recurrent neural networks, which number of neurons is equal to that of decision variables in the programming problems, have more simply architecture complexity than the other neural networks such as Lagrangian network and primal-dual networks.

4.2. Related Definitions and Lemmas

Definition 4.1.[31] Suppose $E \subset \mathbb{R}^n$. $F : x \mapsto F(x)$ is called a set-valued function from $E \hookrightarrow \mathbb{R}^n$, if to each point x of a set E, there corresponds to a nonempty closed set $F(x) \subset \mathbb{R}^n$.

Definition 4.2.[32] A function $\varphi : \mathbb{R}^n \to \mathbb{R}$ is said to be Lipschitz near $x \in \mathbb{R}^n$ if there exist $\varepsilon, \delta > 0$, such that for any $y, z \in \mathbb{R}^n$ satisfy $\|y - x\| < \delta$ and $\|z - x\| < \delta$, we have $|\varphi(y) - \varphi(z)| \le \varepsilon \|y - z\|$, where $\| \cdot \|$ denotes the Euclidean norm. If φ is locally Lipschitz near any point $x \in \mathbb{R}^n$, φ is also said to be locally Lipschitz in $\mathbb{R}^n$.

Definition 4.3. Assume that φ is Lipschitz near x. The generalized directional derivative of φ at x in the direction $v \in \mathbb{R}^n$ is defined as

$$\varphi^0(x, v) = \limsup_{\substack{y \to x \\ \xi \to 0^+}} \frac{\varphi(y + \xi v) - \varphi(y)}{\xi}.$$

The Clarke's generalized gradient of φ is defined as

$$\partial \varphi(x) = \{y \in \mathbb{R}^n : \varphi^0(x, v) \ge y^T v, \forall v \in \mathbb{R}^n\}.$$

As a special case, the derivative of φ at x from the right in the direction v is defined as

$$D\varphi(x, v) = \lim_{\xi \to 0^+} \frac{\varphi(x + \xi v) - \varphi(x)}{\xi}.$$

If $D\varphi(x, v)$ exists for all directions, we say that φ is differentiable from the right at x. We say that the closed convex subset (possibly empty)

$$\partial \varphi(x) = \{\xi \in \mathbb{R}^n : \forall v \in \mathbb{R}^n, \xi^T v \le D\varphi(x, v)\}$$

is the sub-differential of φ at x. The element ξ of $\partial \varphi(x)$ is called the sub-gradient of φ at x. If φ is convex, the sub-differential is equivalent to the Clarke's generalized gradient.

If φ is locally Lipschitz in $\mathbb{R}^n$ and differentiable for almost all (a.a.) $x \in \mathbb{R}^n$ (in the sense of Lebesgue measure), then the Clarke's generalized gradient of φ at $x \in \mathbb{R}^n$ is equivalent to

$$\partial \varphi(x) = K\{\lim_{n \to \infty} \nabla \varphi(x_n) : x_n \to x, x_n \notin \Omega_\varphi \cup \mathcal{N}\},$$

where $K(\cdot)$ denotes the closure of the convex hull of the corresponding set, $\mathcal{N} \subset \mathbb{R}^n$ is an arbitrary set with Lebesgue measure zero, and $\Omega_\varphi \subset \mathbb{R}^n$ is the set of points where φ is not differentiable.

To tackle an optimization problem by using a recurrent neural network, the key lie in the construction of its dynamic equations such that the equilibrium points correspond to the desired optimal solutions. In general, a neural network can be described by the following dynamical equation governed by

$$\frac{dx}{dt} = \psi(x), \quad x(t_0) = x_0. \tag{4.1}$$

If $\psi(x)$ is discontinuous on $\mathbb{R}^n$, it is necessary to explain what is meant by a solution of (4.1). A possible definition is that of Filippov.[31]

Definition 4.4. A set-valued map: $\phi : \mathbb{R}^n \hookrightarrow \mathbb{R}^n$ is defined as

$$\phi(x) = \bigcap_{\delta > 0} \bigcap_{\mu(\mathcal{N})=0} K[\psi(\mathcal{B}(x, \delta) - \mathcal{N})],$$

where $\mathcal{N}$ is an arbitrary set with measure zero, $\mu(\mathcal{N})$ is the Lebesgue measure of set $\mathcal{N}$, and $\mathcal{B}(x, \delta) = \{y \in \mathbb{R}^n : \|y - x\| \leq \delta\}$. A Filippov solution of (4.1) is an absolutely continuous function $x(t)$ defined on an interval $[t_0, t_1](t_0 \leq t_1 \leq +\infty)$, which satisfies $x(t_0) = x_0$ and differential inclusion:

$$\frac{dx}{dt} \in \phi(x), \quad \text{a.a. } t \in [t_0, t_1].$$

Definition 4.5. $\bar{x}$ is said to be an equilibrium point of system (4.1) if

$$0 \in \phi(\bar{x}). \tag{4.2}$$

Definition 4.6. The neural network (4.1) is said to be globally convergent to an equilibrium point if for any trajectory $x(t)$ of the neural network with initial point $x(t_0) \in \mathbb{R}^n$, there exists an equilibrium point $\bar{x}$ such that $\lim_{t \to +\infty} x(t) = \bar{x}$.

4.3. For Linear Programming

Consider the following linear programming (LP) problem as:

$$\begin{aligned}
\text{minimize} \quad & c^T x, \\
\text{subject to} \quad & Ax = b, \\
& l \leq x \leq u,
\end{aligned} \tag{4.3}$$

where $x \in \mathbb{R}^n$ is the vector of decision variables; $A \in \mathbb{R}^{m \times n}$ is a full row-rank matrix (i.e., $\text{rank}(A) = m$, $m < n$); $c, l, u \in \mathbb{R}^n$ and $b \in \mathbb{R}^m$. Here, if $l = -\infty$ or $h = \infty$, the inequality constraints are one-side.

4.3.1. *Model Description and Convergence Results*

According to the Karush-Kuhn-Tucker (KKT) conditions,[2] x^* is an optimal solution of (4.3), if and only if there exist $y^* \in \mathbb{R}^m$ and $z^* \in \mathbb{R}^n$ such that $(x^*, y^*, z^*)^T$ satisfies the following optimality conditions:

$$c + A^T y + z = 0, \tag{4.4}$$

$$Ax = b, \tag{4.5}$$

$$\begin{cases} z_i \geq 0, & \text{if } x_i = u_i, \\ z_i = 0, & \text{if } x_i \in (l_i, u_i), \\ z_i \leq 0, & \text{if } x_i = l_i. \end{cases} \tag{4.6}$$

From (4.4), we have

$$x = x - c - A^T y - z. \tag{4.7}$$

Substituting (4.7) into (4.5), we have

$$Ax - Ac - AA^T y - Az = b.$$

Because A is full row-rank, AA^T is invertible. Then

$$y = (AA^T)^{-1}(Ax - Az - Ac - b). \tag{4.8}$$

Substituting (4.8) into (4.4), we have

$$c + A^T (AA^T)^{-1}(Ax - Az - Ac - b) + z = 0. \tag{4.9}$$

Let $P = A^T (AA^T)^{-1} A$ and $q = -(I - P)c + A^T (AA^T)^{-1} b$. Then (4.9) can be written as

$$Px + (I - P)z - q = 0, \tag{4.10}$$

where I is an identity matrix. The matrix P, called the projection matrix, has the following properties by simple calculation.

Lemma 4.1.

 (*i*) *P is symmetric, $P^2 = P$ and $(I - P)^2 = I - P$;*
 (*ii*) *For any $\zeta > -1$, $I + \zeta P$ is invertible and $(I + \zeta P)(I - P) = I - P$.*

Based on above analysis, a recurrent neural network is proposed to solve (4.3) with its dynamical equation governed by

$$\frac{dx}{dt} = \gamma\{-Px - \sigma(I - P)g(x) + q\}, \tag{4.11}$$

where γ is a positive scaling constant, σ is a nonnegative gain parameter, and $g(x) = (g_1(x_1), g_2(x_2), \ldots, g_n(x_n))^T$ with $g_i(x_i)$ being the following discontinuous activation function

$$g_i(x_i) = \begin{cases} 1, & \text{if } x_i > u_i, \\ [0, 1], & \text{if } x_i = u_i, \\ 0, & \text{if } x_i \in (l_i, u_i), \quad i = 1, 2, \ldots, n, \\ [-1, 0], & \text{if } x_i = l_i, \\ -1, & \text{if } x_i < l_i. \end{cases} \tag{4.12}$$

The following lemma gives a necessary condition for the existence of equilibrium point of the neural network in (4.11)

Lemma 4.2.[16] *If there exists an equilibrium point of neural network* (4.11), *then*

$$\sigma \geq \begin{cases} 0, & \text{if } (I - P)c = 0; \\ \dfrac{c^T(I - P)c}{\|(I - P)c\|_1}, & \text{if } (I - P)c \neq 0, \end{cases}$$

where $\| \cdot \|_1$ *denotes the* l_1*-norm of the corresponding vector in* $\mathbb{R}^n$.

In general, the equilibrium point of neural network (4.11) may not always be an optimal solution of problem (4.3). However, the following lemma reveals the relationship between the optimal solution of problem (4.3) and the equilibrium point of neural network (4.11).

Lemma 4.3.[16]

(i) *Any optimal solution of problem* (4.3), *denoted as* x^*, *is an equilibrium point of neural network* (4.11) *if* $\sigma \geq 0$ *when* $(I - P)c = 0$, *or one of the following conditions holds when* $(I - P)c \neq 0$:

   (a) $\sigma \geq \|(I - P)c\| / \min_{\xi \in X}^+ \|(I - P)\xi\|$, *or*
   (b) $\sigma \geq c^T(I - P)c / \min_{\xi \in X}^+ \{|c^T(I - P)\xi|\}$,

   where $\min_{\xi \in X}^+ \{c^T(I - P)\xi\}$ *denotes the minimum positive value of the corresponding function and* $X = \{\xi = (\xi_1, \ldots \xi_n)^T \in \mathbb{R}^n : \xi_i = -1, 0 \text{ or } 1, i = 1, 2, \ldots, n\}$.

(ii) *Any equilibrium point of neural network (4.11), denoted by $\bar{x}$, is an optimal solution of problem (4.3), if one of the following conditions holds:*

(a) $\bar{x} \in \Omega = \{x \in \mathbb{R}^n : l \le x \le u\}$, *or*
(b) $\bar{x}$ *is unique and σ satisfies the conditions in (i).*

Theorem 4.1. *If the equilibrium point set of neural network (4.11) is not empty, the neural network is globally convergent to an equilibrium point.*

From Lemma 4.3 and Theorem 4.1, the optimal solutions of problem (4.3) can be guaranteed by the following two corollaries.

Corollary 4.1. *The neural network (4.11) is globally convergent to an optimal solution of problem (4.3), if the equilibrium point set of (4.11) is included in the set Ω defined in Lemma 4.3(ii).*

Corollary 4.2. *The neural network (4.11) is globally convergent to an optimal solution of problem (4.3), if it has a unique equilibrium point and $\sigma \ge 0$ when $(I - P)c = 0$ or one of the following conditions holds when $(I - P)c \ne 0$:*

(i) $\xi \ge \|(I - P)c\| / \min^+_{\xi \in X} \|(I - P)\xi\|$, *or*
(ii) $\sigma \ge c^T(I - P)c / \min^+_{\xi \in X}\{|c^T(I - P)\xi|\}$,

where X is defined in Lemma 4.3(i).

4.3.2. *Simulation Results*

Example 4.1. Consider the following linear programming problem:

$$
\begin{aligned}
\text{minimize} \quad & 3x_1 - x_2 + 2x_3 - 4x_4, \\
\text{subject to} \quad & x_1 - 2x_2 - x_3 + x_4 = 4, \\
& x_1 + x_2 + 2x_3 - 3x_4 = 2, \\
& -2 \le x_1, x_2, x_3, x_4 \le 2.
\end{aligned}
$$

Next, the neural network (4.11) is utilized to solve this problem. By Lemma 4.2, a necessary condition for existence of the equilibrium point of (4.11) is $\sigma \ge 0.1112$. According to the conditions in Corollary 4.2, to get the optimal solution, the lower bound of σ is 1. Simulation results are depicted in Fig. 4.1 with $\gamma = 1$ and four different values of σ. It shows that the state trajectories of the neural network globally converge to the unique optimal solution $x^* = (2, -1.6, 2, 0.8)^T$ from 10 random initial points when $\sigma = 0.6, 1.2$ and 4 as shown in Fig. 4.1. However, if $\sigma = 0.2$, the simulation

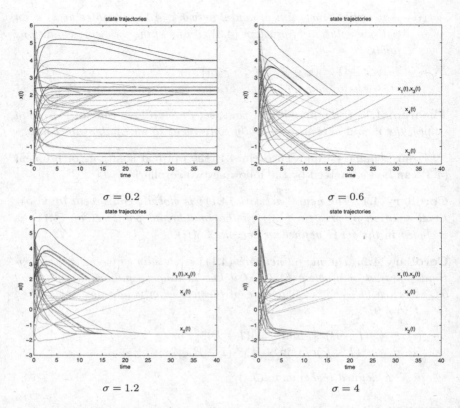

$$\sigma = 0.2 \qquad\qquad\qquad \sigma = 0.6$$

$$\sigma = 1.2 \qquad\qquad\qquad \sigma = 4$$

Fig. 4.1.: Transient behaviors of the neural network (4.11) with four different values of σ in Example 4.1.

results in Fig. 4.1 show that the neural network may not converge to an optimal solution.

4.4. For Quadratic Programming

In this section, we are concerned with a quadratic programming (QP) problem as follows:

$$
\begin{aligned}
\text{minimize} \quad & f(x) = \frac{1}{2}x^T Q x + c^T x, \\
\text{subject to} \quad & Ax = b, \\
& l \le x \le u,
\end{aligned}
\tag{4.13}
$$

where $x \in \mathbb{R}^n$ is the vector of decision variables, $Q \in \mathbb{R}^{n \times n}$ is symmetric but not necessarily positive semidefinite, A, b, c, l and u are defined in (4.3).

4.4.1. *Model Description*

According to the Karush-Kuhn-Tucker (KKT) conditions,[2] if x^* is a optimal solution of (4.13), then there exist $y^* \in \mathbb{R}^m$ and $z^* \in \mathbb{R}^n$ such that $(x^*, y^*, z^*)^T$ satisfies the following optimality conditions:

$$Qx + c + A^T y + z = 0, \tag{4.14}$$

$$Ax = b, \tag{4.15}$$

$$\begin{cases} x_i = u_i, & \text{if } z_i > 0, \\ l_i \leq x_i \leq u_i, & \text{if } z_i = 0, \\ x_i = l_i, & \text{if } z_i < 0. \end{cases} \tag{4.16}$$

If (4.13) is a convex programming problem, the KKT conditions (4.14)-(4.16) are both necessary and sufficient.

From (4.14), we have

$$x = (I - Q)x - c - A^T y - z. \tag{4.17}$$

Substituting (4.17) into (4.15), we have

$$A(I - Q)x - Ac - AA^T y - Az = b.$$

Because A is full row-rank, AA^T is invertible. Then

$$y = (AA^T)^{-1}(A(I - Q)x - Az - Ac - b). \tag{4.18}$$

Substituting (4.18) into (4.14), we have

$$Qx + c + A^T(AA^T)^{-1}(A(I - Q)x - Az - Ac - b) + z = 0. \tag{4.19}$$

Let $P = A^T(AA^T)^{-1}A$ and $s = -c + Pc + A^T(AA^T)^{-1}b$, then, (4.19) can be written as

$$(I - P)z + [(I - P)Q + P]x - s = 0. \tag{4.20}$$

According to Lemma 4.1, $I + \zeta P$ is invertible as $\zeta > -1$. By multiplying $I + \zeta P$ in both sides of (4.20), it follows that

$$(I - P)z + [(I - P)Q + \alpha P]x - q = 0, \tag{4.21}$$

where $\alpha = 1 + \zeta > 0$ is a parameter and $q = (I + \zeta P)s = -c + Pc + \alpha A^T(AA^T)^{-1}b$.

Based on (4.21), the proposed recurrent neural network model for solving (4.13) is described as follows:

- State equation

$$\frac{dz}{dt} = \gamma\{-(I - P)z - [(I - P)Q + \alpha P]h(z) + q\}, \qquad (4.22)$$

- Output equation

$$x = ((I - P)Q + \alpha P)^{-1}(-(I - P)z + q), \qquad (4.23)$$

where γ is a positive scaling constant and $h(z) = (h_1(z_1), \ldots, h_n(z_n))^T$, in which $h_i(z_i)$ is the hard-limiting activation function defined as

$$h_i(z_i) = \begin{cases} u_i, & \text{if } z_i > 0, \\ [l_i, u_i], & \text{if } z_i = 0, \\ l_i, & \text{if } z_i < 0. \end{cases} \qquad (4.24)$$

Here we assume that there exists $\alpha > 0$ such that $(I - P)Q + \alpha P$ is invertible. In façt, if the objective function in (4.13) is strictly convex on the set defined by the equality constraints, the matrix $(I - P)Q + \alpha P$ is always invertible, which can be guaranteed by the results listed in the next subsection.

For any $z \in \mathbb{R}^n$, the output vector x of the neural network satisfies equation (4.21). Since (4.19), (4.20), and (4.21) are equivalent, equation (4.19) also holds. Let $y = (AA^T)^{-1}(A(I - Q)x - Az - Ac - b)$ and substitute it into (4.19), then equation (4.14) holds. From (4.14), we have $z = -Qx - A^T y - c$ and substitute it into $y = (AA^T)^{-1}(A(I - Q)x - Az - Ac - b)$. Then, we have $Ax = b$. That is to say, for any $z \in \mathbb{R}^n$, the output vector x of the neural network always satisfies the equality constraints.

According to above analysis, if x^* is an optimal solution of (4.13), there exists $z^* \in \mathbb{R}^n$ such that z^* is an equilibrium point of system (4.22). Conversely, if the objective function $f(x)$ of (4.13) is convex on the set $S = \{x \in \mathbb{R}^n : Ax = b\}$ and z^* is an equilibrium point of system (4.22), then $x^* = ((I - P)Q + \alpha P)^{-1}(-(I - P)z^* + q)$ is an optimal solution of (4.13).

In the following, some convergence results of neural network (4.22) are listed for solving the problem (4.13). The main results have been published in Ref. 17.

4.4.2. *Convergence Results*

Let $[E]^S$ be the symmetric part of matrix E; i.e., $[E]^S = (E + E^T)/2$. Based on the Lyapunov method, the stability and global convergence of (4.22) can be obtained.

Theorem 4.2.[17] *The state vector $z(t)$ of the neural network defined in (4.22) is stable in the sense of Lyapunov and globally convergent to an equilibrium point if there exists a constant $s \geq \alpha$ such that $2s[(I - P)Q + \alpha P]^S - Q(I - P)Q - \alpha^2 P$ is positive semidefinite.*

As a special case, if $(I - P)Q + \alpha P$ is positive definite, the conditions in Theorem 4.2 hold if $s \geq \max\{\alpha, \lambda_{\max}(Q(I - P)Q + \alpha^2 P)/(2\lambda_{\min}([(I - P)Q + \alpha P]^S))\}$, where $\lambda_{\max}(\cdot)$ and $\lambda_{\min}(\cdot)$ are respectively the maximum and minimum eigenvalues of corresponding matrix. Then the following corollary holds.

Corollary 4.3. *The state vector $z(t)$ of the neural network defined in (4.22) is stable in the sense of Lyapunov and globally convergent to an equilibrium point if $(I - P)Q + \alpha P$ is positive definite.*

Remark 4.1. According to the results in Theorem (4.2) and Corollary (4.3), if (4.13) is a convex quadratic programming problem (i.e., the objective function $f(x)$ is convex on the equality constraints), the output vector $x(t)$ in (4.23) is globally convergent to an optimal solution if only the conditions in Theorem 4.2 or Corollary 4.3 are satisfied by choosing proper α. Moreover, if (4.13) is a strictly convex quadratic programming problem (i.e., the objective function $f(x)$ is strictly convex on the equality constraints), the output vector $x(t)$ in (4.23) is always globally convergent to an optimal solution if only α is large enough. We show the details in the rest part of this subsection.

According to the definition of P, the rank of P is m (i.e., $\text{rank}(P) = m$). Since $P^2 = P$, there exists an orthogonal matrix Γ such that

$$\Gamma^T P \Gamma = \begin{pmatrix} I & O \\ O & O \end{pmatrix},$$

where O is a zero matrix. Suppose

$$\Gamma^T Q \Gamma = \bar{Q} = \begin{pmatrix} \bar{Q}_1 & \bar{Q}_2 \\ \bar{Q}_2^T & \bar{Q}_3 \end{pmatrix},$$

where $\bar{Q}_1 \in \mathbb{R}^{m \times m}$, $\bar{Q}_2 \in \mathbb{R}^{m \times (n-m)}$ and $\bar{Q}_3 \in \mathbb{R}^{(n-m) \times (n-m)}$.

If the matrix Q in (4.13) is positive definite, we can choose a sufficiently large value of the parameter α to guarantee the stability of the neural network according to the following theorem.

Theorem 4.3.[17] *Assume that Q is positive definite. If*

$$\alpha \geq \lambda_{\max}(\bar{Q}_1)/2,$$

then the state vector $z(t)$ of the neural network in (4.22) is globally conver-
gent to an equilibrium point and the output vector $x(t)$ in (4.23) is globally
convergent to an optimal solution of problem (4.13).

Since $\lambda_{\max}(\bar{Q}_1) \leq \lambda_{\max}(\bar{Q}) = \lambda_{\max}(Q)$, the following corollary is more convenient to estimate the lower bound of parameter α.

Corollary 4.4. *Assume that Q is positive definite. If*

$$\alpha \geq \lambda_{\max}(Q)/2,$$

then the state vector $z(t)$ of the neural network in (4.22) is globally conver-
gent to an equilibrium point and the output vector $x(t)$ in (4.23) is globally
convergent to an optimal solution of problem (4.13).

Denote $\text{trace}(Q)$ as the trace of Q; i.e., $\text{trace}(Q) = \sum_{i=1}^{n} q_{ii}$, where q_{ii} is the ith diagonal element of Q. Then, $\lambda_{\max}(Q) \leq \text{trace}(Q)$.

Corollary 4.5. *Assume that Q is positive definite. If*

$$\alpha \geq \text{trace}(Q)/2,$$

then the state vector $z(t)$ of the neural network in (4.22) is globally conver-
gent to an equilibrium point and the output vector $x(t)$ in (4.23) is globally
convergent to an optimal solution of problem (4.13).

If the objective function $f(x)$ of problem (4.13) is not convex every-where, but strictly convex on the set $\mathcal{S} = \{x \in \mathbb{R}^n : Ax = b\}$, the neural network in (4.22) and (4.23) can still solve the problem with proper param-eter α based on the following results.

Theorem 4.4.[17] *Assume that the objective function $f(x)$ of problem (4.13) is strictly convex on the set $\mathcal{S} = \{x \in \mathbb{R}^n : Ax = b\}$. If*

$$\alpha > \lambda_{\max}(\bar{Q}_2 \bar{Q}_3^{-1} \bar{Q}_2^T)/4,$$

then the state vector $z(t)$ of the neural network in (4.22) is globally conver-
gent to an equilibrium point and the output vector $x(t)$ in (4.23) is globally
convergent to an optimal solution of problem (4.13).

Since

$$\begin{aligned}
\lambda_{\max}(\bar{Q}_2 \bar{Q}_3^{-1} \bar{Q}_2^T) &\leq \lambda_{\max}(\bar{Q}_2 \bar{Q}_2^T)\lambda_{\max}(\bar{Q}_3^{-1}) \\
&\leq \lambda_{\max}(\bar{Q}^2)\lambda_{\max}(\bar{Q}^{-1}) \\
&= \lambda_{\max}(Q^2)\lambda_{\max}(Q^{-1}),
\end{aligned}$$

the following corollary holds.

Corollary 4.6. *Assume that the objective function $f(x)$ of problem (4.13) is strictly convex on the set $\mathcal{S} = \{x \in \mathbb{R}^n : Ax = b\}$. If*

$$\alpha > \lambda_{\max}(Q^2)\lambda_{\max}(Q^{-1})/4,$$

then the state vector $z(t)$ of the neural network in (4.22) is globally convergent to an equilibrium point and the output vector $x(t)$ in (4.23) is globally convergent to an optimal solution of problem (4.13).

Because $\lambda_{\max}(Q^2) \leq \text{trace}(Q^2)$, and $\lambda_{\max}(Q^{-1}) = 1/\lambda_{\min}^+(Q)$, where $\lambda_{\min}^+(Q)$ is the minimum positive eigenvalue of Q, we can obtain the following corollary.

Corollary 4.7. *Assume that the objective function $f(x)$ of problem (4.13) is strictly convex on the set $\mathcal{S} = \{x \in \mathbb{R}^n : Ax = b\}$. If*

$$\alpha > \frac{\text{trace}(Q^2)}{4\lambda_{\min}^+(Q)},$$

then the state vector $z(t)$ of the neural network in (4.22) is globally convergent to an equilibrium point and the output vector $x(t)$ in (4.23) is globally convergent to an optimal solution of problem (4.13).

As a special case, when $Q = \eta I$, where $\eta > 0$, according to Corollary 4.6, the output vector $x(t)$ of the neural network is convergent to an optimal solution if $\alpha \geq \eta/4$.

4.4.3. *Simulation Results*

Example 4.2. Consider the following quadratic programming problem:

$$
\begin{aligned}
\text{minimize} \quad & f(x) = -x_1^2 + 2.5x_2^2 + x_3^2 + x_1x_2 + 2x_1x_3 \\
& \quad -2x_2x_3 + 2x_1 - x_2 + x_3, \\
\text{subject to} \quad & x_1 + x_2 - x_3 = -1, \\
& -5 \leq x_1, x_2 \leq 5.
\end{aligned}
\tag{4.25}
$$

As

$$
Q = \begin{pmatrix} -1 & 0.5 & 1 \\ 0.5 & 2.5 & -1 \\ 1 & -1 & 1 \end{pmatrix}
$$

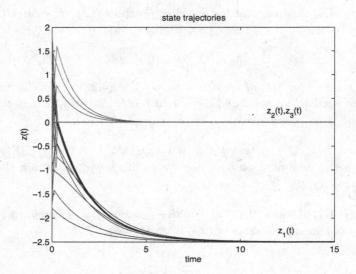

Fig. 4.2.: Transient behaviors of the state variables of the neural network in Example 4.2.

is not positive definite, the objective function is not convex everywhere. However, if we substitute $x_3 = x_1 + x_2 + 1$ into the objective function, then $\tilde{f}(x_1, x_2) = 2x_1^2 + 1.5x_2^2 + 3x_1x_2 + 7x_1 + 2$ is strictly convex. Here, $\lambda_{\max}(Q^2) = 9$ and $\lambda_{\max}(Q^{-1}) = 0.9122$. According to Corollary 4.6, let us choose $\alpha = 2.5 > \lambda_{\max}(Q^2)\lambda_{\max}(Q^{-1})/4 = 2.0524$ and $\gamma = 1$ in the neural network model. Figures 4.2 and 4.3 respectively depict the transient behaviors of state variable $z(t)$ and output variable $x(t)$ with 10 random initial points, from which we can see that the state variables are globally convergent to an equilibrium point $z^* = (-2.5, 0, 0)^T$ and the output variables are globally convergent to the optimal solution $x^* = (-5, 5, 1)^T$.

4.5. For Non-Smooth Convex Optimization Subject to Linear Equality Constraints

In this section, we consider the following nonlinear programming (NP) problem:

$$\begin{aligned} \text{minimize} \quad & f(x), \\ \text{subject to} \quad & Ax = b, \end{aligned} \tag{4.26}$$

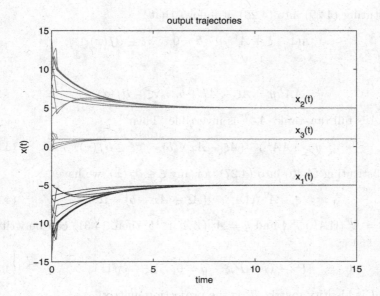

Fig. 4.3.: Transient behaviors of the output variables of the neural network in Example 4.2.

where $x \in \mathbb{R}^n$, $f(x) : \mathbb{R}^n \to \mathbb{R}$ is convex continuous function but not smooth (i.e., not continuously differentiable), A and b are defined in (4.3).

4.5.1. *Model Description and Convergence Results*

Based on the saddle point theorem[2] and the results in Ref. 33, x^* is an optimal solution of problem (4.26) if and only if there exists $y^* \in \mathbb{R}^m$ such that $(x^*, y^*)^T$ satisfies the following equations

$$0 \in \partial f(x) - A^T y, \tag{4.27}$$
$$0 = Ax - b, \tag{4.28}$$

where $\partial f(x)$ is the sub-gradient of $f(x)$.

Next, according to (4.27) and (4.28), the recurrent neural network model will be induced.

From (4.27), for any $\xi \in \partial f(x)$, we have

$$x = x - \xi + A^T y. \tag{4.29}$$

Substituting (4.29) into (4.28), it follows that

$$A(x - \xi + A^T y) - b = 0, \quad \forall \xi \in \partial f(x).$$

That is

$$AA^T y = A\xi - Ax + b, \quad \forall \xi \in \partial f(x).$$

Since A is full raw-rank, AA^T is invertible. Then

$$y = (AA^T)^{-1}(A\xi - Ax + b), \quad \forall \xi \in \partial f(x). \tag{4.30}$$

Substituting (4.30) into (4.27), for any $\xi \in \partial f(x)$, we have

$$\xi - A^T(AA^T)^{-1}(A\xi - Ax + b) = 0. \tag{4.31}$$

Let $P = A^T(AA^T)^{-1}A$ and $q = A^T(AA^T)^{-1}b$, then, (4.31) can be written as

$$Px + (I - P)\xi - q = 0, \quad \forall \xi \in \partial f(x), \tag{4.32}$$

where I is identity matrix, P is the projection matrix.

Based on Eq. (4.32), the proposed recurrent neural network model is described by the following differential inclusion:

$$\frac{dx}{dt} \in \gamma[-Px - (I - P)\partial f(x) + q], \tag{4.33}$$

where γ is a positive scaling constant.

From above analysis, x^* is an optimal solution of problem (4.26) if and only if it is an equilibrium point of neural network (4.33).

According to Lyapunov method, the stability and global convergence of the neural network can be guaranteed shown as the following theorem.[34]

Theorem 4.5. *The neural network* (4.33) *is stable in the sense of Lyapunov and globally convergent to an optimal solution of problem* (4.26).

4.5.2. *Constrained Least Absolute Deviation*

As a special case, we consider the following constrained least absolute deviation (CLAD) problem:

$$\begin{array}{ll} \text{minimize} & \|Cx - d\|_1, \\ \text{subject to} & Ax = b, \end{array} \tag{4.34}$$

where $C \in \mathbb{R}^{p \times n}$, $d \in \mathbb{R}^p$, A and b are defined in (4.3).

The neural network in (4.33) for solving the CLAD problem can be written as

$$\frac{dx}{dt} = \gamma[-Px - (I - P)C^T h(Cx - d) + q], \qquad (4.35)$$

where $h(\cdot)$ is the discontinuous activation function defined in (4.24) with $l_i = -1$ and $u_i = 1$ $(i = 1, 2, \ldots, p)$.

From the results in Theorem (4.5), the neural network (4.40) is always convergent to the solution of the CLAD problem. We illustrate it in the following simulation example.

Example 4.3. Nonlinear Curve-Fitting Problem.

Let us consider a constrained nonlinear least absolute deviation curve-fitting problem: Find the parameters of the combination of exponential and polynomial function $y(x) = a_4 e^x + a_3 x^3 + a_2 x^2 + a_1 x + a_0$, which fits the data given in Table 4.1 and subjects to the equalities $y(0.8) = -3.2$ and $y(4.6) = -3.4$. This problem can be formulated as (4.34) with $x = (x_1, x_2, x_3, x_4, x_5)^T = (a_4, a_3, a_2, a_1, a_0)^T$ and

$$C = \begin{pmatrix} 1 & 1.649 & 2.718 & 4.482 & 7.389 & 12.183 & 20.086 & 33.116 & 54.598 & 90.017 \\ 0 & 0.125 & 1 & 3.375 & 8 & 15.625 & 27 & 42.875 & 64 & 91.125 \\ 0 & 0.25 & 1 & 2.25 & 4 & 6.25 & 9 & 12.25 & 16 & 20.25 \\ 0 & 0.5 & 1 & 1.5 & 2 & 2.5 & 3 & 3.5 & 4 & 4.5 \\ 1 & 1 & 1 & 1 & 1 & 1 & 1 & 1 & 1 & 1 \end{pmatrix}^T,$$

$$d = \begin{pmatrix} -2.6 & -3.6 & -3.2 & -3.6 & -4.5 & -5.2 & -2.2 & -7.2 & -6.6 & -8.2 \end{pmatrix}^T,$$

$$A = \begin{pmatrix} 2.2255 & 0.512 & 0.64 & 0.8 & 1 \\ 99.484 & 97.336 & 21.16 & 4.6 & 1 \end{pmatrix}, b = \begin{pmatrix} -3.2 \\ -3.4 \end{pmatrix}.$$

The neural network in (4.40) is used for solving this problem, the simulation results are shown in Fig. 4.4 with $\gamma = 1$ and 10 random initial points, from which we can see that the neural network is globally convergent to the optimal solution $x^* = (0.2685, -0.2946, 0.3378, -1.2427, -2.8687)^T$ or $(a_4, a_3, a_2, a_1, a_0) = (0.2685, -0.2946, 0.3378, -1.2427, -2.8687)$. The curve fitting is drawn in Fig. 4.5 for l_1-norm (solid line) and l_2-norm (dashed line). It shows that least absolute (LA) has better fitting performance than least square (LS).

Table 4.1.: Nonlinear fitting data for Example 4.3.

x	0	0.5	1	1.5	2	2.5	3	3.5	4	4.5
y	-2.6	-3.6	-3.2	-3.6	-4.5	-5.2	-2.2	-7.2	-6.6	-8.2

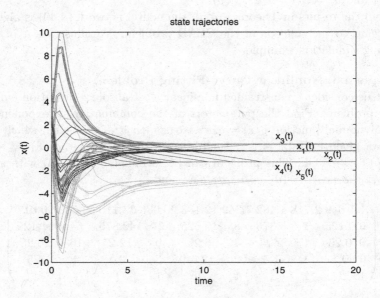

Fig. 4.4.: Transient behavior of neural network (4.33) in Example 4.3.

4.6. Application to k-Winners-Take-All

The winner-take-all (WTA) operation is to identify the largest value from a collection of input signals. The k-winners-take-all (kWTA) operation selects the k largest inputs out of n inputs ($1 \leq k \leq n$), is a generalization of WTA operation. It is well known that the kWTA operation has important applications in science and engineering problems, such as signal processing, associative memories, machine learning, etc. As the number of inputs increases and/or the selection process should be operated in real time, parallel algorithms and hardware implementation are desirable. For these reasons, there have been many attempts to design very large scale

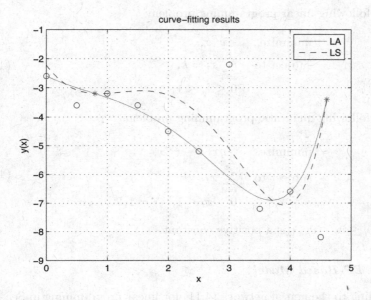

Fig. 4.5.: Comparison of the two nonlinear curve fitting methods between the LA and LS in Example 4.3.

integrated (VLSI) circuits to perform kWTA operations. In the literature, many WTA and kWTA networks have been proposed.[24,35,36]

Generally, the kWTA operation can be defined as the following function

$$x_i = f(v_i) = \begin{cases} 1, \text{ if } v_i \in \{k \text{ largest elements of } v\}, \\ 0, \text{ otherwise}, \end{cases} \quad (4.36)$$

where $v = (v_1, v_2, \ldots, v_n)^T$ is the input vector and $x = (x_1, x_2, \ldots, x_n)^T$ is the output vector.

The solution to the kWTA problem can be determined by solving the following linear integer programming problem:

$$\begin{aligned} \text{minimize} \quad & -\sum_{i=1}^{n} v_i x_i, \\ \text{subject to} \quad & \sum_{i=1}^{n} x_i = k, \\ & x_i \in \{0, 1\}, \quad i = 1, 2, \ldots, n. \end{aligned} \quad (4.37)$$

According to Ref. 37, if the kth and $(k+1)$th largest elements of v (denoted as $\bar{v}_k$ and $\bar{v}_{k+1}$ respectively) are different, problem (4.37) is equivalent to the following linear programming problem:

$$
\begin{aligned}
&\text{minimize} \quad -v^T x, \\
&\text{subject to} \quad \sum_{i=1}^{n} x_i = k, \\
&\hphantom{\text{subject to} \quad} 0 \le x_i \le 1, \quad i = 1, 2, \ldots, n,
\end{aligned}
\tag{4.38}
$$

or the following quadratic programming problem:

$$
\begin{aligned}
&\text{minimize} \quad \frac{1}{2}\eta x^T x - v^T x, \\
&\text{subject to} \quad \sum_{i=1}^{n} x_i = k, \\
&\hphantom{\text{subject to} \quad} 0 \le x_i \le 1, \quad i = 1, 2, \ldots, n,
\end{aligned}
\tag{4.39}
$$

where $\eta \le \bar{v}_k - \bar{v}_{k+1}$ is a positive constant.

4.6.1. *LP-Based Model*

According to the neural network (4.11) for linear programming in Sec. 4.3, the dynamic equation of the LP-based kWTA network model is described as follows:

$$
\epsilon \frac{dx}{dt} = -Px - \sigma(I - P)g(x) + q,
\tag{4.40}
$$

where $P = ee^T/n$, $q = v - Pv + ke/n$, $e = (1, 1, \ldots, 1)^T \in \mathbb{R}^n$, ϵ is a positive constant, σ is a nonnegative gain parameter, and $g(x) = (g(x_1), g(x_2), \ldots, g(x_n))^T$ is the activation function defined in (4.12) with $l_i = 0$ and $u_i = 1$ $(i = 1, 2, \ldots, n)$.

The following two corollaries are directly from the results in Sec. 4.3.

Corollary 4.8. *The network* (4.40) *can perform the kWTA operation if the equilibrium point set* $\Omega^e \subset \{x \in \mathbb{R}^n : 0 \le x \le 1\}$.

Corollary 4.9. *The network* (4.40) *can perform the kWTA operation if it has a unique equilibrium point and $\sigma \ge 0$ when $(I - ee^T/n)u = 0$ or one of the following conditions holds when $(I - ee^T/n)u \ne 0$:*

(i) $\sigma \ge n\sqrt{\dfrac{\sum_{i=1}^{n}(v_i - \sum_{j=1}^{n} v_j/n)^2}{n(n-1)}}$, *or*

(ii) $\sigma \ge \dfrac{\sqrt{\sum_{i=1}^{n}(v_i - \sum_{j=1}^{n} v_j/n)^2}}{\min_{\xi_i \in \{-1,0,1\}}^{+}\left\{\left|\sum_{i=1}^{n}(v_i - \sum_{j=1}^{n} v_j/n)\xi_i\right|\right\}}.$

4.6.2. *QP-Based Model*

According to the neural network in (4.22) and (4.23) for quadratic programming in Sec. 4.4, the QP-based kWTA network model is described as follows:

• State equation

$$\epsilon \frac{dz}{dt} = -(I - P)z - [\eta I + (1 - \eta)P]h(z) + q, \qquad (4.41)$$

• Output equation

$$x = -\frac{1}{\eta}(I - P)z + \frac{q}{\eta} + \frac{k(\eta - 1)}{n\eta}e, \qquad (4.42)$$

where P, q and ϵ are defined in (4.40), $h(z) = (h(z_1), h(z_2), \ldots, h(z_n))^T$ is the activation function defined in (4.24) with $l_i = 0$ and $u_i = 1$ ($i = 1, 2, \ldots, n$).

From the results in Sec. 4.4, we can get the following two corollaries.

Corollary 4.10. *The system* (4.41) *with any* $\eta > 0$ *is stable in the sense of Lyapunov and any trajectory is globally convergent to an equilibrium point.*

Corollary 4.11. $x^* = -(I - P)z^*/\eta + q/\eta + (\eta - 1)ke/(n\eta)$ *is an optimal solution of* k*WTA problem* (4.39), *where* z^* *is an equilibrium point of system* (4.41).

From the results of Corollaries 4.10 and 4.11, the network in (4.41) and (4.42) is globally convergent and the output vector is globally convergent to an optimal solution of problem (4.39). Thus, if the kth and $(k+1)$th largest elements of the input signals $v_i(i = 1, 2, \ldots, n)$ are different, the network in (4.41) and (4.42) is capable of guaranteeing the kWTA operation.

Here, we have proposed two kWTA networks, but they have different architectures and properties. The network in (4.40) has lower model complexity than the network in (4.41) and (4.42). However, when problem size n is large, the value of σ in (4.40) needs to be large. The convergence time decreases when the problem scale n increases for network (4.40). However, for network (4.41) and (4.42), the parameter η needs to be sufficiently small when the kth and $(k + 1)$th largest element of u are contiguous in regard of the problem size. Some comparisons can be found in Ref. 37.

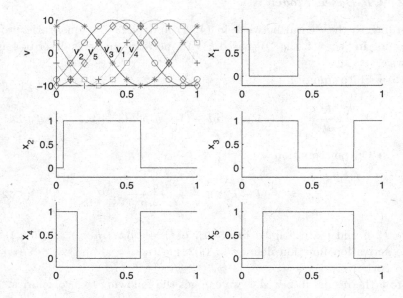

Fig. 4.6.: Inputs and outputs of the two kWTA networks in Example 4.4.

4.6.3. *Simulation Results*

Example 4.4. Let us consider a set of five sinusoidal input signals with the following instantaneous values $v_p(t) = 10\cos[2\pi(t + 0.3p)]$ $(p = 1, 2, 3, 4, 5)$ and $k = 3$. The five input signals and the transient outputs of the three kWTA networks are depicted in Fig. 4.6, in which $\epsilon = 0.01$, $\sigma = 20$ and $\eta = 0.001$. The simulation results show that the kWTA networks can generate the three largest signals in real time.

4.7. Concluding Remarks

Recurrent neural networks are computational models for solving optimization problems. Ever since the seminal work of Hopfield and Tank on neurodynamic optimization, numerous recurrent neural networks have been developed for optimization. The recurrent neural networks have been improved with desirable properties such as exact optimality, global convergence, and simple architecture.

This chapter is devoted to novel recurrent neural network approaches to linear, quadratic and non-smooth optimization problems. Several recurrent neural networks with discontinuous activation functions are discussed with detailed model description, design procedures, and convergence and complexity analysis. For linear and convex quadratic optimization problems, two one-layer recurrent neural networks have been proposed and they are shown to be effective for kWTA application.

Acknowledgments

The work described in the chapter was supported by the Research Grants Council of the Hong Kong Special Administrative Region, China, under Grants CUHK417209E and CUHK417608E.

References

1. D. Bertsekas, *Constrained Optimization and Lagrange Multiplier Methods.* (New York: Academic, 1982).
2. M. Bazaraa, H. Sherali, and C. Shetty, *Nonlinear Programming: Theory and Algorithms (2nd Ed.).* (New York: John Wiley, 1993).
3. E. Bonabeau, M. Dorigo, and G. Theraulaz, *Swarm Intelligence: From Natural to Artificial Systems.* (Oxford University, USA, 1999).
4. D. Tank and J. Hopfield, Simple neural optimization networks: An a/d converter, signal decision circuit, and a linear programming circuit, *IEEE Transactions on Circuits and Systems.* **33**(5), 533–541, (1986).
5. M. Kennedy and L. Chua, Neural networks for nonlinear programming, *IEEE Transactions on Circuits and Systems.* **35**(5), 554–562, (1988).
6. A. Bouzerdoum and T. Pattison, Neural network for quadratic optimization with bound constraints, *IEEE Transactions on Neural Networks.* **4**(2), 293–304, (1993).
7. X. Wu, Y. Xia, J. Li, and W. Chen, A high-performance neural network for solving linear and quadratic programming problems, *IEEE Transactions on Neural Networks.* **7**(3), 643–651, (1996).
8. Y. Xia and J. Wang, Neural network for solving linear programming problems with bounded variables, *IEEE Transactions on Neural Networks.* **6**(2), 515–519, (1995).
9. J. Wang, Analysis and design of a recurrent neural network for linear programming, *IEEE Transactions on Circuits and Systems-I.* **40**(9), 613–618, (1993).
10. J. Wang, A deterministic annealing neural network for convex programming, *Neural Networks.* **7**(4), 629–641, (1994).
11. Y. Xia and J. Wang, A general methodology for designing globally convergent

     optimization neural networks, *IEEE Transactions on Neural Networks.* **9**(6), 1331–1343, (1998).

12. Y. Xia and J. Wang, Global exponential stability of recurrent neural networks for solving optimization and related problems, *IEEE Transactions on Neural Networks.* **11**(4), 1017–1022, (2000).

13. Y. Xia, H. Leung, and J. Wang, A projection neural network and its application to constrained optimization problems, *IEEE Transactions Circuits and Systems-I.* **49**(4), 447–458, (2002).

14. Y. Xia and J. Wang, A general projection neural network for solving monotone variational inequalities and related optimization problems, *IEEE Transactions on Neural Networks.* **15**(2), 318–328, (2004).

15. M. Forti, P. Nistri, and M. Quincampoix, Generalized neural network for nonsmooth nonlinear programming problems, *IEEE Transactions on Circuits and Systems-I.* **51**(9), 1741–1754, (2004).

16. Q. Liu and J. Wang, A one-layer recurrent neural network with a discontinuous activation function for linear programming, *Neural Computation.* **20**(5), 1366–1383, (2008).

17. Q. Liu and J. Wang, A one-layer recurrent neural network with a discontinuous hard-limiting activation function for quadratic programming, *IEEE Transactions on Neural Networks.* **19**(4), 558–570, (2008).

18. S. Zhang and A. Constantinides, Lagrange programming neural networks, *IEEE Transactions on Circuits and Systems-II.* **39**(7), 441–452, (1992).

19. Y. Xia and J. Wang. Primal neural networks for solving convex quadratic programs. In *International Joint Conference on Neural Networks (IJCNN 1999)*, vol. 1, pp. 582–587, (1999).

20. Y. Xia, A new neural network for solving linear and quadratic programming problems, *IEEE Transactions on Neural Networks.* **7**(6), 1544–1548, (1996).

21. Y. Xia, A new neural network for solving linear programming and quadratic programming problems, *Neural Networks.* **9**(6), 1544–1547, (1996).

22. Y. Xia and J. Wang, A dual neural network for kinematic control of redundant robot manipulators, *IEEE Transactions on Systems, Man and Cybernetics-B.* **31**(1), 147–154, (2001).

23. Y. Zhang, J. Wang, and Y. Xia, A dual neural network for redundancy resolution of kinematically redundant manipulators subject to joint limits and joint velocity limits, *IEEE Transactions on Neural Networks.* **14**(3), 658–667, (2003).

24. S. Liu and J. Wang, A simplified dual neural network for quadratic programming with its kwta application, *IEEE Transactions on Neural Networks.* **17**(6), 1500–1510, (2006).

25. T. Friesz, D. Bernstein, N. Mehta, R. Tobin, and S. Ganjalizadeh, Day-to-day dynamic network disequilibria and idealized traveler information systems, *Operations Research.* **42**(6), 1120–1136, (1994).

26. Y. Xia and J. Wang, On the stability of globally projected dynamical systems, *Journal of Optimization Theory and Applications.* **106**(1), 129–150, (2000).

27. Y. Xia, An extended projection neural network for constrained optimization, *Neural Computation.* **16**, 863–883, (2004).

28. Q. Liu, J. Cao, and Y. Xia, A delayed neural network for solving linear projection equations and its analysis, *IEEE Transactions on Neural Networks.* **16**(4), 834–843, (2005).

29. Y. Yang and J. Cao, Solving quadratic programming problems by delayed projection neural network, *IEEE Transactions on Neural Networks.* **17**(6), 1630–1634, (2006).

30. Q. Liu, J. Wang, and J. Cao. A delayed lagrangian network for solving quadratic programming problems with equality constraints. In *Lecture Notes in Computer Science (ISNN2006)*, vol. 3971, pp. 369–378. Springer, (2006).

31. A. Filippov, *Differential Equations with Discontinuous Righthand Sides.* (Mathematics and its applications (Soviet series). Boston: Kluwer Academic Publishers, 1988).

32. F. Clarke, *Optimization and Nnonsmooth Analysis.* (New York: Wiley, 1983).

33. Q. Liu and J. Wang. A recurrent neural network for non-smooth convex programming subject to linear equality and bound constraints. In *Proc. 13th Int. Conference on Neural Information Processing*, pp. 1004–1013. Springer LNCS 4233, (2006).

34. Q. Liu and J. Wang. A one-layer recurrent neural network for non-smooth convex optimization subject to linear equality constraints. In *Proc. 15th Int. Conference on Neural Information Processing*, Auckland, New Zealand (25–28 November, 2008).

35. A. Yuille and D. Geiger. Winner-take-all networks. In *The Handbook of Brain Theory and Neural Networks (2nd ed.)*, pp. 1228–1231. MIT Press Cambridge, MA, (2003).

36. C. Marinov and J. Hopfield, Stable computational dynamics for a class of circuits with o(n) interconnections capable of kwta and rank extractions, *IEEE Transactions on Circuits and Systems-I.* **52**(5), 949–959, (2005).

37. Q. Liu and J. Wang, Two k-winners-take-all networks with discontinuous activation functions, *Neural Networks.* **21**(2-3), 406–413, (2008).

Chapter 5

Automated Power Quality Disturbance Classification Using Evolvable Neural Network

B. K. Panigrahi[1,*], A. Mohapatra[2], P. Ray[1] and S. Das[3]

[1]*Department of Electrical Engineering,
Indian Institute of Technology, New Delhi, India*
*[*]*bijayaketan.panigrahi@gmail.com*

[2]*Department of Electrical Engineering,
College of Engineering, BPUT, Orissa, India*

[3]*Department of Electronics and Communication Engineering,
Jadavpur University, West Bengal, India*

This chapter presents the application of Wavelet Transform (WT), one of the most frequently used signal processing techniques for the extraction of important information in different frequency sub-bands from the non-stationary Power Quality (PQ) signals. The features derived by WT are used for the classification purpose to identify the nature of the PQ disturbances. We have used the Multi Layer Feed Forward Neural Network (FNN) for the classification. Traditional learning algorithm like back propagation learning is used for the training of the FNN. Simultaneously an attempt has been made for the design of an evolvable neural network classifier using one of the most promising Swarm Intelligence algorithms known as Particle Swarm Optimization (PSO). The basic PSO algorithm was modified to make it adaptive and the proposed algorithm is known as adaptive PSO (APSO). The performance of the NN for the classification is compared for both back propagation based learning and APSO based learning.

5.1. Introduction

In the present scenario of deregulated electric market, Power Quality (PQ)[1] study has become one of the key issues for electric utilities and their customers. As a result, power quality study is gaining interest day by day among researchers around the globe. Among the PQ disturbances sag, swell and momentary interruption are generally treated as the steady state phenomena with varying amplitude of the system root mean square (rms) voltage around the nominal system voltage. Notch and spike are very short duration disturbances and may be repetitive and mostly caused due to the power electronic loads. Transients are the short duration events caused by line switching, capacitor bank switching etc. These PQ events results in malfunctions, instabilities, short lifetime, failure of electrical equipments and so on. In electric power network system faults may cause voltage sag or momentary interruption whereas switching off of large load or energization of large capacitor bank may lead to voltage swell. On the other hand use of solid state switching devices, non-linear and power electronically switched loads such as rectifiers or inverters may cause harmonic distortion and notching in the voltage and current. Use of arc furnaces may lead to flickers. Ferroresonance, transformer energization, capacitor switching may cause transients and lightning strikes may lead to spikes.

In a realistic distribution system, in order to improve power quality these disturbances need to be identified before appropriate mitigating action can be taken. Therefore, in this chapter 11 types of disturbances are considered. To detect, solve and mitigate the PQ problem, many utilities perform PQ monitoring for their industrial and key customers. In the deregulated market, the PQ monitoring would be an effective means for providing better customer services as well as reinforcing competitiveness among the utilities.

Generally PQ monitoring is carried out by capturing the disturbance or event. Disturbance waveforms are recorded continuously using power monitoring instruments, producing yearly data files in the gigabytes range. Unfortunately, existing methods to analyze and identify power disturbance are laborious and tedious since the methods are based on visual inspection of the waveform. As a result, power quality engineers are inundated with enormous amount of data to inspect. Moreover, they may loose the important information while monitoring. Hence, a robust method for automatic classification of disturbances is highly demanded. Recent advances in signal analysis and pattern recognition techniques have led to the development of new methods for characterizing and identifying various power quality disturbances. In order to identify the type of PQ disturbance more effectively, several authors have presented different methodologies based on the combination of wavelet transform (WT)[2] and artificial neural network (ANN).[3-9] Using the features derived through WT and subsequently training with an ANN, it is possible to extract important information about the disturbance signal and to determine what type of disturbance has caused a power quality (PQ) problem to occur. Wavelet exhibits its notable capabilities in detection and localization of power quality disturbances.

This chapter presents automatic classification of different PQ disturbances using the features extracted from the raw PQ signal using WT and a NN based classifier. The other important investigation carried out in this paper is to show the effectiveness of one of the most promising Swarm Intelligence (SI) algorithm known as Particle Swarm Optimization (PSO) for the design of an evolvable neural network classifier and its application to the classification of non-stationary power quality disturbances. Firstly, a brief description of the algorithm and the general procedures are described in detail. The proposed methodology adopts an adaptive inertia weight based PSO strategy and hybrid with a back propagation (BP) algorithm. The PSO parameters are carefully designed to optimize the neural network, avoiding premature convergence. The results are compared with the traditional classification algorithm like back propagation learning based feed forward multi layer perception neural network classifier.

5.2. Wavelet Transform (WT)

The Discrete Wavelet Transform (DWT) is a special case of the WT that provides a compact representation of a signal in time and frequency that can be computed efficiently. The DWT is calculated based on two fundamental equations: the scaling function $\phi(t)$, and the wavelet function $\psi(t)$, where

$$\phi(t) = \sqrt{2} \sum_{i=1} h_k \phi(2t - k) \tag{5.1}$$

$$\psi(t) = \sqrt{2} \sum_{i=1} g_k \phi(2t - k) \tag{5.2}$$

These functions are two-scale difference equations based on a chosen scaling function (mother wavelet), with properties that satisfy the following conditions

$$\sum_{k=1}^{N} h_k = \sqrt{2} \tag{5.3}$$

$$\sum_{k=1}^{N} h_k h_{k+2l} = 1 \qquad if \ l = 0$$

$$= 0 \qquad if \ l\epsilon z, \ l \neq z \tag{5.4}$$

The discrete sequences h_K and g_k represent discrete filters that solve each equation, where $g_k = (-1)^k h_{N-1-K}$. The scaling and wavelet functions are the prototype of a class of orthonormal basis functions of the form

$$\phi_{j,k}(t) = 2^{j/2}\phi(2^j t - k); \quad j, k \epsilon z \tag{5.5}$$

$$\psi_{j,k}(t) = 2^{j/2}\psi(2^j t - k); \quad j, k \epsilon z \tag{5.6}$$

where the parameter j controls the dilation or compression of the function in time scale and amplitude. The parameter k controls the translation of the function in time. Z is the set of integers. Once a wavelet system is created, it can be used to expand a function $f(t)$ in terms of the basis functions

$$f(t) = \sum_{l \epsilon z} c(l)\phi_l(t) + \sum_{j=0}^{J-1} \sum_{k=0}^{\infty} d(j, k)\psi_{j,k}(t) \tag{5.7}$$

where, the coefficients c(l) and d(j,k) are calculated by inner product as

$$c(l) = \langle \phi_l \mid f \rangle = \int f(t)\phi_l(t)dt \tag{5.8}$$

$$d(j, k) = \langle \psi_{j,k} \mid f \rangle = \int f(t)\psi_{j,k}(t)dt \tag{5.9}$$

The expansion coefficients c(l) represent the approximation of the original signal f(t) with a resolution of one point per every 2^J points of the original signal. The expansion coefficients d(j,k) represent details of the original signal at different levels of resolution. c(l) and d(j,k) terms can be calculated by direct convolution of f(t) samples with the coefficients h_k and g_k, which are unique to the specific mother wavelet chosen.

The WT can be implemented with a specially designed pair of FIR filters called a quadrature mirror filters (QMFs) pair. QMFs are distinctive because the frequency responses of the two FIR filters separate the high- and low-frequency components of the input signal. The dividing point is usually halfway between 0 Hz and half the data sampling rate (the Nyquist frequency). The outputs of the QMF filter pair are decimated (or desampled) by a factor of two. The low-frequency (low-pass) filter output is fed into another identical QMF filter pair. This operation can be repeated recursively as a tree or pyramid algorithm, yielding a group of signals that divides the spectrum of the original signal into octave bands with successively coarser measurements in time as the width of each spectral band

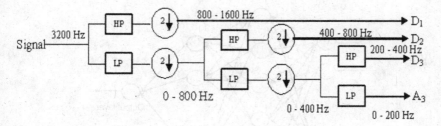

Fig. 5.1.: Wavelet multi-resolution algorithm.

narrows and decreases in frequency. The tree or pyramid algorithm can be applied to the WT by using the wavelet coefficients as the filter coefficients of the QMF filter pairs as shown in the Fig. 5.1. In WT multi-resolution algorithm, same wavelet coefficients are used in both low-pass (LP) and high-pass (HP) filters. The LP filter coefficients are associated with the scaling function, and the HP filter is associated with the wavelet function. Figure 1 shows the tree algorithm of a multi resolution WT for a discrete signal sampled at 3200 Hz. The outputs of the LP filters are called the approximations (A), and the outputs of the HP filters are called the details (D). In wavelets applications, different basis functions have been proposed and selected. Each basis function has its feasibility depending on the application requirements. Daubechies wavelet family is one of the most suitable wavelet families in analyzing power system transients In the present work, the db4 wavelet has been used as the wavelet basis function for power quality disturbance detection and classification.

5.3. Brief Overview of Neural Network Classifiers

In this chapter the Multi Layer Feed Forward (MLFF) Neural Network (NN) is considered for effective classification of PQ disturbances. MLFF is the most popular network architectures used in most of the research applications in engineering, medicine, mathematical modeling, etc. In this architecture, the weighted sum of the inputs and bias term are passed to activation level through a transfer function to produce the output, and the units are arranged in a layered feed-forward topology called Feed Forward Neural Network (FFNN). The schematic representation of MLFF with n inputs, m hidden units and one output unit along with the bias term of the input unit and hidden unit is shown in Fig. 5.2. Backpropagation algorithm for Multilayer Feed forward Neural Network[10,11] is one of most

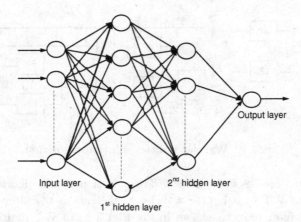

Fig. 5.2.: Structure of feed forward neural network.

popular neural network which can be used for various applications including classification.

The Backpropagation algorithm is the classical technique for supervised training. It works by measuring the output error, calculating the gradient of this error, and adjusting the ANN weights and biases in the descending gradient direction. Hence, BP is a gradient-descent local search procedure. Thus it may lead to stuck in local optima in a complex error landscapes. Generally back propagation algorithm uses the mean squared error of the ANN for a set of input patterns. The value of the error depends on the weights of the network, which were randomize at the initial stage. The basic BP algorithm calculates the gradient of the error for all the patterns and updates the weights by moving them along the gradient-descendent direction. This can be expressed as $\Delta w = -\eta \Delta E$, where the parameter $\eta > 0$ is the learning rate that controls the learning speed.

Despite the popularity of the gradient descent technique as an optimization tool for neural network training, it also suffers from several drawbacks. For instance, the performance of the network learning is strictly dependent on the shape of the error surface, values of the initial connection weights, learning rate etc. A common error surface may have many local minima making it multimodal, or it may be non-differentiable, there by causing the gradient descent algorithm to stick at local minima while moving across the error surface. On the other side, evolutionary computation techniques like Genetic Algorithm (GA) and Particle Swarm Optimization (PSO) offers an efficient search method for a complex problem space may be multomodal, and / or non-differentiable. With regard to the above-mentioned problems of the gradient descent a complete substitution of them by a GA / PSO might be advantageous. Recently some investigations into neural network training using genetic algorithms[12–15] and PSO[16] to form an evolvable neural network have been reported. In this chapter we have used a new variant of PSO, known as Adaptive PSO (APSO) as an evolutionary computing tool to design an evolvable neural network where during the neural network training process the evolution of connection weights takes place by PSO. With this strategy, PSO can then be used effectively in the evolution to find a near-optimal set of connection weights globally without computing gradient information. This paper attempts to classify 11 types of power quality disturbances using an evolvable NN, trained by APSO.

5.4. Overview of Particle Swarm Optimization

Eberhart and Shi introduce a concept for the optimization of nonlinear functions using particle swarm methodology.[17–19] The performance of particle swarm optimization using an inertia weight is compared with performance using a constriction factor is also explained. Developments and resources in the particle swarm algorithm are reviewed in Refs. 20–22. PSO method conducts search using a population of particles, corresponding to individuals. Each particle in the swarm represents a candidate solution to the problem. It starts with a random initialization of a population of individuals in the search space and works on the social behavior of the particles in the swarm like bird flocking, fish schooling and the swarm theory. Therefore, it finds the global optimum by simply adjusting the trajectory of each individual towards its own best location and towards the best particle of the swarm at each generation of evolution. However, the trajectory of each individual in the search space is adjusted by dynamically altering the velocity of each particle, according to the flying experience of its own and the other particles in the search space.

Advantages of PSO

1) PSO is easy to implement and there are few parameters to adjust.

2) Unlike GA, PSO has no evolution operators such as crossover and mutation.

3) In GAs chromosomes share information, so that the whole population moves like one group, but in PSO only Gbest gives out information to others. It is more robust than that of GA.

4) PSO can be more efficient than GAs; that is, PSO often finds the solutions with fewer objective function evaluations than are required by GAs.

5) PSO uses payoff (performance index or objective function) information to guide the search in the problem space.

6) Unlike GA and other heuristic algorithms, PSO has the flexibility to control the balance between the global and local exploration of the search space. This unique feature of PSO overcomes the premature convergence problem and enhances the search capability.

PSO Algorithm

The position and the velocity of the ith particle in the d-dimensional search space can be represented as $X_i = [x_{i1}, x_{i2}, ..., x_{id}]^T$ and $V_i = [v_{i1}, v_{i2}, ..., v_{id}]^T$, respectively. Each particle has its own best position (*pbest*), $P_i(t) = [p_{i1}(t), p_{i2}(t), ..., p_{id}(t)]^T$ corresponding to the personal best objective value obtained so far at generation 't'. The global best particle (*Gbest*) is denoted by $P_g(t) = [p_{g1}(t), p_{g2}(t), ..., p_{gd}(t)]^T$, which represents the best particle found so far at generation 't' in the entire swarm. The new velocity of each particle is calculated as follows:

$$V_{ij}(t+1) = \omega V_{ij}(t) + c_1 r_1 [p_{ij}(t) - x_{ij}(t)] + c_2 r_2 [p_{gi}(t) - x_{ij}(t)]$$
$$j = 1, 2, ..., d; \ i = 1, 2, ..., n \qquad (5.10)$$

where c_1 and c_2 are constants named acceleration coefficients corresponding to the cognitive and social behavior, ω is called the inertia factor, n is the population size, r_1 and r_2 are two independent random numbers uniformly distributed in the range of [0, 1]. Thus, the position of each particle at each generation is updated according to the following equation:

$$x_{ij}(t+1) = x_{ij}(t) + V_{ij}(t+1)$$
$$j = 1, 2, ..., d; \ i = 1, 2, ..., n \qquad (5.11)$$

Equation (5.10) shows that the new velocity is updated according to its previous velocity and to the distance of its current position from both its best historical position and the global best position of the swarm. Generally, the value of each component in Vi can be clamped to the range $[V_{imin}, V_{imax}]$ to control excessive roaming of particles outside the search space $[X_{imin}, X_{imax}]$. Then the particle flies towards a new position according to (5.11). The process is repeated until a user-defined stopping criterion is reached.

Adaptive PSO Algorithm

In simple PSO method the inertia weight is made constant for all the particles in a single generation. But the most important parameter that moves the current position towards the optimum position is inertia weight ω. In order to increase the search ability, the algorithm should be redefined in the manner that the movement of swarm should be controlled by the objective function. In our Adaptive PSO, the particle position is adjusted such that the highly fitted particle (best particle) moves slowly when compared to the low fitted particle. This can be achieved by selecting different ω values for each particle according to their rank, between ω_{min} and ω_{max} as in the following form.

$$\omega_i = \omega_{min} + \frac{(\omega_{max} - \omega_{min}) * Rank_i}{Total population} \qquad (5.12)$$

So from (5.12) it can be interpreted that the best particle takes the first rank and inertia weight for that particle is set to minimum value and for the lowest fitted particle it will take the maximum of inertia weight and makes that particle to move with a high velocity. The velocity of each particle is updated using (5.13), and if any updated velocity goes beyond Vmax, then limit it to Vmax using (5.14).

$$V_{ij}(t+1) = \omega_i V_{ij}(t) + c_1 r_1 [p_{ij}(t) - x_{ij}(t)] + c_2 r_2 [p_{gj}(t) - x_{ij}(t)] \quad (5.13)$$

$$V_{ij}(t+1) = sign(V_{ij}(t+1)) * (min|V_{ij}(t+1)|, V_{jmax})$$

$$j = 1, 2, ..., d; \ i = 1, 2, ..., n \qquad (5.14)$$

The new particle position is obtained by using the (5.15) and if any particle position goes beyond the range specified, then it is adjusted to its boundary using (5.16).

$$x_{ij}(t+1) = x_{ij}(t) + V_{ij}(t+1) \qquad j = 1, 2, ..., d; \ i = 1, 2, ..., n \quad (5.15)$$

$$x_{ij}(t+1) = max(x_{ij}(t+1), \ range_{jmin})$$

$$x_{ij}(t+1) = min(x_{ij}(t+1), range_{jmax}) \qquad (5.16)$$

The concept of re-initialization is introduced to the proposed APSO algorithm after a specific number of generations if there is no improvement in the convergence of the algorithm. The population of the proposed APSO at the end of the above mentioned specific generation is re-initialized with new randomly generated individuals. The number of this new individuals is selected from k least fit individuals of the original population, where 'k' is the percentage of total population to be changed. This effect of population re-initialization is in a sense similar to the mutation operator in GA.[16] This effect is favorable when the algorithm prematurely converges to a local optimum and further improvement is not noticeable. This re-initialization of population is performed after checking the changes in 'Fbest' value in each and every specific number of generations.

The procedure of Adaptive PSO

Step 1: Get the input parameters like range [min max] for each variables, c_1, c_2, samples of the signal, Iteration counter=0, V_{max}, ω_{min} and ω_{max}.

Step 2: Initialize n number of population of particles of dimension d with random positions and velocities.

Step 3: Increment Iteration counter by one.

Step 4: Evaluate the fitness function of all particles in the population, find particles best position *pbest* of each particle and update its objective value. Similarly find the global best position among all particles and update its objective value.

Step 5: If stopping criterion is met go to step (11). Otherwise continue.

Step 6: Evaluate the inertia factor according to (5.12), so that each particles movement is directly controlled by its fitness value.

Step 7: Update the velocity using (5.13) and correct it if $V_{new} > V_{max}$ using (5.16).

Step 8: Update the position of each particle according to (5.15) and if new position goes out of range set it to the boundary value using (5.16).

Step 9: The Elites are inserted in the first position of the new population in order to maintain the best particle found so far.

Step 10: For every 5 generations, this F_{Best}, new value (at the end of these 5 generations) is compared with the F_{Best}, old value (at the beginning of these 5 generations), if there is no noticeable change then re-initialize k % of the population. Go to step (3).

Step 11: Output the G_{best} particle and its objective value.

5.5. Signal Generation, Feature Extraction and Classification

The signals are simulated using MATLAB.[26] The sampling frequency chosen is 3.2 kHz. The following 11 types of PQ disturbances have been simulated in the MATLAB environment.

C1 $\rightarrow$ Normal

C2 $\rightarrow$ Pure Sag

C3 $\rightarrow$ Pure Swell

C4 $\rightarrow$ Momentary Interruption (MI)

C5 $\rightarrow$ Harmonics

C6 $\rightarrow$ Sag with Harmonic

C7 $\rightarrow$ Swell with Harmonic

C8 $\rightarrow$ Flicker

C9 $\rightarrow$ Notch

C10 $\rightarrow$ Spike

C11 $\rightarrow$ Transient

The parametric model of the disturbances are reported in Table 5.1. Using the parametric model of each event we generated totally 1100 different cases (100 from each class) according to the range of parameter variation in each class. The sag and swell are generated with different amplitudes (A as

Table 5.1.

PQ disturbances	Model	Parameters variation
Normal	$x(t) = A sin(\omega t)$	$A = 1$
Voltage Sag	$x(t) = A(1 - \alpha(u(t-t1) - u(t-t2)))sin(\omega t)$ $t1 < t2,\ u(t) = 1\ if\ t \geq 0,\ = 0\ if\ t < 0$	$0.05 \leq \alpha \leq 0.85\ T \leq (t2 - t1) \leq 7T$
Voltage Swell	$x(t) = A(1 + \alpha(u(t-t1) - u(t-t2)))sin(\omega t)$ $t1 < t2,\ u(t) = 1\ if\ t \geq 0,\ = 0\ if\ \ t < 0$	$0.05 \leq \alpha \leq 0.8\ T \leq (t2 - t1) \leq 7T$
Harmonics	$x(t) = A(\alpha_1 sin(\omega t) + \alpha_2 sin(2\omega t) + \alpha_3 sin(3\omega t) + \alpha_5 sin(5\omega t) + \alpha_7 sin(7\omega t))$	$\alpha_1 = 1.0,\ \alpha_2 - \alpha_7 = (0.0 - 0.3)$
Flicker	$x(t) = A(1 + \alpha_f sin(\beta_f \omega t))sin(\omega t)$	$0.01 \leq \alpha_f \leq 0.25,\ 2Hz \leq f_f \leq 8Hz$
Oscillatory Transient	$x(t) = A[sin(\omega t) + e^{-\gamma(t - t_1)} sin(\omega_{tr}(t - t_1)]$	$-2 \leq b \leq 2,\ 50 \leq \gamma \leq 100,\ 500Hz \leq f_{tr} \leq 1500Hz$
Sag with Harmonics	$x(t) = A(1 - \alpha(u(t - t1) - u(t - t2)))$ $(\alpha_1 sin(\omega t) + \alpha_2 sin(2\omega t) + \alpha3 sin(3\omega t) + \alpha_5 sin(5\omega t) + \alpha_7 sin(7\omega t))$	$\alpha_1 = 1.0,\ 0.0 \leq \alpha_2,\ \alpha_3,$ $\alpha_5\ and\ \alpha_7 \leq 0.3,\ 0.05 \leq \alpha \leq 0.85,\ T \leq (t2 - t1) \leq 7T$
Swell with Harmonics	$x(t) = A(1 + \alpha(u(t - t1) - u(t - t2)))$ $(\alpha_1 sin(\omega t) + \alpha_2 sin(2\omega t) + \alpha3 sin(3\omega t) + \alpha_5 sin(5\omega t) + \alpha_7 sin(7\omega t))$	$\alpha_1 = 1.0,\ 0.0 \leq \alpha_2,\ \alpha_3,$ $\alpha_5\ and\ \alpha_7 \leq 0.3,\ 0.05 \leq \alpha \leq 0.8,\ T \leq (t2 - t1) \leq 7T$
Interruption	$x(t) = A(1 - \alpha(u(t-t1) - u(t-t2)))sin(\omega t)$	$0.88 \leq \alpha \leq 1.0,\ T \leq t2 - t1 \leq 7T$

per Table 5.1) and with different duration, harmonics are generated with different harmonic components superimposed with the fundamental one. Transients are generated by varying the location of the transient in the signal, transient duration, frequency and its amplitude. Notch and spike are generated by varying their location, depth and duration.

After the signal is generated, the signal is processed through WT. The energy at each decomposition level of the WT is calculated using the following equation

$$ED_i = \sum_{j=1}^{N} |D_{ij}|^2 \qquad i = 1, 2, 3, ..., l$$

$$ED_1 = \sum_{j=1}^{N} |A_{ij}|^2 \qquad\qquad (5.17)$$

where i=1,2,...,l is the wavelet decomposition level from level 1 to level l. N is the number of coefficients of detail or approximate at each decomposition level. Thus for a 'l' level decomposition using WT, the feature vector adopted is of length 'l' and is denoted by

Feature = [ED_1 ED_2 ED_3 .... ED_n]. In the proposed work the PQ signal is decomposed upto 13 level, thereby making the feature vector dimension 13.

Based on the feature extraction by the WT, a 13 dimensional feature sets for training and testing are constructed. All the data sets of features for various classes are applied to NN for automatic classification of PQ events. Each neural network is trained with 100 events of each class and 100 events of each class are considered for testing. The NN is trained with the traditional BP learning as described in Sec. 5.3 as well as the weights of the NN are evolved through the optimization process using the APSO as described in Sec. 5.4. Mean Squared Error is selected as the search objective of the proposed optimization approach. The goal is to minimize the error through APSO. The block diagram of the proposed approach is shown in Fig. 5.3.

5.6. Results and Discussion

The learning convergence of the NN trained by classical back propagation learning and the APSO based learning for 500 epochs and 1000 epochs are shown in Figs. 5.4 and 5.5 respectively. As observed in both the cases, the APSO based learning results lesser MSE as compared to that of BP based

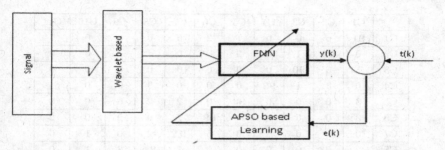

Fig. 5.3.: Basic block diagram for the APSO based evolvable Neural Network.

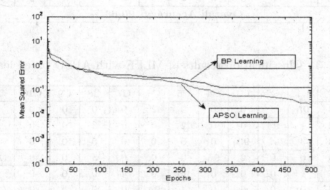

Fig. 5.4.: Comparison of learning by BP and APSO for 500 epochs.

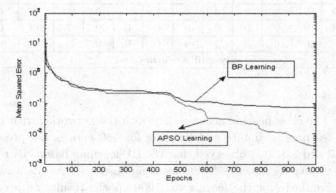

Fig. 5.5.: Comparison of learning by BP and APSO for 1000 epochs.

Table 5.2.: Classification results of MLFF with back propagation learning.

	C1	C2	C3	C4	C5	C6	C7	C8	C9	C10	C11
C1	91	6	0	0	1	2	0	0	0	0	0
C2	0	93	1	0	2	0	0	4	0	0	0
C3	5	5	90	0	0	0	0	0	0	0	0
C4	0	3	1	92	0	0	0	3	0	0	1
C5	3	0	0	0	93	0	2	2	0	0	0
C6	3	0	0	0	4	89	2	0	0	0	2
C7	0	1	0	0	0	5	91	2	0	1	0
C8	2	0	0	0	0	0	7	87	3	1	0
C9	0	0	0	0	5	0	4	0	91	0	0
C10	0	2	0	2	0	0	0	1	2	90	3
C11	1	3	0	0	0	0	0	0	4	2	90
Overall Accuracy: 90.63 %										

Table 5.3.: Classification results of MLFF with APSO based learning.

	C1	C2	C3	C4	C5	C6	C7	C8	C9	C10	C11
C1	95	3	0	0	0	2	0	0	0	0	0
C2	1	94	0	0	2	0	0	3	0	0	0
C3	3	4	93	0	0	0	0	0	0	0	0
C4	0	0	2	95	0	0	0	3	0	0	0
C5	1	0	0	0	95	0	4	0	0	0	0
C6	0	0	0	0	4	94	0	0	0	0	2
C7	0	0	0	0	0	3	94	0	0	3	0
C8	2	0	0	0	0	0	1	92	5	0	0
C9	0	0	0	0	2	0	4	0	94	0	0
C10	0	0	0	0	0	0	0	0	7	93	0
C11	0	0	0	0	0	0	0	0	0	5	95
Overall Accuracy: 94%										

learning. The classification accuracy for both the networks, with the same common features as inputs and learning for 500 epochs are presented in Tables 5.2 and 5.3. It is observed that the BP learning based NN results an accuracy of 90.63% and the APSO learning based NN results an accuracy of 94%. Similarly for the learnig for 1000 epochs results an accuracy of 91.2% for BP base NN and 96.42% for the APSO based NN.

5.7. Conclusions

This paper presents automatic classification of different Power Quality (PQ) disturbances using Wavelet Transform as an signal processing tool to extract important information from the raw disturbance signal and an evolvable Neural Network (NN) based classifier as a pattern recognition tool. The features extracted by processing the PQ disturbance signal using WT are used as the inputs for the NN for training by both back propagation technique as well as the APSO, and subsequently it is tested for an effective classification. It is observed that the evolvable NN provides a better classification accuracy for the classification of PQ disturbances.

References

1. Math H. J. Boolen, *Understanding Power Quality Problems: Voltage Sags and Interruption*, IEEE Press, (2000).
2. S. G. Mallat, "A theory of multiresolution signal decomposition: the wavelet representation," *IEEE Transactions on Pattern Analysis and Machine Intelligence*, 11(7), 674-693, (1989).
3. S. Santoso, E. J. Powers, and W. Grady, "Power quality disturbance identification using wavelet transformers and artificial neutral network," *The 1996 International Conference on Harmonic and Quality of Power*, Las Vegas, NV, U.S.A, 615-618, (1996).
4. S. Santoso., E. J. Powers, W. M. Grady, and P. Hofmann, "Power quality assessment via wavelet transform analysis," *IEEE Trans. on Power Delivery*, 11(2), 924-930, (1996).
5. P. Pillay and A. Bhattacharjee, "Application of wavelets to model short-term power system disturbances," *IEEE Trans. on Power Systems*, 11(4), 2031-2037, (1996).
6. L. Angrisani, P. Daponte, M. D'Apuzzo, and A. Testa, "A measurement method based on the wavelet transform for power quality analysis," *IEEE Transactions on Power Delivery*, 13(4), 990-998, (1998).
7. A. M. Gaouda, M. M. A. Salama, M. K. Sultan, and A.Y. Chikhani, "Power Quality Detection and Classification Using Wavelet-Multiresolution Signal Decomposition," *IEEE Trans. on Power Delivery*, 14(4), 1469-1476, (1999).
8. C. H. Lin and C. H. Wang, "Adaptive Wavelet Networks for Power-Quality Detection and Discrimination in a Power System," *IEEE Trans. on Power Delivery*, 21(3), 1106-1113, (2006).
9. Z. L. Gaing, "Wavelet-Based Neural Network for Power Disturbance Recognition and Classification," *IEEE Trans. on Power Delivery*, 19(4), 1560-1568, (2004).
10. S. Haykin, *Neural Networks: A Comprehensive Foundation*, Pearson Prentice Hall, Singapore, (2005).
11. B. Yegnanarayana, *Artificial Neural Networks*, Prentice Hall, India, (2000).

12. L. C. K. Hui, K.-Y. Lam, and C. W. Chea, "Global Optimisation in Neural Network Training," *Neural Computing & Application,* 5, 58-64, (1997).

13. A. Georgieva and I. Jordanov, "Supervised Neural Network Training with a Hybrid Global Optimization Technique," *2006 International Joint Conference on Neural Networks,* Vancouver, BC, Canada, doi. 0-7803-9490-9/06, (2006).

14. M. Mclnerney and A. P. Dhawan, "Use of genetic algorithms with back propagation in training of feed-forward neural networks," *Proceedings of IEEE International Conference On Neural Networks,* 2, 203-208, (1993).

15. A. Abraham, C. G. S. Tigan, "Ensemble of hybrid neural network learning approaches for designing pharmaceutical drugs," *Neural Comput & Applic,* 16(3), 307-316, (2007).

16. S. Mohaghegi, Y. del Valle, G. K. Venayagamoorthy and R. G. Harley, "A Comparison of PSO and Backpropagation for Training RBF Neural Networks for Identification of a Power System with Statcom," *IEEE Conference,* Id 0-7803-8916-6/05.

17. R. C. Eberhart and Y. Shi, "Comparison between genetic algorithms and particle swarm optimization," *IProc. IEEE Int. Conf. Evol. Comput.,* 611-616, (1998).

18. R. C. Eberhart and Y. Shi, "Particle swarm optimization: developments, applications and resources," *Proc. Congress on Evolutionary Computation. Piscataway: IEEE, Soul,* 81-86, (2001).

19. Y. Shi and R. C. Eberhart, "Fuzzy Adaptive Particle Swarm Optimization," *Proceedings of Evolutionary Computation,* 1, 101-106, (2001).

20. L. Wang, Q. Kang, H. Xiao, and Q. Wu, "A modified adaptive particle swarm optimization algorithm," *IEEE International Conference on Industrial Technology, 2005. ICIT 2005,* 209-214, (2005).

21. D. Li, L. Gao, J. Zhang, and Y. Li, "Power System Reactive Power Optimization Based on Adaptive Particle Swarm Optimization Algorithm," *Proceedings of the 6th World Congress on Intelligent Control and Automation,*China, 7572-7576, (2006).

22. S. L. Ho, S. Yang, G. Ni, and H. C. Wong, "A Particle Swarm Optimization Method With Enhanced Global Search Ability for Design Optimizations of Electromagnetic Devices," *IEEE Trans. on Magnetics,* 42(4), 1107-1110, (2006).

23. V. K. Koumousis and C. P. Katsaras, "A sawtooth genetic algorithm combining the effects of variable population size and reinitialization to enhance performance," *IEEE Trans. on Evolutionary computation,* 10(1), 19-28, (2006).

24. *MATLAB, The Math Works,* (www.mathworks.com).

Chapter 6

Condition Monitoring and Fault Diagnosis Using Intelligent Techniques

Sandeep Kumar Yadav*, Vrijendra Singh[†] and P. K. Kalra[‡]

Department of Electrical Engineering,
Indian Institute of Technology, Kanpur-208016,
2103, Computer Center-I, India
Indian Institute of Information Technology,
Allahabad, India

Condition monitoring, fault diagnosis as well as Quality Assurance have become of paramount importance in various industrial sectors such as automobile-industries, aeronautics industries, power generating units as well as manufacturing units. Interestingly, all these sectors can use generic solution approach to achieve a greater autonomy with quality assurance using non-parametric classification and prediction methodologies. This chapter provides a generic solution using intelligent methodologies based on statistical techniques, neural networks as well as other machine learning approaches. The methods were justified by using a real life engine data taken from one of the Two Wheeler Industry. The following applications has been investigated at length: Classifying good and bad single cylinder engines based on audio data collected during the running of the engine at various speeds and predicting possible type of the defect in engines. The chapter has been concluded by futuristic approach and vide applicability of the proposed techniques.

6.1. Introduction

Quality assurance related applications in the automobile industry, power generating units as well as manufacturing units lend themselves to similar generic solutions. If relevant plant data can be generated for various operating conditions, these data can be used for purposes of making inferences and predictions. There are several parametric and non-parametric techniques

---

*IIT Kanpur, e-mail: sandeepy@iitk.ac.in
[†]IIIT Allahabad, e-mail: vrij@iiita.ac.in
[‡]IIT Kanpur, e-mail: kalra@iitk.ac.in

that can be used to analyze such data, like Data Mining, Artificial Neural Networks, Principal Component Analysis and Independent Component Analysis. These techniques are already proven for feature extraction, classification and prediction and hence can be used in classification of engines into good and not good classes.

In any two wheeler company, currently, trained operators working on the testing stands at the end of the assembly line, listen to the sound of the engine under a specific operating condition. The operator tries to identify any deviant noise in the engine sound, and based on that he classifies the engine in to two classes Good/ Not Good. If the operator classifies an engine as Not Good, he then goes on to make suggestions about the possible defect in the engine. This human testing process is subjective and prone to errors on account of fatigue, boredom, judgment errors etc. This testing process of the engines can be automated using the statistical as well as Neural Networks based software proposed by us.

Efficiency demands more and more equipment reliability. Productivity must be continually improved to keep costs down and profits up. No one technology is able to perform everything effectively. It is much better to integrate a set of technologies that will best meet the challenges of a particular plant.

The rest of the chapter will focus on the classification of the engine into two classes of Good and Not Good in order to facilitate quality control and improvement in the engine assembly lines.

6.2. Methodology

We can apply any mathematical technique to the data generated from the plant. So, in the current problem objective we have taken engine audio signal. Artificial Neural Networks (ANN),[2] Support Vector Machine (SVM), Decision Trees and Statistical Models[1] have already proven themselves robust for encouraging results in the classification of sensors appropriately. The proposed methods are all model-free approaches and can be extended to any sensors/engines in general. Thus, once a model for single cylinder engine is formed, it can be extended with a little effort to multi-cylinder engine.

For developing a system that is capable of classifying the single-cylinder engines into good/ not good classes, following steps are useful:

(1) Hardware specification, System Setup and Audio Data Generation

(2) Data pre-processing / De-noising/ Data Reduction
(3) Classification of engines into "Defective" and "Not Defective" ones through pattern recognition
(4) In the case of "Defective" engines, the identification of possible defect

6.2.1. *Hardware Specification, System Setup and Audio Data Generation*

The audio signals can be captured using a set of directional microphones in a specific position with the help of data acquisition card. The complete experimental arrangement for measurement, including the Industrial microphones (PCB ICP-130D20), two cylinder internal combustion engine, data acquisition card (cDAQ-9172) from National Instruments are shown in the Fig. 6.1:

This system has been connected to a PC to store the digitized audio signal. The audio data are the amplitude values of sound of engine for small

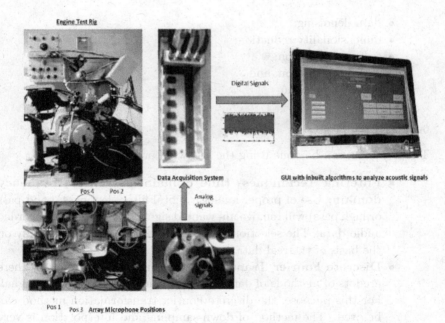

Fig. 6.1.: Experimental setup for data collection.

durations. The time-duration was taken 1 minute after a few experiments. The audio signals of the engines having different defects has been stored in a database with defect signatures. Some of the possible defect-classes are specified as below:

(1) Tappet noise;
(2) Primary gear whine;
(3) Primary gear damage;
(4) Camshaft defects;
(5) Good.

6.2.2. *Data Pre-Processing*

The data obtained from any plant will be always noisy. The noise may be of the sensors or ambient noise from the surrounding environment. All the suggested methods work only on the basis of data supplied. If the data supplied in not "Quality Data", the chance of getting wrong results will increase. Thus, it is very necessary to pre-process the data obtained from the sensors in the real life environment. The pre-processing can include:

- data denoising;
- dimensionality reduction;
- feature extraction;
- signal segregation; and
- data compression.

6.2.2.1. *Data Denoising*

The denoising can be done using the following proven techniques:

- **Filtering Techniques- time domain as well as frequency domain:** Use of proper analog/ digital filter (low pass, band pass or high pass) will remove un- wanted signal/noise from the recorded audio data. The selection of the proper filter can be done only on the basis of the real data.
- **Discrete Fourier Transformation:** in frequency domain- there are lots of methods of denoising in frequency domain of the signal. For this purpose, the discrete Fourier transformation method can be used. The method of down-sampling and interpolation is very useful for data compression.

6.2.2.2. *Dimensionality Reduction*

In available data, many input variables are considered due to unavailability of proper knowledge regarding the percentage of contribution of each variable in systems behavior. The large number of input variables results in a large sized network, which in turn indicates an increase in cost and complexity in the hardware implementation.

Therefore, it is very important to reduce the number of input variables by a method which determines the percentage of contribution of each input variables in systems input-output behavior.

Principal Component Analysis is one of the major techniques for dimensionality reduction. It is based on second order statistical measures and uses eigen-values and eigen-vector decomposition.

Another approach is based on the radial basis network. In which any radial basis function network can be used in a successive manner to reduce the dimensionality. It acts like a moving average algorithm. It can also be treated as a filter to denoise the signal.

6.2.2.3. *Feature Extraction & Signal Segregation*

Different classes will certainly have different features and the system learns only with the help of features present in the data. Thus, Feature extraction is very important when we talk about classification. Principal component analysis and Independent component analysis are the popular methods to extract features. In a real life environment, Recorded sounds will normally include ambient sounds too.

Independent component analysis (ICA) is a method of separation of different sources from a set of mixtures. ICA transforms the data into another domain where the signals become statistically as independent from each other as possible. In the current problem, the audio recordings will be a mixture of engine sound and the surrounding unwanted sounds which can be considered as the noise signal. So, Independent Component Analysis is one of the best approaches to denoise/ separate the mixture signals to get Quality Audio data. ICA is based on the higher order statistical measures.

6.2.3. *Data Classification Techniques*

In signal processing, the one dimensional signal is the audio signal, the two dimensional signal is the image signal and the three dimensional signal is the video signal. Signals store information and this information can be represented in terms of the data. The term DATA can be defined as the measurements of some quantity/variable. Thus, the algorithm based on data will work for one, two or three dimensional signals. The term data will now be used for audio, image as well as video signals. The following data based techniques can be used for classification of the engines:

- Artificial Neural Networks
- Support Vector Machine
- Information Theoretic Model-Decision Trees
- Statistical Models

6.2.3.1. *Artificial Neural Networks*

In Artificial neural networks (ANN), one can create a model according to the available data. The mapping from a known set of input space to known output space can be done using ANN techniques, even when we dont have any physical formula for input-output relationship. In ANN, first we have to supply some data to train a created ANN model. The training means the learning of the system with known input data with corresponding outputs. Once the learning of the model is done, the model is tested with some known data, previously not supplied to the model. This process is known as the testing or validation of the model. This validated model can then be used to predict for a given unknown set of inputs. The process of training and testing is a one time job, while the prediction can be done for several times. The learning is purely based on the supplied data. If proper data is supplied to the model it will perform as desired otherwise one can get wrong predictions. The ANN learning can be categorized into Supervised and Unsupervised case.

Supervised Learning

In supervised learning, we have to supply the input patterns as well as corresponding output patterns. In the case of supervised classification, there are two phases to construct a classifier - the learning/training phase and the prediction phase. In the learning phase, the training set is used to decide how the parameters ought to be weighted and combined in order to separate the various classes of objects. In the prediction phase, the weights determined in the training set are applied to a set of objects that do not have known classes in order to determine what their classes are likely to be. Following algorithms can be used in the classification of the engine:

1. Back-Propagation Algorithms

The conventional neural network model which is based on the McCulloch Pitt's neuron model has been extensively used for the engineering applications. A number of learning algorithms have been formulated to reduce the learning complexity of the ANN and to achieve faster convergence. Back-propagation is an example of such an algorithm. Back-propagation algorithm works into two passes: the forward pass and the backward pass. In forward pass it predicts the output for a given set of inputs and compares this output with the desired one to find out the training error. In backward pass, this estimated error is back-propagated through the network in order to update weight of the neural network model. The weight represents the strength of connection. This process is iterative until the training error is converged or change in the weight stop. The conversion of the error means getting predicted output very near (or same) to the desired output. The multi-layered neural network model will be look like as follows: These nodes known as neurons are the mathematical function units. The data flows in the direction of the arrows. The input node will be equal to the input variables/ attributes identified for classification while the output node will be equal to the desired outputs. The intermediate layers known as hidden layers are subjected to optimization of the neural network models for the given real data sets.

2. Conjugate Gradient Algorithms

The Conjugate Gradient algorithm is based upon a class of optimization techniques well known in numerical analysis as the Conjugate Gradient Methods. Con- jugate Gradient uses second order information from the neural network but requires only O(N) memory usage, where 'N' is the number of weights in the network. Con-jugate Gradient yields a speed up of at least an order of magnitude relative to back-propagation. Conjugate Gradient is fully automated including no user dependent parameters.

Un-Supervised Learning:

In un-supervised learning, there is no class labels defined. Generally, in unsupervised case the data is clustered into different groups. Each cluster stores the data of similar characteristics/ information. Selection of few samples from each of the groups can be used to create a small database. If

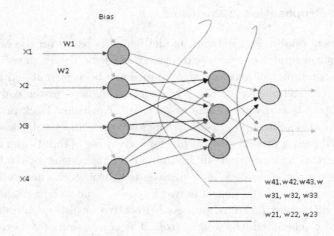

Fig. 6.2.: Artificial neural network architecture.

this database will be used in supervised case, it will result in faster learning of the model. The following clustering algorithms can be used:

- K-means algorithm
- K-mediod algorithm
- Gustafson-Kessel algorithm

All the above specified algorithms works by an explicit minimization of the objective function (generally, the distance between the patterns) and since the cluster centers are points of minimum in a distribution the algorithm has a natural proclivity to find them.

6.2.3.2. *Support Vectors Machine (SVM)*

Support vector machines (SVMs) are a set of related supervised learning methods used for classification and regression. Their common factor is the use of a technique known as the "kernel trick" to apply linear classification techniques to non-linear classification problems. It is composed of the following two modules:

- A general purpose learning machine
- A problem specific kernel function

Suppose we want to classify some data points into two classes. Often we are interested in classifying data as part of a machine-learning process. These data points may not necessarily be points in R2 but may be multidimensional Rp (statistics notation) or Rn (computer science notation) points. We are interested in whether we can separate them by a hyperplane. As we examine a hyperplane, this form of classification is known as linear classification.We also want to choose a hyperplane that separates the data points "neatly", with maximum distance to the closest data point from both classes - this distance is called the margin. We desire this property since if we add another data point to the points we already have; we can more accurately classify the new point since the separation between the two classes is greater. Now, if such a hyperplane exists, the hyperplane is clearly of interest and is known as the maximum-margin hyperplane or the optimal hyperplane, as are the vectors that are closest to this hyperplane, which are called the support vectors.

Conclusively, SVM transform the data into a richer feature space including non-linear features where the data can be linearly classified.

6.2.3.3. *Information Theoretic Model-Decision Trees*

The decision trees method is to create and explore a tree based on the information theoretical approach in order to extract rules. A decision tree describes a tree structure wherein leaves represent classifications and branches represent conjunctions of features that lead to those classifications. A decision tree can be learned by splitting the source set into subsets based on an attribute value test. This process is repeated on each derived subset in a recursive manner. The recursion is completed when splitting is either non-feasible, or a singular classification can be applied to each element of the derived subset. A random forest classifier uses a number of decision trees in order to improve the classification rate. The following decision trees can be used:

- Iterative Dichotomiser 3 Algorithm (ID3);
- C4.5 (Extended ID3) Algorithm.

6.2.3.4. *Statistical Models*

There are different statistical measures that can be used to categorize/classify the data. The cross-correlation measure will describe about the relation between the two samples that how much they are correlated or not correlated. Other distance based methods can also be explored.

6.2.4. *Signal Segregation using Independent Component Analysis*

ICA is a linear transformation method, in which data is transformed into a newer coordinate system, where the axes are non-orthogonal to each other. The directions of the axes of the new coordinate system are determined by both the second and higher order statistics of the original data. ICA tries to linearly transform the data in such a way so that the resulting variables are as statistically independent from each other as possible. In general, ICA can be defined as follows:

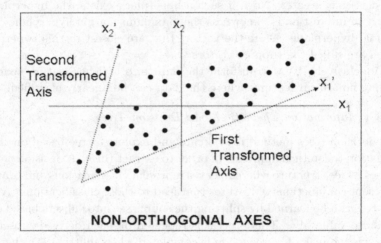

Fig. 6.3.: The original axes are X_1 and X_2. After ICA transformation the new axes are X_1' and X_2', which may or may not be orthogonal to each other.

ICA

Let $X : \Phi \rightarrow R^m$ be a random vector. A measurable mapping $f : R^m \rightarrow R^n$ is called an Independent Component Analysis (ICA) of X if $Y = f \circ X$ is independent. The components Y_i of Y are said to be the Independent Components (ICs) of X.

In case if $m = n$, ICA is known as square ICA and in this case f is then assumed to be invertible.

Linear ICA

Linear ICA or noise - free ICA

Let $X : \Phi \to R^m$ be a random vector. A full rank $n \times m$ matrix W is called a linear independent component analysis (Linear ICA) of X if $Y = WX$ is independent or the components $Y_i's$ are as independent from each other as possible.

Thus, in the case of square linear ICA, W will be an $m \times m$ matrix. ICA of X is always a PCA of X. The vice - versa is not necessarily true. For deterministic or Gaussian signals, PCA of X is equivalent to ICA of X. In linear ICA, there are two possible type of cases:

1. Noise - free ICA

In noise - free ICA the observed data Y is represented as,

$Y = AX$, where A is the unknown mixing matrix.

2. Noisy ICA

In noisy ICA, the observed data Y is represented as,

$Y = AX + n$, where n is the noise.

Non-Linear ICA

In non-linear ICA the observed data is represented as

$$Y = f(AX); \qquad Noise\ free\ ICA \qquad (6.1)$$

$$Y = f(AX + n); \qquad Noisy\ ICA \qquad (6.2)$$

Where f is an arbitrary non -linear function. Thus linear ICA is a special case of non - linear ICA where f is a linear function.

All the above mentioned methods can be used for extracting the signatures of a particular type of the fault and can be used separately as well as in combination to classify the single cylinder engine into Good and Not Good classes.

6.3. Experimental Details

For the current problem, we have acquired the audio signals from a two wheeler single cylinder engine. The property of the acquired signals have been found as following:

- Data File Format: wav file
- Data Size: 1 x 1100001 having about 2.5 MB file size
- Sampling frequency 100000
- Channels: 4 for different position

6.3.1. *Pre-Processing*

The collected data is very large in size. So, during pre-processing it has been empirically found to take the initial transition phase data only. For this, the signal has been trimmed with the engine start to **1,00,000** data to preserve the features. After that, we applied two steps filtering using two Radial Basis Networks (RBN) with varying input size. These filters are also acting as a data compressor to reduce the overall system cost. The Network Architecture of Radial Basis Network is as following:

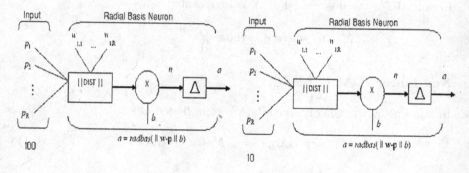

Fig. 6.4.: RBF pre processor.

Pseudocode for RBF Filtering

(1) *Read the Raw data (.wav file)*
(2) *Select the first 100000 x 1 data*

(3) *For the first RBF Network, rearrange the selected into 100x1000 matrix i.e. with window size 100. (Then first filter (RBN1) will produce 1000x1 size output).*
(4) *Rearrange the output data as obtained after first RBF net- work, into 10x100 size matrix i.e. with window size 10.*
(5) *Use another RBF network to get 100x1 as final output.*
(6) *This final output is assumed to be the featured data which has also been compressed from 100000 to 100.*

6.3.2. *Method 1: Artificial Neural Network Setup for Engine Classification*

Simple Feed forward error back propagation algorithm with momentum has been applied for the classification of the engine. The normalised data has been presented to the neural network for training. The normalization range was 0.1 to 0.9. **Network architecture for training**:

- Input layer: 100 nodes
- Hidden layer: 14 nodes
- Output layer: 1 node
- Learning rate: 0.02
- Momentum: 0.8

Binary flagging has been done for target. Good engine noise, cylinder head noise, cam chain noise, tappet noise & PGD noise flagged with class-1, 2, 3, 4 & 5 respectively. The network was trained with around 2500 epochs. The training and testing results are as follows:

Table 6.1.: Training and testing file information.

Engine Information		Position (No. of file)				total	Files used for training
type	class/flags	P1	P2	P3	P4		
Good	1	161	216	216	177	770	238
CHN	2	35	35	35	19	124	113
CCN	3	73	72	72	21	238	214
TN	4	61	60	60	00	181	163
PGD	5	39	40	40	05	124	111

S. K. Yadav, V. Singh and P. K. Kalra

Table 6.2.: Matching between operator observation & classifier prediction.

Eng	Position	Operator observation	Similar prediction as operator	Prediction matching with operator observation
Good	1	216	216	100%
CHN	1	35	14	40%
CCN	1	71	58	81.69%
TN	1	61	57	93.44%
PGD	1	40	25	62.50%
Good	2	216	216	100%
CHN	2	35	8	22.86%
CCN	2	72	52	72.22%
TN	2	61	59	96.72%
PGD	2	40	24	60%
Good	3	216	216	100%
CHN	3	35	8	22.86%
CCN	3	72	58	80.56%
TN	3	60	52	86.67%
PGD	3	40	27	67.50
Good	4	177	177	100%
CHN	4	19	11	57.89%
CCN	4	19	16	84.21
TN	4	0	NA	NA
PGD	4	5	3	60%

So, the net result can be written by taking average as:

GOOD	100%
CHN	36%
CCN	80%
TN	92%
PGD	63%

6.3.3. *Method 2: Artificial Neural Networks based PCA for feature extraction and ANN for classification*

Pseudo-code for PCA based algorithm

(1) Read the Raw data (.wav file)
(2) Select the first 100000 x 1 data
(3) Apply the generalize habbian algorithm[1] on this reduced data and extract first 100 features from every position.
(4) create six network for each of the class.
(5) do classification through Feed forward error backpropagation algorithm.

By applying the above method, the results on the testing data was found very less (less than 20% in all cases).

6.3.4. *Method 3: Feature Extraction using Wigner Willey Transformation and Classification using Decision Trees*

The steps are as follows:

- Time-Frequency Analysis by Wigner-Ville transformation.[2] The Wigner distribution P(x, p) is defined as:

$$P(x,p) = \tfrac{1}{\pi h} \int_{-\infty}^{\infty} \Psi^*(x+y)\Psi(x-y)e^{2piy/\hbar}dy$$

where Ψ is the wave function and x and p are position and momentum but could be any conjugate variable pair. (i.e. real and imaginary parts of the frequency and time of a signal). It is symmetric in x and p:

$$P(x,p) = \tfrac{1}{\pi h} \int_{-\infty}^{\infty} \Phi^*(p+q)\Phi(p-q)e^{-2piy/\hbar}dq$$

where Φ is the Fourier transform of Ψ.
- Decision tree for classification.

The block diagram of Method 3 has been shown in Fig. 6.5.

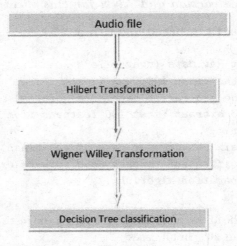

Fig. 6.5.: RBF pre processor.

The classification was done by using the decision tree function available in matlab. The decision tree will look like as following:

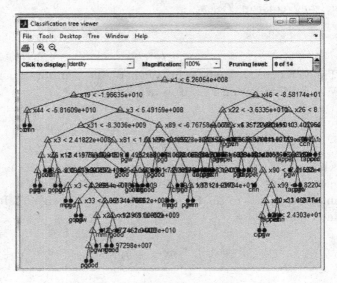

Fig. 6.6.: Decision tree classification.

The position wise training and testing results are as following:

Table 6.3.: Position 1 result.

Training Data	Training Result	Testing Data	Testing Result
30 %	89.05%	70%	19.62%
50%	87.90%	50%	22.11%
70%	87.58%	30%	18.25%

Table 6.4.: Position 2 result.

Training Data	Training Result	Testing Data	Testing Result
30%	87.62%	70%	17.43%
50%	91.25%	50%	11.70%
70%	92.38%	30%	23.54

Table 6.5.: Position 3 result.

Training Data	Training Result	Testing Data	Testing Result
30%	94.36%	70%	36.12%
50%	93.39%	50%	36.79%
70%	92.62%	30%	32.56%

Table 6.6.: Position 4 result.

Training Data	Training Result	Testing Data	Testing Result
30%	91.18%	70%	23.48%
50%	89.09%	50%	26.85%
70%	89.41%	30%	22.19%

Table 6.7.: Combined Training and Position wise testing result.

Training Data (%)	Training Result (%)	Test Result(%) Position 1	Test Result (%) Position 2	Test Result (%) Position 3	Test Result (%) Position 4
30	90.67	18.08	23.07	21.37	25.14
50	91.69	21.28	22.68	25.2	23.42
70	92.10	24.27	20.39	21.28	15.59

Section-3 (Results for Position-1 only)

Data under consideration: training (80%), testing (20%)

Fault Class	Testing Result
GOOD	80 %
CHN	83%
CCN	86%
TAPPET	90%
PGD	78%
PGW	91%

6.3.5. *Method 4: NSUR and ICA based Classification*

(1) NSUR (Number of Samples in Unit Rotation) calculation:
(2) RPM + Sampling Frequency, based signal balancing:
(3) Sampling Frequency (SF) = 100000
(4) RPM varies as per data recording

For balancing the data, number of samples varies with variation in RPM, so number of samples in unit rotation may play a suitable role in reducing the intra-class variation in data. Number of Samples in Unit Rotation (NSUR) can be calculated as:

$$NSUR = (SF * 60) / RPM$$

Pseudo-code for NSUR and ICA Based Classification

(1) *Apply ICA on four positions audio signals*
(2) *Select one IC with maximum Eigen value*
(3) *Calculate NSUR on extracted signal.*
(4) *Collect 100 column vectors, each of NSUR length*
(5) *Repeat the entire process to some selected files for each class*
(6) *Apply Decision tree for classification*

6.3.6. *Method 5: ICA and FFT based Classification*

Pseudo-code for ICA and FFT Based Classification

(1) *Apply ICA on four positions audio signals*
(2) *Select one IC with maximum Eigen value*
(3) *Apply FFT, and trim its imaginary part up to 100000*
(4) *Calculate Entropy*
(5) *Evaluate Range for entropy variation*
(6) *Take Decision*

Results were tested on data with 2500 rpm & 3500 rpm. Results for 2500 rpm data (50 % data was involve in Range decision & 50 % were tested) The results are as follows:

Table 6.8.: Classification result for 3500 RPM data.

Result For 3500 RPM		
Set 1	50% data for training	85.74%
	50 % data for testing	26.68%
Set 2	50% data for training	66.32%
	50 % data for testing	27.84%

Table 6.9.: Classification result for 2500 RPM data.

Class	Correct/Total	Testing Results
Primary Gear (PGD + PGW)	65/68	95%
Tappet	39/39	100%
Cam Chain Noise	34/35	97%
Cylinder Head Noise	18/21	85%
Good	00/15	00%

6.4. Discussion and Conclusion

As described above, we have applied several methods independently as well as in combination. The results varies algorithm to algorithm. Some algorithms are giving good result for some of the classes. And, independently, none of them are able to give more than 80% results to all. But, in combination, they are giving much better result. It has also been found that in the case of 2500 RPM data, the results are highly encouraging. So, in real life situation, it is better to use these algorithms in combination and one can

ICA + NSUR + Decision tree classification		
		Classification Accuracy
SET-1	238 data for training	66.55%
	3500 data for testing	43.71%
SET-2	245 data for training	67.76%
	1530 data for testing	63.01%
SET-3	250 data for training	67.60%
	1530 data for testing	54.72%
SET-4	259 data for training	63.32%
	1530 data for testing	54.71%
SET-5	260 data for training	63.08%
	1530 data for testing	54.71%
SET-6	354 data for training	53.67%
	1530 data for testing	41.70%
SET-7	520 data for training	44.23%
	1530 data for testing	52.03%

Fig. 6.7.: Classification result of NSUR and ICA algorithm.

think of a hierarchical kind of architecture for a successful implementation and application.

References

1. T. D. Sanger, Optimal unsupervised learning in a single-layer linear feedforward neural network, *Neural Networks.* **2**(6), 459–473, (1989). doi: 10.1016/0893-6080(89)90044-0.
2. H. J. Carmichael, *Statistical Methods in Quantum Optics I: Master Equations and Fokker-Planck Equations.* (Springer-Verlag, 2002).

Chapter 7

Hue-Preserving Color Image Enhancement Using Particle Swarm Optimization

A. Ghosh[1] and A. Gorai[2]

Center for Soft Computing Research,
Indian Statistical Institute,
203 B. T. Road, Kolkata 700108, India
[1] *ash@isical.ac.in*

Machine Intelligence Unit,
Indian Statistical Institute,
203 B. T. Road, Kolkata 700108, India
[2] *apurbagorai@gmail.com*

Image enhancement techniques are used to improve image quality by maximizing the information content in the input image using intensity transformation functions. For color image enhancement, gray-level enhancement techniques may be applied directly on R, G, B color planes separately, but in this case colors of the enhanced image is distorted as R, G, B planes are enhanced in different scales. So, perception of the original color in the image will be changed, which means hue of the input image is altered. Sometimes enhancement is done in HSI/LHS color space by processing intensity and saturation components keeping the hue unaltered to preserve the original color perception. While performing reverse transformation from these color spaces to RGB space, it usually creates gamut problem. In this article a PSO based hue preserving gamut problem free color image enhancement technique is proposed. The process is as follows. Image enhancement is considered as an optimization problem and particle swarm optimization (PSO) is used to solve it. The three color components are used to construct an intensity image. The contrast of the intensity image is enhanced by a parameterized transformation function, in which parameters are optimized by PSO, based on an objective function. The transformation function uses local and global information of the input image and the objective function considers the entropy and edge information to measure the image quality. We tried to achieve the best enhanced image according to the objective criterion. The enhanced R, G and B components are obtained

157

from enhanced intensity image by scaling. Scaling may produce gamut problem for few pixels in the enhanced color image. Pixels with the gamut problem are transformed to HSI color space and rescaling is done to the saturation component to remove the gamut problem. Afterwards the pixel is again transformed to RGB space. The algorithm is tested on several color images and the results are compared with two other popular color image enhancement techniques like hue-preserving color image enhancement without gamut problem (HPCIE) and a genetic algorithm based approach to color image enhancement (GACIE). Visual analysis, detail and background variance of the resulted images are reported. It has been found that the proposed method produce better results than the other two compared methods.

7.1. Introduction

Image enhancement, one of the important image processing techniques, can be treated as transforming one image to another to improve the interpretability or perception of it for human viewers, or to extract finer details of images which may provide better input for other automated image analysis systems. With the rapid increase in the usage of color images, it has become a necessity to develop tools and algorithms for color image enhancement.

Histogram transformation is considered as one of the fundamental processes for contrast enhancement of gray level images. Some other methods are the variants of histogram equalization.[1] Linear contrast stretching[1] employs a linear transformation that maps the gray-levels in a given image to fill the dynamic range of gray levels. Pseudocoloring is an enhancement technique that artificially "color" the gray-scale image based on a color mapping, with the extensive interactive trials required to determine an acceptable mapping.[1] These methods indirectly modify the image histogram. Consequently, they reveal maximum information contained in the image.

Generalizing gray scale image enhancement to color image enhancement is not a trivial task. Several factors, such as selection of a color model, characteristics of the human visual system, and color contrast sensitivity, must be considered for color image enhancement. Contrast enhancement of color images is a well studied problem. Boccignone and Picariello[2] have suggested a multi-scale approach to contrast enhancement using a nonlinear scale-space representation of image generated by anisotropic diffusion. Another multi-scale contrast enhancement technique is developed by Toet[3] through non-linear pyramid recombination. Color images can be enhanced by separating the image into chromaticity and intensity components.[4] A scheme for local contrast enhancement of gray-scale images using multi-scale morphology is proposed in Ref. 5. Strickland *et al.*[6] have proposed a scheme for color enhancement based on saturation. A method for color equalization with its application to color images is developed by Bockstein.[7] A scheme for color image enhancement employing genetic algorithm is proposed by Shyu *et al.*[8] Tang *et al.*[4] have proposed a method of enhancing color images via chromaticity diffusion. Oakley *et al.*[9] have proposed an enhancement scheme for color images under poor visibility conditions. A hue preserving and gamut problem free color image enhancement is proposed by Naik and Murthy[10] considering RGB and CMY color spaces.

Evolutionary algorithms have been previously used to perform gray image enhancement.[11–14] In Ref. 11, the authors applied a global contrast enhancement technique using genetic programming (GP)[15] to adapt the color map in the image so as to fit the demands of human interpreter. In Ref. 12 a real coded genetic algorithm (GA) is used with a subjective evaluation criterion to globally adapt the gray-level intensity transformation in the image. Combination of different transformation functions with different parameters are used to produce the enhanced image by GA in Ref. 14.

Particle swarm optimization is stochastic optimization technique modeled after the simulation of the social behavior of insects and bird flocks. This technique has parallel positive feedback mechanisms having advantages of parallelism, robustness and can easily be combined/hybridized with other methods. The parallelism and robust global optimum solution searching ability of the swarm based algorithms make them potential candidates for solving difficult problems. This technique is easy to understand and has been successfully applied to solve a wide range of optimization problems. Thus, due to its simplicity and efficiency in navigating large search spaces for optimal solutions, these techniques are used to develop efficient, robust and flexible algorithms to solve some of the difficult problems in the field of color image processing. Color image enhancement is a difficult and challenging task for present researchers. Therefore usefulness of PSO can be used to solve the problem of color image enhancement. At the same time, in comparison to GA, PSO does not require selection, crossover and mutation operations (for details of PSO refer to Ref. 16) and also takes less time to converge. This optimization technique estimates better parameters set in continuous search space than other optimization techniques. As we have a small continuous search space to estimates the parameters, PSO is chosen to perform the task.

In this paper we have performed color image enhancement by particle swarm optimization (PSO), a population based optimization technique. In the proposed method an intensity image is created from the input color image. Enhancement is done to this intensity image using a parameterized transformation function in which parameters are optimized using PSO considering an edge and entropy based objective function. Scaling factor α is calculated from the enhanced intensity image and enhanced color image is formed by scaling. If any pixel faces gamut problem during scaling then it is corrected considering HSI color space. The enhanced color images by PSO are found to be better compared to other automatic color image enhancement techniques. Both objective and subjective evaluations are performed on the output images which speaks about the goodness of PSO. As the proposed methodology deals with the gamut problem and uses an optimization algorithm to perform the image enhancement task, we have chosen two such algorithms for comparison in which, one considers gamut problem and other one uses GA.

The rest of the paper is organized as follows: In Sec. 2, image enhancement (enhancement/evaluation function) is described. In Sec. 3, proposed methodology (basic PSO, proposed methodology for color image enhancement using PSO, removal of gamut problem, parameter setting) is discussed. In Sec. 4, two methods for comparison are briefly described. In Sec. 5, results and discussion are put, and finally in Sec. 6, conclusion of the work are made.

7.2. Image Enhancement

For image enhancement task, a transformation function is required which takes the intensity value of each pixel from the input image and generates a new intensity value for the corresponding pixel to produce the enhanced image. To evaluate the quality of the enhanced image automatically, an evaluation function is needed which tells us about the quality of the enhanced image. In this section we describe the function used for the proposed work.

7.2.1. Enhancement Function

Image enhancement, done in spatial domain, uses a transformation function which generates a new intensity value for each pixel of the $M \times N$ original image to generate the enhanced image, where M denotes the number of columns and N denotes the number of rows. This can be denoted by

$$g(i,j) = T[f(i,j)], \tag{7.1}$$

where $f(i,j)$ is the gray value of the (i,j)th pixel of the input image and $g(i,j)$ is the gray value of the (i,j)th pixel of the enhanced image; T is the transformation function. Local enhancement method apply transformation on a pixel considering intensity distribution among its neighboring pixels.[17] Adaptive histogram equalization (AHE) is one such local enhancement method which gives good result on medical images.[1] However AHE is quite expensive. The method used in this paper is less time consuming and is similar to statistical scaling presented in Ref. 1. The function used here is designed in such a way that it takes both global as well as local information to produce the enhanced image. Local information is extracted from a user defined window of size $n \times n$. The transformation T is defined as:

$$g(i,j) = K(i,j)[f(i,j) - c \times m(i,j)] + m(i,j)^a. \tag{7.2}$$

In Eq. (7.2) a, and c are two parameters, $m(i,j)$ is the local mean of the (i,j)th pixel of the input image over an $n \times n$ window and $K(i,j)$ is the enhancement function which takes both local and global information into account. Expression for local mean function is given as:

$$m(i,j) = \frac{1}{n \times n} \sum_{x=0}^{n-1} \sum_{y=0}^{n-1} f(x,y). \tag{7.3}$$

One form of the enhancement function $K(i,j)$ used in this work is

$$K(i,j) = \frac{k.D}{\sigma(i,j) + b}. \tag{7.4}$$

where k, and b are two parameters, D is the global mean and $\sigma(i,j)$ is the local standard deviation of (i,j)th pixel of the input image over an $n \times n$ window, and are defined as:

$$D = \frac{1}{M \times N} \sum_{i=0}^{M-1} \sum_{j=0}^{N-1} f(i,j), \tag{7.5}$$

$$\sigma(i,j) = \sqrt{\frac{1}{n \times n} \sum_{x=0}^{n-1} \sum_{y=0}^{n-1} (f(x,y) - m(i,j))^2}. \tag{7.6}$$

Thus the transformation function looks like:

$$g(i,j) = \frac{k.D}{\sigma(i,j)+b}[f(i,j) - c \times m(i,j)] + m(i,j)^a. \qquad (7.7)$$

By this transformation (Eq. (7.7)), contrast of the image is stretched considering local mean as the center of stretch. Four parameters are used in the transformation function, namely a, b, c, and k to produce large variation in the processed image.

7.2.2. *Enhancement Evaluation Criterion*

To evaluate the quality of an enhanced image without human intervention, we need an objective function which will speak all about the image quality. Many objective functions are presented in literature.[18–20] In this study the objective function is formed by combining three performance measures, namely entropy value, sum of edge intensities and number of edgels (edge pixels). It is observed that compared to the original image good contrast enhanced image has more number of edgels[1] and enhanced version should have a higher intensity of the edges.[13] But these two are not sufficient to test an enhanced image and that is why one more measure has been taken i.e. entropy value of the image. Entropy value reveals the information content in the image. If the distribution of the intensities are uniform, then we can say that histogram is equalized and the entropy of the image is more. The objective function considered here is:

$$F(I_e) = log(log(E(I_s))) \times \frac{n\_edgels(I_s)}{M \times N} \times H(I_e). \qquad (7.8)$$

In the above equation I_e is the enhanced image of I_o (the original image) produced by the transformation function defined in Eq. (7.7). The edges or edgels can be detected by different edge detector algorithms such as Sobel,[1] Laplacian,[1] Canny[21] etc. In this study Sobel[1] is used as an automatic edge detector.[22] After using Sobel edge operator we produce an edge image I_s on the enhanced image I_e as:

$$I_s(i,j) = \sqrt{\delta m_{I_e}(i,j)^2 + \delta n_{I_e}(i,j)^2}; \qquad (7.9)$$

$\delta m_{I_e}(i,j) = g_{I_e}(i+1,j-1) + 2g_{I_e}(i+1,j) + g_{I_e}(i+1,j+1) - g_{I_e}(i-1,j-1) - 2g_{I_e}(i-1,j) - g_{I_e}(i-1,j+1)$

and

$\delta n_{I_e}(i,j) = g_{I_e}(i-1,j+1) + 2g_{I_e}(i,j+1) + g_{I_e}(i+1,j+1) - g_{I_e}(i-1,j-1) - 2g_{I_e}(i,j-1) - g_{I_e}(i+1,j-1)$.

$E(I_s)$ is the sum of $M \times N$ pixel intensities of Sobel edge image I_s. *n_edgels* is the number of pixels, whose intensity value is above a threshold in the Sobel edge image. Threshold value is calculated using global thresholding[1] approach on the sobel edge image I_s. Entropy value is calculated on the enhanced image I_e as:

$$H(I_e) = - \sum_{i=0}^{L-1} e_i, \qquad L = 256 \tag{7.10}$$

where $e_i = h_i log_2(h_i)$ if $h_i \neq 0$ otherwise $e_i = 0$, and h_i is the probability of occurrence of ith intensity value of image I_e.

7.3. Proposed Methodology

7.3.1. *Theory of PSO*

PSO is an optimization algorithm proposed by Kennedy and Eberhart in 1995.[23] This optimization algorithm is a multiagent based search strategy[24,25] modeled on the social behavior of organisms such as bird flocking and fish schooling. PSO as an optimization tool, provides a population-based search procedure in which individuals called particles change their position with time. In a PSO system, particles fly around in a multi-dimensional search space. During flight, each particle adjusts its position according to its own experience, and the experience of its neighboring particles, making use of the best position encountered by itself and its neighbors. Thus, as in GAs[26] and memetic algorithms,[27] a PSO system combines local search with global search, attempting to balance exploration and exploitation.

In PSO, each single solution is a "particle". All the particles have fitness values which are evaluated by the objective function to be optimized, and have velocities which direct the flying of the particles. The particles fly through the search space by following the personal and global best positions.

The swarm is initialized with a group of particles randomly and it then searches for optima by updating through iterations. In every iteration, each particle is updated by following two "best" values. The first one is the best position of each particle achieved so far. This value is known as *pbest* solution. Another one is the best position tracked by any particle among all generations of the swarm. This best value is known as *gbest* solution. These two best values are responsible to drive the particles to move to new better positions.

After finding the two best values, a particle updates its velocity and position with the help of the following equations:

$$v_i^{t+1} = W^t.v_i^t + c_1.r_1.(pbest_i^t - X_i^t) + c_2.r_2.(gbest^t - X_i^t), \qquad (7.11)$$

and

$$X_i^{t+1} = X_i^t + v_i^{t+1}; \qquad (7.12)$$

where X_i^t and v_i^t denote the position and velocity of ith particle at time t, W^t is the inertia weight at tth instant of time, c_1 and c_2 are positive acceleration constants, and r_1 and r_2 are random values in the range [0,1], sampled from a uniform distribution. $pbest_i$ is the best solution of ith individual particle over its flight path, $gbest$ is the best particle obtained over all generations so far.

Algorithm 7.1 (PSO algorithm)

Create and initialize P number of d dimensional particles;
Repeat
for Each Particle $i = 1$ to P **do**
  set the personal best position
  if $F(P_i) > F(pbest_i)$ **then**
    $pbest_i = P_i$
  end if
  set the global best position
  if $F(P_i) > F(gbest)$ **then**
    $gbest = P_i$
  end if
end for
for Each Particle $i = 1$ to P **do**
  update the velocity using Eq. (7.11)
  update the position using Eq. (7.12)
end for
until stopping condition is true;

7.3.2. *Proposed Methodology*

In this work, first we have constructed an intensity image from the input color image with the help of I component of HSI color space [*Where*

$I = (R + G + B)/3$]. For the proposed algorithm, we have considered this intensity image as input. Now enhancement is done on this intensity image.

To produce an enhanced image from an input image, a parameterized transformation function defined in Eq. (7.7) is used, which incorporates both global and local information of the input image. The transformation function contains four parameters namely, a, b, c, and k. Different values of these parameters produce different enhanced images. These four parameters have their defined ranges (mentioned in the parameter setting section). Now our aim is to find out the set of values for these four parameters which can produce the best result (according to the fitness function value) with *PSO*.

In this algorithm P number of particles are initialized, each with four parameters a, b, c, and k with random values within their specified range. It means position vector of each particle X has four components a, b, c, and k. Now using these parameter values, each particle generates an enhanced image using the intensity transformation function defined in Eq. (7.7). Transformation function is applied on each pixel in the input image taking the parameter values from each particle, and generate a modified intensity value for that pixel. So each generation will produce P number of enhanced image. Quality of each enhanced image is calculated by an objective function (fitness function) defined in Eq. (7.8) which is termed as fitness of the particle. Fitness value of all the enhanced images generated by all the particles are calculated. From these fitness values *pbest* and *gbest* locations are found. *Pbest* solution is the best location (solution) of a particular particle that it has visited so far, it is also refereed to as cognitive component and *gbest* solution is the best location of any particle among all generation, also refereed as social component. In PSO the most attractive property is that *pbest* and *gbest* are highly responsible to drive each particle (solution) to the direction of best location (Eq. (7.11) and (7.12)).

In each step (iteration) P number of new locations are generated. From every generation *pbest* and *gbest* are found out according to their fitness values. With the help of these two best values, componentwise new velocity of each particle is calculated to get the new solution using (Eqs. (7.11) and (7.12)). In this way new positions of particles are created for generations. When the process is completed the enhanced image is created by the *gbest* position of the particles, as it provides the maximum fitness value.

The enhanced image produced by the *gbest* location is the enhanced version of the intensity image. A pixel in a color image has a color vector with three components R, G, and B. Enhancement procedure applied on three different planes (R, G, and B) modifies the three components of the color vector by three different scales. This leads to change in hue of the color vector which is undesirable. A way of making this transformation hue preserving is to have the same scale for each of the three components of the color vector, known as scaling. As it is known, scaling and shifting are hue preserving transformations,[10] so from the enhanced image we can find out an α scaling factor. $\alpha(i,j)=f(i,j)/g(i,j)$, where $\alpha(i,j),f(i,j)$, and $g(i,j)$ are the scaling factor, intensity value of the input intensity image and intensity value of the enhanced image of (i,j)th pixel respectively. Now we construct the enhanced color image with the help of the scaling factor α, as $(eR, eG, eB)(i,j) = (\alpha R, \alpha G, \alpha B)(i,j)$, here eR stands for enhanced red value.

It can easily be seen that for few pixels, enhanced R, G, B values produced after scaling by α may not lie in the range [0-255], which is popularly known as gamut problem. We have solved the problem taking help of HSI color space for those pixels which have faced gamut problem. Detail description of this technique is given in the next section:

7.3.3. Removal of Gamut Problem

When α scaling factor is used to get the enhanced R, G, B values, few pixels faced gamut problem. We have analyzed the problem considering HSI color space. Figure 7.1 shows the HSI color model with other details. Let R, G, and B values of (i,j)th pixel be such that it provides intensity component 0.5, indicated as 'O' on the intensity axis, (OP) the saturation value and 'H' the hue. After enhancement 'O' is shifted to 'O'' on the intensity axis. As we are only processing the intensity component so hue and saturation will remain unaltered. From Fig. 7.1 it is clear that at 'O'' saturation value will be $(O'P')$, where point P' is outside the color gamut. Thus keeping the same saturation value, if R, G, B values of (i,j)th pixel are modified then it will produce gamut problem. At 'O'', it is allowed to have maximum saturation of $(O'P'')$. Our interest here is to find out the value of $(O'P'')$ to overcome the gamut problem.

To do this, first let us consider $\triangle ORP$, where $\angle ORP = 60°$ as $\triangle ORY$ is an equilateral triangle, $\angle ROP = H$, $\angle RPO = h = (180-(H+60))°$, and $OR=1$. Therefore $OP=(\sin 60°)/(\sin h)=S$. Now we can consider $\triangle WOP$ with two sides 'OW' and 'OP' known. At the same time $\triangle WOP$ is a right angle triangle with $\angle WOP = 90°$. Now we can easily find out the length of $(O'P'')$ by ratio rule in case of right angled triangle, where $(O'P'') =$

Algorithm 7.2 (PSO based color image enhancement)

Create Intensity image 'I' of the input color image, where (I=(R+G+B)/3).

Provide 'I' as the input image to the algorithm.

Create P number of d dimensional particles.

for Each Particle $i = 1$ to P **do**

  Initialize parameters a, b, c, k and velocities randomly.

end for

while (Termination condition $\neq$ true) **do**

  for each Particle $i = 1$ to P **do**

    Generate enhanced image I_e using Eq. (7.7).

    Calculate objective functional value $F()$ using Eq. (7.8).

    //Set *pbest* as personal best of ith particle achieved so far.

    if $F((I_e)_i) > F(pbest_i)$ **then**

      $pbest_i = P_i$

      // P_i is the ith particle

    end if

    //Set g_{best} as global best achieved so far among all generation.

    if $F((I_e)_i) > F(gbest)$ **then**

      $gbest = P_i$

    end if

  end for

  for Each Particle $i = 1$ to P **do**

    Update the velocity using Eq. (7.11).

    Update the position using Eq. (7.12).

  end for

end while

Enhanced image is formed by *gbest* location of the particle.

Scaling factor α (where $\alpha = I_e(i,j)/I(i,j)$) is calculated for each pixel.

(eR,eG,eB)(i,j)=(αR,αG,αB,)(i,j).

if eR,eG,eB are not in the range [0-255] **then**

  Remove the gamut problem as described.

end if

$S-((OO'/OW) \times S)$. Thus we get the enhanced hue value of (i,j)th pixel as '$O'P'''$'. We are now able to calculate the RGB values of the corresponding HSI value by the HSI to RGB conversion[1] using H (hue), newly calculated S (saturation) i.e. $(O'P'')$ and 'O'' as intensity. Processing the saturation

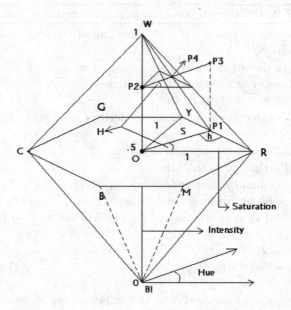

Fig. 7.1.: HSI color model with other details regarding removal of gamut problem.

component in this way helps us to overcome gamut problem. Figure 7.1 shows the solution for only 60° hue, considering $\triangle ORY$, similarly we can solve the problem for any hue considering other concerned triangles.

7.3.4. *Parameter Setting*

The result of PSO algorithm is very much parameter dependent. Fine tuning of the parameters can provide better results than other optimization algorithms. Parameter W used in Eq. (7.11) is called the inertia weight. Maximum and minimum value for W is set to two and zero respectively, which is the same for all particles. The process starts with maximum inertia value and gradually reduces to minimum. Therefore initially inertia component is big and explore larger areas in the solution space, but gradually inertia component becomes small as time 't' passes ($t = iteration$ $number$) and exploit better solutions in the solution space. Inertia value

W is calculated as:

$$W^t = W_{max} - \frac{W_{max} - W_{min}}{t_{max}} \times t. \qquad (7.13)$$

Parameters c_1, and c_2 of Eq. (7.11) are positive acceleration constants, a random number in [0,2]. These parameters are fixed for each particle throughout its life. We have fixed it at 1.3 for both c_1, and c_2. Parameters r_1 and r_2 are random numbers in [0,1] and varies for each component of the particles in every generation.

In this study there are four problem specific parameters, a, b, c, and k of Eq. (7.7). The range of these parameters are the chosen as in Ref. 17. $a \in [0, 1.5]$, $b \in [0, 0.5]$, $c \in [0, 1]$ and $k \in [0.5, 1.5]$. We have changed the range for parameter b as the supplied range did not produce good result. It has been observed that small value for b stretch the intensity by a large amount. So, after normalizing the intensity values between [0,255] to display the image, it becomes very discritized and the originality of the image is lost. So, we increase the range to overcome this specific problem, and the range is fixed to $[1, (D/2)]$, where D is the global mean of the original image. Initial velocities for parameters 'a', 'c' and 'k' are set randomly between [0, 0.5] and for 'b', it is between [0, 10].

7.4. Methods Compared With

In this article a gamut problem free color image enhancement algorithm is proposed in which we have done the enhancement using particle swarm optimization considering enhancement as an optimization problem. To compare our results, we have selected two different color image enhancement algorithms, in which one deals with hue preservation and gamut problem and the other one uses genetic algorithm to enhance the color images. Brief description of these algorithms are given in the next section.

7.4.1. *Hue-Preserving Color Image Enhancement Without Gamut Problem (HPCIE)*

Hue preserving color image enhancement without gamut problem is proposed by Naik and Murthy.[10] In this method α is taken as a function of $l_{\tilde{x}}$, where $l_{\tilde{x}} = x_1 + x_2 + x_3$ [where, $x_1 = R/255$, $x_2 = G/255$ and $x_3 = B/255$]. Then the transformation will be of the form

$$x'_k = \alpha(l_{\tilde{x}})x_k \qquad \forall \tilde{x} \in I, k = 1, 2, 3. \qquad (7.14)$$

Initially, they defined $\alpha(l_{\tilde{x}}) = f(l_{\tilde{x}})/l_{\tilde{x}}$, where $f(l_{\tilde{x}})$ is a S-type[10] nonlinear transformation used in contrast enhancement for gray scale images. In this case $l_{\tilde{x}}$, and $f(l_{\tilde{x}}) \in [0,3]$ as x_1, x_2, and $x_3 \in [0,1]$. Therefore $\alpha(l_{\tilde{x}})$ is a ratio of $f(l_{\tilde{x}})$ and $l_{\tilde{x}}$, when $f(l_{\tilde{x}}) > l_{\tilde{x}}$, value of $\alpha(l_{\tilde{x}})$ will be greater than 1. In such a case value of x'_k may exceed 1 and thus resulting in gamut problem. A possible solution proposed in this paper is to convert those pixels to CMY color space and process in the same way taking C, M and Y values instead of R, G and B values. Doing so, ratio will be less than 1 and gamut problem can be rectified.

Actually $\alpha > 1$ does not produce gamut problem always. So, in case of standard S-type contrast enhancement function all the pixels belonging to the upper half of S-type function are transformed to CMY color space although all of them may not face gamut problem. It means few pixels take unnecessary transformation from RGB to CMY and subsequently CMY to RGB color space. At the same time it produces the same type of enhancement based on the enhancement function. It is not automatically adjusted according to the image type.

7.4.2. *A Genetic Algorithm Approach to Color Image Enhancement (GACIE)*

A genetic algorithm approach to color image enhancement is proposed by Shyu and Leou.[8] In this study, LHS color space is considered and a GA based color image enhancement is proposed, in which color image enhancement is formulated as an optimization problem. In the proposed approach, a set of generalized transformations for color image enhancement is formed by a linear weighted combination of four types of nonlinear transformations. These four transformations are applied on the L component and beta transformation is applied on the saturation component of the LHS color space. The fitness function for GAs is formed by four performance measures, namely, the AC power measure, the compactness measure, the Brrenner's measure, and the information-noise change measure. Then GAs are used to determine the "optimal" set of parameters for the generalized transforms with the largest fitness function value. The transformed R, G, B values may not be in the range [0,255] in the enhanced image. To bring them in the range, all pixel values are normalized with the help of max(R, G, B) value. After normalization enhanced color image is displayed.

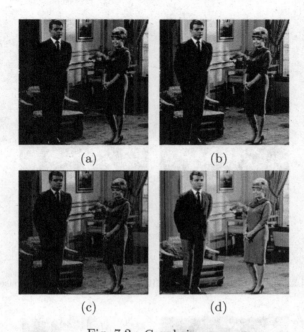

Fig. 7.2.: Couple image.
(a) Original Image, (b) HPCIE, (c) GACIE, and (d) Proposed PSO based Method.

7.5. Results and Discussion

The proposed method is tested on many color images. Here we put results of six color images. Results of the proposed method is compared with two other methods, namely (i) Hue-preserving color image enhancement without gamut problem ($HPCIE$) and (ii) a genetic algorithm approach to color image enhancement ($GACIE$). The results of the proposed method and GACIE are taken the best produced results based on the respective fitness value out of several runs. The results of all these algorithms are evaluated using the same evaluation function and the results are put in Table 7.2. The description of the input images and *Detail Variance (DV)*, *Background Variance (BV)*, and *fitness* are given in the Table 7.1. Description of DV and BV are given below.

Objective Evaluation: The objective evaluation criterion considered here is the Detail Variance (DV) and Background Variance (BV).[28] DV and BV values are calculated by computing the local variance considering the neighbors of each pixel over an $n \times n$ window ($n = 3$) of the intensity

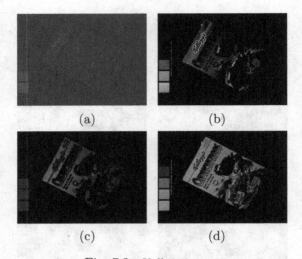

Fig. 7.3.: Kellogs image.
(a) Original Image, (b) HPCIE, (c) GACIE, and (d) Proposed PSO based Method.

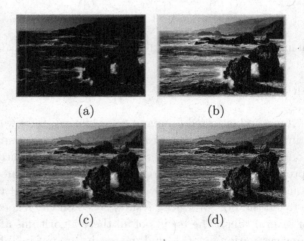

Fig. 7.4.: Sea image.
(a) Original Image, (b) HPCIE, (c) GACIE, and (d) Proposed PSO based Method.

image of the input color image and enhanced color image. The pixel is classified as a foreground one when the variance is more than a threshold (1.5 in the present case), otherwise it is classified as a background pixel. The averaged variance of all pixels included in the foreground class is DV,

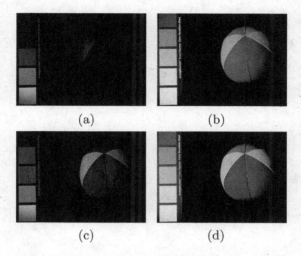

(a)   (b)

(c)   (d)

Fig. 7.5.: Ball image.
(a) Original Image, (b) HPCIE, (c) GACIE, and (d) Proposed PSO based Method.

Table 7.1.: Details of the original images.

Image	Size(M x N)	P/t/w	Fitness	DV	BV
Couple	256 x 256	30/20/5	0.394272	8.939399	1.189810
Kellogs	637 x 468	30/20/3	0.031078	6.838592	0.849381
Sea	400 x 266	30/20/5	0.526847	10.058320	1.056729
Ball	637 x 468	30/20/3	0.086996	7.165190	0.843959
Flower	637 x 468	30/20/3	0.033159	6.840421	1.046732
Duck	576 x 768	40/20/7	0.792870	10.126090	1.664255

P, t, and w in the third column of Table 7.1 signify the number of particles, maximum number of generations and window size taken to extract the local information, correspondingly.

and the averaged variance of all pixels included in the background class is BV. An image is said to be enhanced if the DV value of the resulted image increases, while BV value is not changed much compared to the input image. From Table 7.2 it is very clear that the proposed technique is giving better results than other techniques according to the *fitness*, DV, and BV values.

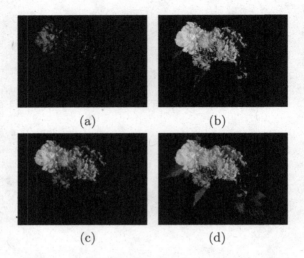

(a)   (b)

(c)   (d)

Fig. 7.6.: Flower image.
(a) Original Image, (b) HPCIE, (c) GACIE, and (d) Proposed PSO based Method.

Visual analysis of the enhanced images: Results of six enhanced color images have been displayed. If we visually analyze the images then we see that for the *Couple* image in Fig. 7.2, the *PSO* based enhancement result is better than other two methods. Darker portion under the table and chair is more visible, and the legs of the chair are quite clear compared to the other two methods. In the *Kellogs* image in Fig. 7.3, all the methods provide good results, as detail information on the books are visible in all the resulted images. Most interesting thing is that if we minutely see the images, then we can say that only the proposed method shows the prominent stripes on the background clearly which is not so for the other two methods. In the *Sea* image in Fig. 7.4, the result produced by HPCIE percept good visually but if we see carefully then we can analyze that shadow formed by the stones are darker than other resulted images. Both Fig. 7.4(b) and 7.4(c) give less variation of pixel values where water is striking the stones and looks very much brighter whereas in Fig. 7.4(d) it provides every details of the image. Next two images *Ball* and *Flower* in Figs. 7.5 and 7.6 are the same type of low contrast image. In these two images GACIE and the proposed method provided better results than HPCIE. Though HPCIE enhanced the object region properly, but the stripes on the background are not extracted at all which is done by GACIE and the proposed method as well. In the flower image the proposed method provides inner details of the

(a)   (b)

(c)   (d)

Fig. 7.7.: Duck image.
(a) Original Image, (b) HPCIE, (c) GACIE, and (d) Proposed PSO based Method.

background which can be visualized from Fig. 7.7(d). If we consider the *Duck* image in Fig. 7.7, HPCIE produced a smoother, brighter and a good contrast image but the darker regions are not well enhanced. Details on the stone are still hidden in dark and black regions of the duck are not very prominent. Folds on the wings of the duck are not very clear to viewers. Same is true for GACIE based result also. All these details are very much clear in the enhanced image by the proposed method. Feather details and ground details are also very prominent here. From Table 7.2, we can draw conclusion on the results of all the methods and can say that for all the six images the proposed method provides better results in all respect i.e., *fitness*, *DV* and *BV* values than the other two methods.

Table 7.2.: Fitness, detail variance and background variance of the enhanced images for different methods w.r.t. input image.

Image	Measure	Input Image	HPCIE	GACIE	PSOCIE
Couple	Fitness	0.394272	0.876685	0.458671	**1.357177**
	DV	8.939399	10.423670	9.282858	**13.040270**
	BV	1.18981	1.745627	**1.452154**	1.497653
Kellogs	Fitness	0.031078	0.374782	0.148802	**0.573081**
	DV	6.838592	7.374213	5.681680	**14.055740**
	BV	0.849381	1.952481	1.862559	**1.519606**
Sea	Fitness	0.526847	1.316842	0.918073	**1.413641**
	DV	10.05832	14.248320	12.280040	**15.528380**
	BV	1.056729	1.003202	1.158371	**1.056729**
Ball	Fitness	0.086996	0.184661	0.884720	**1.791936**
	DV	7.16519	8.509681	14.683910	**17.437290**
	BV	0.843959	1.292823	**1.142411**	1.301100
Flower	Fitness	0.033159	0.261068	0.080697	**2.942127**
	DV	6.840421	7.431453	7.861288	**21.982720**
	BV	1.046732	**1.584418**	1.777610	1.728160
Duck	Fitness	0.79287	1.178543	0.846922	**1.571635**
	DV	10.12609	12.449450	10.412370	**13.453540**
	BV	1.664255	1.340164	**1.682339**	1.146161

7.6. Conclusion

In this paper we have proposed a PSO based automatic color image enhancement technique. In the proposed method enhancement is done on the intensity image. Scaling factor α is calculated from the enhanced intensity image by which scaling is done to produce enhanced color image. While scaling, if any pixel faces gamut problem then it is corrected considering HSI color space. Results of the proposed technique are compared with two other techniques, like hue-preserving color image enhancement without gamut problem and a genetic algorithm approach to color image enhancement. For all the six images shown in this paper, it is observed that the proposed technique is giving better results compared to other methods mentioned above. In HPCIE technique pixels may get transformed to CMY color space without having gamut problem also. It also produces the same enhancement to all images based on the transformation function. This technique is not adoptive with image type. The proposed technique takes care of these points. In PSO, the most important property is that it can produce better results with proper tuning of parameters. It is also true for GA based image enhancement, but it needs more number of chromosomes and needs more generations to obtain the optimal solutions compared to PSO. PSO also does not need crossover and mutation operations like GA, which helps to reduce its complexity. At present there are many variants of PSO, we can try our proposed algorithm using these variants to improve the result further. We can also try with multi objective particle swarm optimization to improve the enhancement quality of color images considering other relevant objective functions.

Acknowledgment

Authors would like to thank the Department of Science and Technology, Government of India and University of Trento, Italy, the sponsors of the ITPAR program and the Center for Soft Computing Research, Indian Statistical Institute, Kolkata for providing computational facilities.

References

1. R. Gonzales and R. Woods, *Digital Image Processing.* (Prentice Hall, 2008).
2. G. Boccignone and A. Picariello. Multiscale contrast enhancement of medical images. In *Proceedings ICASSP'97*, Munich, Germany, (April 21-24, 1997).
3. A. Toet, A hierarchical morphological image decomposition, *Pattern Recognition Letter.* **11**(4), 267–274, (1990).
4. B. Tang, G. Sapiro, and V. Caselles, Color image enhancement via chromaticity diffusion, *IEEE Transaction on Image Processing.* **10**(5), 701–707, (2001).
5. S. Mukhopadhyay and B. Chanda, Multiscale morphological approach to local contrast enhancement, *Signal Processing.* **80**(4), 685–696, (2000).
6. R. N. Strickland, C. S. Kim, and W. F. McDonnell, Digital color image enhancement based on the saturation component, *Optical Engineering.* **26**, 609–616, (1987).
7. I. M. Bockstein, Color equalization method and its application to color image processing, *Journal of Opt. Soc. Am.* **A3**(5), 735–737, (1986).
8. M. Shyu and J. Leon, A genetic algorithm approach to color image enhancement, *Pattern Recognition.* **31**(7), 871–8880, (1998).
9. J. P. Oakley and B. L. Satherley, Improving image quality in poor visibility conditions using a physical model for contrast degradation, *IEEE Transaction on Image Processing.* **7**, 167–179, (1998).
10. S. K. Naik and C. A. Murthy, Hue-preserving color image enhancement without gamut problem, *IEEE Transaction on Image Processing.* **12**(12), 1591–1598, (2003).
11. R. Poli and S. Cagnoni, Evolution of pseudo-colouring algorithms for image enhancement with interactive genetic programming, *Univ. Birmingham.* **CSRP-97-5**, (1997).

12. C. Munteanu and V. Lazarescu. Evolutionary contrast stretching and detail enhancement of satellite images. In *In Proc. Mendel*, pp. 94–99, (Berno, Czech Rep. 1999).

13. F. Saitoh. Image contrast enhancement using genetic algorithm. In *Proc. IEEE SMC*, pp. 8–14, Tokyo, Japan, (1999).

14. S. K. Pal, D. Bhandari, and M. K. Kundu, Genetic algorithms for optimal image enhancement, *Pattern Recognition Letter.* **15**, 261–271, (1994).

15. T. Bäck, D. Fogel, and Z. Michalewicz, *Handbook of Evolutionary Computation.* (Oxford Univ. Press, London, U. K., 1997).

16. A. P. Englebrecht, *Computational Intelligence: An Intorduction.* (John Wiley and Sons, 2007).

17. C. Munteanu and A. Rosa, Gray-scale enhancement as an automatic process driven by evolution, *IEEE Transaction on Systems,Man and Cybernatics-Part B:Cybernetics.* **34**(2), 1292–1298, (2004).

18. S. Jingquan, F. Mengyin, and Z. Chanjian, An image enhancement algorithm based on chaotic optimization, *Computer Engineering and applications.* **27**, 4–6, (2003).

19. G. Runqiu, L. Junfeng, and L. Xiaochun, The infrared image enhancement and the correlative technique based on the parallel genetic algorithm, *XIDIAN Univ.J.* **31**, 6–8, (2004).

20. T. Xiaodong and L. Zhong, Compare and analysis of enhancement methods of sonar image, *Ship Electronics Eng.J.* **26**, 154–157, (2006).

21. A. K. Jain, *Fundamentals of digital image processing.* (Prentice-Hall, Inc., Upper Saddle River, NJ, USA, 1989). ISBN 0-13-336165-9.

22. P. Rosin. Edges: saliency measures and automatic thresholding. In *Geoscience and Remote Sensing Symposium, 1995. IGARSS '95. 'Quantitative Remote Sensing for Science and Applications', International*, Vol. 1, pp. 93–95, (1995).

23. R. C. Eberhart and J. Kennedy. A new optimizer using particle swarm theory. In *Proceedings of the sixth International Symposium on Micromachine and Human Science*, pp. 39–43, Nagoya, Japan, (1995).

24. M. Clerc and J. Kennedy, The particle swarm-explosion, stability, and convergence in a multidimensional complex space, *IEEE Transaction on Evol. Comput.* **6**(1), 58–73, (2002).

25. K. E. Parsopoulos and M. N. Vrahatis, Recent approaches to global optimization problens through particle swarm optimization, *Neural Comput.* **1**, 235–306, (2002).

26. D. E. Goldberg, *Genetic Algorithms in Search, Optimization and Machine Learning.* (Addison-Wesley Longman Publishing Co., Inc., Boston, MA, USA, 1989). ISBN 0201157675.

27. N. Krasnogor. *Studies on the Theory and Design Space of Memetic Algorithms.* PhD thesis, University of the West of England, (2002). Supervisor: Dr. J.E. Smith.

28. G. Ramponi, N. Strobel, S. K. Mitra, and T. H. Yu, Nonlinear unsharp masking methods for image contrast enhancement, *Journal of Electronic Imaging.* **5**(3), 353–366, (1996).

Chapter 8

Efficient Classifier Design with Hybrid Polynomial Neural Network

B. B. Misra, P. K. Dash and G. Panda

Department of Information Technology,
Silicon Institute of Technology,
Bhubaneswar-751024, ORISSA, India
bijanmisra@ieee.org

Department of Electrical Engineering,
Silicon Institute of Technology,
Bhubaneswar-751024, ORISSA, India
pkdash_india@yahoo.co.in

School of Electrical Sciences, Indian Institute of Technology,
Bhubaneswar-751013, ORISSA, India
ganapati.panda@gmail.com

This chapter presents the procedure for designing classifier model using polynomial neural network (PNN). The PNN is a flexible neural architecture. The number of layers of the PNN is not fixed in advance but it is developed on the fly. This network is also considered as a self-organizing network. The essence of the design procedure of PNN dwells on the group methods of data handling (GMDH) technique. Each node of PNN exhibits a high level of flexibility and realizes a polynomial mapping between input and output variables. This chapter presents how artificial neural network (ANN) is incorporated with PNN to enhance the performance substantially. During the study of PNN, it is observed that a huge amount of time is consumed during training the classifier and the developed model is complex in nature. To alleviate these drawbacks the growth of PNN layers is restricted but the approximation capability is maintained by incorporating ANN. Bench mark databases of different domain have been taken to test the performance of this hybrid model. The simulation results of PNN and its hybrid models are compared, which shows that the performance gain obtained in the hybrid model is much better in comparison to the PNN model.

179

8.1. Introduction

Modeling is an important task in control, identification, forecasting and classification problems.[1] Usually limited information is available about the underlying physical laws of complex systems such as engineering, ecological and economic systems. To make structural identification mathematical tools with certain assumptions about their features are needed. Generally these systems consist of subsystems with unknown interrelationships. The behavior of these subsystems is not important as the exhibitory behavior of the system, which is highlighted by analyzing, transforming and manipulating the input and output features.

Very often a complex system is characterized as a black box. The information available about such systems are only the number and nature of input and output variables. In such systems the main concentration remains only on the study of the relationship between the input-output rather than on the internal mechanism of the system. In black box approach, the knowledge about the system is obtained only from the data, where deductive or inductive sorting methods are applied to identify and process the knowledge. For simple problems where the theory of the system is well established and physical laws are applied to identify the physical model like many of the engineering problems, deductive techniques are applied, whereas for complex problems like the ecological and economic problems, the inductive techniques are used. The statistical methods and regression analysis may be considered as the basic approaches for modeling, where a priori knowledge about the laws governing the data as well as their properties is assumed. Such assumptions very often lead to inaccurate model design.

The task of modeling has been investigated by a significantly large number of researchers using back propagation learning for neural networks over three decades. Even though a small number of assumptions in comparison to statistical methods is made but still a significant amount of a priori information about the model's structure, such as the number of hidden layers and neurons, their activation function, quality and quantity of input arguments are needed. In such cases the knowledge about the theory of neural networks and also the rules for the translation of this knowledge into the languages of neural networks[2] are essential. Finding an ideal model is difficult in majority of cases as determination of the network architecture is driven by heuristic approaches. Sarle[3] had compared the behavior of neural networks in data analysis to the statistical methods and claimed that it is not appropriate to be viewed as competitors since there are many overlaps between them.

The inductive approach which is based on the principles of self-organization overcome the subjectiveness of neural network. The inductive approach is similar to neural networks but is unbounded in nature, where the independent variables of the system are shifted in a random way and activated so that the best match to the dependent variables is ultimately selected.[4] In these procedures there is a gradual increase of complexity and the optimum model is found with respect to the popular incompleteness theorem issued by Godel in 1931, i.e., it is impossible to obtain a unique model of an object on the basis of empirical data without using an external complement. The existence of a single optimum model is based on the principle of self-organization which states that when a model's complexity gradually increases, certain selection criteria or objective functions that hold the property of external complement pass through a minimum.[5]

The experts have a limited role to play in case of inductive learning algorithms. In this case the communication with machines is made in a generalized language of integrate signals like selection criterion or objective functions.[6] But in deductive methods the dominant role belongs to the experts and the computer works like large calculators. Further these inductive approaches can be compared to the evolutionary methods,[7] where a number of solutions are created and an external criterion plays the role of finding the fittest.

The polynomial neural network (PNN)[8] is an inductive self organized data driven method. In this chapter the basic concepts of PNN, its architecture, and design procedure are presented. How PNN can be used to solve the task of classification and to develop different classifier models are also dealt in this chapter. Different bench mark databases have been used in this study and the simulation results are presented for reference. For clarity in understanding an example of the various models developed by this technique is also included. It is observed that PNN consumes a large amount of time to develop a classifier model. To improvise the time requirement a hybrid model of PNN, which uses the technique of PNN and ANN to develop the reduced and comprehensible polynomial neural network (RCPNN)[9] is also selected for classification task. This model reduces the time requirement substantially as well as enhances the classification accuracy to a remarkable extent.

Rest of the chapter is organized as follows. Section 2 covers the architecture and procedure for designing a classifier model using PNN. Section 3 discusses the fundamentals and recent developments of particle swarm optimization. Section 4 describes the reduced and comprehensible polynomial neural network for classifier design. Finally the conclusion of investigation is provided in Sec. 5.

8.2. Classification using PNN

The GMDH-type PNN belongs to a kind of inductive self-organization data driven approaches. It requires small data samples which can optimize the structure of the models objectively.

The relationship between input-output variables can be approximated by Volterra functional series, the discrete form of which is Kolmogorov-Gabor polynomial,[10] i.e.,

$$y = C_0 + \sum_{k_1} C_{k_1} x_{k_1} + \sum_{k_1 k_2} C_{k_1 k_2} x_{k_1} x_{k_2} + \sum_{k_1 k_2 k_3} C_{k_1 k_2 k_3} x_{k_1} x_{k_2} x_{k_3} \cdots \quad (8.1)$$

where C_k denotes the coefficients or weights of the Kolmogorov-Gabor polynomial and x vector is the input variable. Any stationary random sequence of observations can be approximated by Eq. (8.1) and then solved by either Gaussian or adaptive methods. This polynomial is intractable if the number of input variables of the database increases; also it takes more computation time to solve all necessary normal equations when the input variables are large.

The GMDH algorithm developed by Ivakhnenko[2,10] is a form of Kolmogorov-Gabor polynomial. He has proved that a second order polynomial with two input variables i.e.,

$$y = a_0 + a_1 x_1 + a_2 x_2 + a_3 x_1 x_2 + a_4 x_1^2 + a_5 x_2^2 \quad (8.2)$$

can able to construct the complete Kolmogorov-Gabor polynomial through an iterative procedure.

The GMDH-type PNN method belongs to the category of heuristic self-organization methods where the approaches like black box, connectionism and induction can be applied.[11] It is a fact that the black-box method is a principal approach to analyze systems on the basis of input-output samples and the method of connection as well as induction which are thought of as representation of complex functions through network of elementary functions. Thus the PNN algorithm has the ability to trace out all input-output relationships through an entire system which is essentially very complex.

The PNN technique has been used extensively in many real life applications. Example of some of the applications of PNN to different fields are prediction of plasma etching,[12,13] signal processing,[14] exchange rate forecasting,[15] time series prediction,[16-19] GPS based power line fault detection,[20] identification of nonlinear system,[21,22] modeling the flux linkage characteristics,[23] prediction of NOx emission,[24] classification,[8,9,25-30] robot kinematics learning,[31] forecasting,[32] determining strength of concrete,[33] face detection,[34] economized screw water chillers,[35] piles shaft capacity,[36] software engineering,[37] variable valve-timing spark-ignition engine,[38] prediction of explosive processes,[39,40] heat transfer,[41] machine error compensation,[42] etc.

8.2.1. *PNN Architecture*

The GMDH-type PNNs[8,12,25,43-54] are multi-layered models consisting of the neurons/active units/partial descriptions (PDs), whose transfer function is a short-term polynomial described in Eq. (8.2). At the first layer $L = 1$, an algorithm, using all possible combinations of two inputs out of m variables, generates the first population of PDs. Hence the total number of PDs in first layer is $k = N(N - 1)/2$ and the output of each PD in layer $L = 1$ is computed by applying the Eq. (8.2). Let the output of first layer be denoted as $y_1^1, y_2^1, \cdots, y_k^1$. The coefficient vector of the PDs is determined by the least square estimation approach. An overall architecture of the PNN is presented in Fig. 8.1.

The details of the development of the PNN model is discussed here. Let the input and output data for training is represented in a matrix form as:

$$(X_i, y_i) = (x_{1i}, x_{2i}, \cdots, x_{Ni}, y_i) \tag{8.3}$$

where $i = 1, 2, 3, \cdots, n$. Hence the input and output relationship of the above data by PNN algorithm can be described in the following manner:

$$y = f(x_1, x_2, x_3, \cdots, x_N) \tag{8.4}$$

where N is the number of features in the database.

Now the number of PDs, k in each layer depends on the number of input features N as in Eq. (8.5).

$$k = \frac{N(N - 1)}{2} \tag{8.5}$$

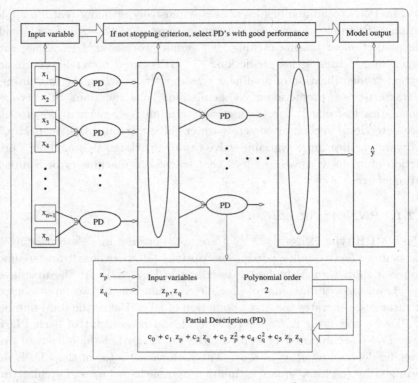

Fig. 8.1.: An overall architecture of the PNN.

After obtaining the values of all the coefficients from Eq. (8.2) based on the training database the target is estimate as:

$$\widehat{y}_i = (C_{j1} + C_{j2}x_{ip} + C_{j3}x_{iq} + C_{j4}x_{ip}x_{iq} + C_{j5}x_{ip}^2 + C_{j5}x_{iq}^2) \qquad (8.6)$$

where $1 \le j \le k$, $k = \frac{N(N-1)}{2}$, $\forall i, 1 \le i \le n$

If the error level is not up to expectation, the PDs for the next layer of PNN are generated based on the output of the previous layer.

$$Z_j = (C_{j1} + C_{j2}x_{ip} + C_{j3}x_{iq} + C_{j4}x_{ip}x_{iq} + C_{j5}x_{ip}^2 + C_{j5}x_{iq}^2) \qquad (8.7)$$

The same procedure is applied to obtain

$$d_i = y_i - (C_{j1} + C_{j2}z_{ip} + C_{j3}z_{iq} + C_{j4}z_{ip}z_{iq} + C_{j5}z_{ip}^2 + C_{j5}z_{iq}^2) \qquad (8.8)$$

This process is repeated until a tolerance of $(d_i \le \epsilon)$ error is reached.

Table 8.1.: Regression polynomial structure.

Order	Number of Inputs		
	1	2	3
1	Linear	Bilinear	Trilinear
2	Quadratic	Biquadratic	Triquadratic
3	Cubic	Bicubic	Tricubic

8.2.2. *Design Procedure of PNN Model*

This subsection deals with the design procedure of the GMDH-type PNN.

Initially the input variables are defined as x_i, $i = 1, 2, \cdots, n$, related to output variables y. Depending on the type of data, normalization of input data is carried out when required.

The database $(X_i, y_i) = (x_{1i}, x_{2i}, \cdots, x_{Ni}, y_i), i = 1, 2, \cdots, n$, is divided into two parts, training and testing. Let the size of training and test databases are denoted by n_{tr} and n_{te} respectively, where $n = n_{tr} + n_{te}$. The training database is used to construct the PNN model, including an estimation of the coefficients of the PD of nodes situated in each layer of the PNN. Next the testing database is used to evaluate the designed PNN model.

The different types of regression polynomials used to develop the PDs related to PNN structure are presented in Table 8.1.

In particular, the input variables of a node are selected from N input variables $x_1, x_2, \cdots, x_N$. The total number of PDs located at the current layer differs according to the number of the selected input variables from the nodes of the preceding layer. This results in $k = \frac{N!}{(N-r)!r!}$ nodes, where r is the number of the chosen input variables to each PD. The choice of the input variables and the order of a PD itself help selecting the best model with respect to the characteristics of the data, model design strategy, nonlinearity and predictive capability. For example, a PD computed on a basis of two input variables and second order polynomial assumes the form of the quadratic regression polynomial.

$$z_m = c_0 + c_1 x_p + c_2 x_q + c_3 x_p^2 + c_4 x_q^2 + c_5 x_p x_q \tag{8.9}$$

where $m = 1, 2, 3, \cdots, \frac{N(N-1)}{2}$. In the above expression the coefficients $(c_0, c_1, c_2, c_3, c_4, c_5)$ are estimated using the training data subset. As the model is linear with respect to the parameters, a standard least square

Table 8.2.: Types of polynomials used for PNN.

	Polynomial
Bilinear	$c_0 + c_1 x_1 + c_2 x_2$
Biquadratic	$c_0 + c_1 x_1 + c_2 x_2 + c_3 x_1^2 + c_4 x_2^2 + c_5 x_1 x_2$
Biquibic	$c_0 + c_1 x_1 + c_2 x_2 + c_3 x_1^2 + c_4 x_2^2 + c_5 x_1 x_2 + c_6 x_1^3 + c_7 x_2^3 +$ $c_8 x_1^2 x_2 + c_9 x_1 x_2^2$
Trilinear	$c_0 + c_1 x_1 + c_2 x_2 + c_3 x_3$
Triquadraic	$c_0 + c_1 x_1 + c_2 x_2 + c_3 x_3 + c_4 x_1^2 + c_5 x_2^2 + c_6 x_3^2 + c_7 x_1 x_2 +$ $c_8 x_2 x_3 + c_9 x_3 x_1$
Tricubic	$c_0 + c_1 x_1 + c_2 x_2 + c_3 x_3 + c_4 x_1^2 + c_5 x_2^2 + c_6 x_3^2 + c_7 x_1 x_2 +$ $c_8 x_2 x_3 + c_9 x_3 x_1 + c_{10} x_1^3 + c_{11} x_2^3 + c_{12} x_3^3 + c_{13} x_1 x_2 x_3$

method is used for optimization task. This results in $\frac{N(N-1)}{2}$ nodes or PDs.

Different types of polynomials considered for development of the PDs in PNN are listed in Table 8.2.

The coefficients of the PDs are estimated as follows. The vector of coefficients C_i is derived by minimizing the mean squared error between y_i and z_{mi}.

$$E = \frac{1}{n_{tr}} \sum_{i=0}^{n_{tr}} (y_i - z_{mi})^2 \qquad (8.10)$$

Using the training data subset, this gives rise to the set of linear equations

$$Y = X_i C_i \qquad (8.11)$$

Evidently, the coefficients of the PD of nodes in each layer are expressed in the form

$$C_i = (X_i^T X_i) X_i^T Y \qquad (8.12)$$

where
$Y = \begin{bmatrix} y_1 \ y_2 \ \cdots \ y_n \end{bmatrix}$,
$X_i = \begin{bmatrix} X_{1i} \ X_{2i} \ \cdots \ X_{ki} \ \cdots \ X_{n_{tr}i} \end{bmatrix}^T$,
$X_{ki}^T = \begin{bmatrix} 1 \ x_{ki1} \ x_{ki2} \ \cdots \ x_{kin} \ \cdots \ x_{ki1}^m \ x_{ki2}^m \ \cdots \ x_{kin}^m \end{bmatrix}$,
$C_i = \begin{bmatrix} c_{0i} \ c_{1i} \ c_{2i} \ \cdots \ c_{n'i} \end{bmatrix}^T$,
i: node number,
k: data number,
n_{tr}: number of training data subset,

N: number of selected input variables,

m: maximum order, and

n': number of estimated coefficients.

This procedure is implemented repeatedly for all nodes of the layer and also for all layers of PNN starting from the input layer and moving to the output layer.

Each PD is estimated and evaluated using both training and test databases. Then these values are compared and the PDs which give the best predictive performance for the output variable are chosen. Usually a predetermined number W of PDs or a pre specified cutoff value of the performance is used to retain the PDs for the next generation.

The number of PDs is guided by the following selection methods:

Method 1: This method uses the threshold criterion θ_m to select the node with the best performance in each layer. Each PD is evaluated using the training database and stored in a new array Z. Those PDs which offer best performance among others are retained. A new PD is preserved (retained) if the following condition holds:

$$E_j < \theta_m = E^* + \delta \tag{8.13}$$

Where E_j is a minimal identification error of the current layer, θ_m stands for a threshold value while E^* is a minimal identification error of the previous layer. Furthermore δ is a positive constant whose value is specified by the designer of the model.

Method 2: A total of $\frac{N!}{(N-r)!r!}$ PDs according to combinations of nodes in each layer are determined. Each PD whose parameters were estimated using the training data subset is evaluated by computing an identification error (MSE) using the testing database. Some PDs are chosen which are characterized by the best performance. A pre-defined number W of PDs which offer better predictive capability are preserved for optimal operation of the next iteration in the PNN algorithm. The output of the preserved PDs (called survivors), serve as input to the next iteration;

There are two situations for choice of PDs in the next layer.

(1) If $\frac{N!}{(N-r)!r!} < W$, then the number of PDs retained for the next layer is equal to $\frac{N!}{(N-r)!r!}$.

(2) If $\frac{N!}{(N-r)!r!} \geq W$, then for the next layer, the number of the retained PDs is equal to W.

One of the following three termination methods are used for the PNN algorithm

(1) The stopping condition may be

$$E_j \geq E_i \qquad (8.14)$$

Where E_j is a minimal identification error of the current layer where as E_i denotes a minimal identification error that occurred at the previous layer. The inequality at (8.14) indicates that an optimal PNN model has been accomplished at the previous layer, and the modeling can be terminated.

(2) The PNN algorithm terminates when the number of iterations predetermined by the designer is reached.

(3) Terminating criterion may be combination of the above two. A predetermined number of iterations is fixed as a stopping criterion, however if the condition at (8.14) is achieved before reaching predetermined number of iteration, then modeling is terminated.

When setting up a stopping criterion, one should be prudent in achieving a balance between model accuracy and an overall computational complexity associated with the development of the model.

If the stopping criterion is not satisfied, then the model has to be expanded. The outputs of the preserved PDs serve as new inputs to the next layer. This is captured by the expression $x_{1i} = z_{1i}, x_{2i} = z_{2i}, \cdots, x_{wi} = z_{wi}$.

The general framework of the PNN design procedure is presented here.

(1) *Determine system's input variables.*
(2) *Partition the given database into training and test set as per the standard rule.*
(3) *Determine the number of input variables and the order of the polynomial forming a partial description (PD) of data.*
(4) *Estimate the coefficients of the PDs.*
(5) *Select PDs with the best classification accuracy*
(6) *If stopping criterion not satisfied, determine new input variables for the next layer and go to step 3.*

8.2.3. *Experimental Studies with PNN*

For the experimental studies of PNN algorithm six databases have been considered: IRIS, BUPA, HAYES, WBC, WINE, and PIMA to explore

Table 8.3.: Databases.

Database	Records	Attributes	Classes	Type
IRIS	150	4	3	Numeric
HAYES	132	5	3	Mixed
WBC	699	9	2	Numeric
BUPA	345	6	2	Numeric
WINE	178	13	3	Numeric
PIMA	768	8	2	Numeric

the efficiency of this algorithm in biomedical and natural science domains.[8] Details of the databases are presented in Table 8.3.

The layers of PNN models are grown as per the algorithm described before. The residual error between the estimated output and the actual output is calculated at each layer. If the error level is within the tolerable limit then the growth of the model is stopped and the final model is derived taking into account only those PDs which contribute to obtain the best result. Otherwise the next layer is grown. It is observed that the error level decreases rapidly at the first layers of PNN network and relatively slower near to optimal number of layers, and further increasing the number of layers causes increasing the value of error level because of over-fitting.[55–57] Thus in the simulation the number of layers in the model increases one-by-one until the stopping rule i.e., the tolerable error level is met at the layer r. Subsequently a desired PNN model of nearly optimal complexity from (r-1)th layer is taken. Hence only those PDs that contribute to the improved result are preserved. From the simulation it is seen that the output of best two PDs of previous layer not necessarily yields the best result in the next layer. Hence, the more number of PDs preserved in a layer increases the possibility of obtaining better result in the next layer. But due to memory and time constraints, the number of PDs being preserved in any layer are restricted to a maximum of 100 only and the maximum number of PDs that are generated for the next layer is 4950.

The PNN model is tested with the test database. The 10-fold cross validation strategy is adopted for testing. Several simulations on basic PNN model for different databases are carried out to obtain improved classification accuracy. However, it is found that by adopting the basic model desired performance level can not be achieved. In this work a modified version of basic PNN model is used. In modified PNN model the original

Table 8.4.: Average classification accuracy with PNN on 10-fold cross validation.

Database	% of correct classification
IRIS	98.69
HAYES	80.92
WBC	96.01
BUPA	72.48
WINE	98.71
PIMA	77.22

inputs are also considered in every successive layer for forming the PDs. In such case the number of PDs preserved in any layer is restricted to the maximum of $(100 - N)$, where N is the number of features in the database. The percentage of correct classification for each database using the modified PNN model is presented in Table 8.4.

The fitness function is defined as $\frac{1}{1+e^2}$, where e is the error of a PD is evaluated for all the PDs in each layer. As an illustration, the fitness values obtained for different PDs in a layer are plotted and the fitness curves obtained for different layers of IRIS database are shown in Fig. 8.2.

The architecture of modified PNN model for IRIS database is depicted in Fig. 8.3 for reference. The mathematical model for IRIS database is also presented here for reference.

PNN mathematical model for IRIS database

$PD_1^1 = [-1.7572, 1.4898, -2.4317, 0.3308, -0.15092, -0.018686]$
$\times poly(x_1, x_2)$,
$PD_2^1 = [-0.96751, -0.25663, 0.65004, -0.13994, 0.044319, 0.10067]$
$\times poly(x_1, x_3)$,
$PD_3^1 = [1.633, -1.0407, 1.4825, -0.051333, 0.09108, -0.067678]$
$\times poly(x_1, x_4)$,
$PD_{16}^2 = [2.0965, 1.4134, -0.84284, 0.046201, -0.57289, 0.079276]$
$\times poly(PD_2^1, x_3)$,
$PD_{28}^2 = [2.2683, 1.7305, -1.2497, -0.22651, 0.15639, 0.15237]$
$\times poly(PD_3^1, x_2)$,
$PD_{38}^2 = [-1.8284, -0.58075, 0.42213, 0.081458, -0.25631, 0.015491]$
$\times poly(PD_1^1, x_3)$,

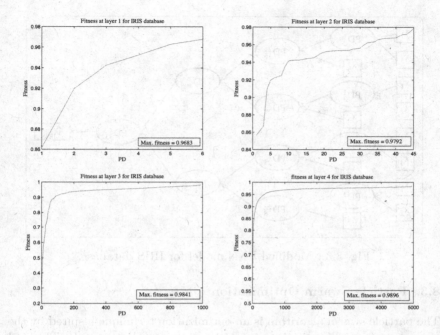

Fig. 8.2.: Fitness achieved in different layers of PNN for IRIS database.

$$PD_{496}^3 = [-0.054529, 0.54087, 0.45923, -1.1876, 0.6309, 0.60812]$$
$$\times poly(PD_{16}^2, PD_{28}^2),$$
$$PD_{557}^3 = [-0.043133, 0.59432, 0.41454, -0.75286, 0.39684, 0.3952]$$
$$\times poly(PD_{28}^2, PD_{38}^2),$$
$$Y_{IRIS} = [0.0057053, 1.158, -0.15925, 5.3642, -2.208, -3.156]$$
$$\times poly(PD_{496}^3, PD_{557}^3).$$

where

(1) function $poly(a_1, a_2)$
$$\{$$
$$return\ [1, a_1, a_2, a_1 * a_2, a_1^2, a_2^2]^T;$$
$$\}$$
(2) PD_j^i is the output of layer i and jth partial description.
(3) y is the estimated output of the model

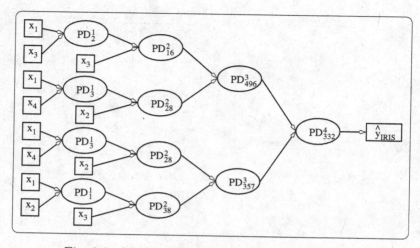

Fig. 8.3.: Modified PNN model for IRIS database.

8.3. Particle Swarm Optimization

The particle swarm algorithm is an optimization technique inspired by the metaphor of social interaction observed among insects or animals. The kind of social interaction modeled within a PSO is used to guide a population of individuals (called particles) moving towards the most promising area of the search space.[58-63] The PSO was developed and first introduced as a stochastic optimization algorithm by Eberhart and Kennedy.[64] During this period, PSO gained increasing popularity due to its effectiveness in performing difficult optimization tasks. Among other applications, it has been applied to tackle multi objective problems,[65] minimax problems,[66,67] integer programming problems,[68] noisy and continuously changing environments,[69-71] errors-in-variables problems,[72] existence of function zeros,[73] power systems[74-80] parameter learning of neural networks (NNs),[81,82] control,[83-86] prediction,[87-89] modeling[90-92] and numerous engineering applications.[93-104]

In a PSO algorithm, each particle is a candidate solution equivalent to a point in a d-dimensional space, so the ith particle can be represented as $x_i = (x_{i,1}, x_{i,2}, ..., x_{i,d})$. Each particle "flies" through the search space, depending on two important factors, $p_i = (p_{i,1}, p_{i,2}, ..., p_{i,d})$, the best position found so far by the current particle and $p_g = (p_{g1}, p_{g2}, ..., p_{gd})$, the global best position identified from the entire population (or within a neighborhood).[60]

The rate of position change of the ith particle is given by its velocity $v_i = (v_{i,1}, v_{i,2}, ..., v_{i,d})$. Equation (8.15) updates the velocity for each particle in the next iteration step, whereas Eq. (8.16) updates each particle's position in the search space:[105]

$$v_{i,d}(t) = \tau(v_{i,d}(t-1) + \phi_1(p_{i,d} - x_{i,d}(t-1)) + \phi_2(p_{gd} - x_{i,d}(t-1))) \quad (8.15)$$

$$x_{i,d}(t) = x_{i,d}(t-1) + v_{i,d}(t) \quad (8.16)$$

where

$$\tau = \frac{2}{(|2 - \phi - \sqrt{\phi^2 - 4\phi}|)}, \phi = \phi_1 + \phi_2, \phi > 4.0 \quad (8.17)$$

τ is referred to as the constriction coefficient.

Two common approaches of choosing p_g are known as gbest and lbest methods. In the gbest approach, the position of each particle in the search space is influenced by the best-fit particle in the entire population; whereas the lbest approach only allows each particle to be influenced by a fitter particle chosen from its neighborhood. Kennedy and Mendes studied PSOs with various population topologies,[106] and have shown that certain population structures provide superior performance over certain optimization functions.

Further, the role of the inertia weight ϕ, in Eq. (8.17), is considered critical for the PSO's convergence behaviour. Improved performance can be achieved through the application of an inertia weight applied to the previous velocity:

$$v_{i,d}(t) = \phi v_{i,d}(t-1) + \phi_1(p_{i,d} - x_{i,d}(t-1)) + \phi_2(p_{gd} - x_{i,d}(t-1)) \quad (8.18)$$

The inertia weight is employed to control the impact of the previous history of velocities on the current one. Accordingly, the parameter ϕ regulates the trade-off between the global (wide-ranging) and local (nearby) exploration abilities of the swarm. A large inertia weight facilitates global exploration (searching new areas), while a small one tends to facilitate local exploration, i.e., fine-tuning the current search area. A suitable value for the inertia weight ϕ usually provides balance between global and local exploration abilities and consequently results in a reduction of the number of iterations required to locate the optimum solution. Initially, the inertia weight is kept constant. However, experimental results indicate that it is better to initially set the inertia weight to a large value, in order to promote global exploration of the search space, and gradually decrease it to get more refined solutions. Thus, an initial value around 1.2 and a gradual decline towards 0 can be considered as a good choice for ϕ.

During the last decade the basic PSO algorithm has been modified and new concepts have been introduced to it such as fully informed particle swarm (FIPS),[106–108] binary particle swarm,[94,109–114] Bare-bones PSO,[115,116] hybrids and adaptive particle swarms,[58,108,113,117–126] PSOs with diversity control,[127–130] etc. Few of them improvise the general performance, and the rest improved performance of particular kinds of problems.

8.4. Classification with Reduced and Comprehensible Polynomial Neural Network

While simulating the PNN model it is observed that the number of partial descriptions generated in each layer grows rapidly. Consequently a lot of time is consumed in generating the PDs. However, the PDs providing poor performance are rejected. Still a substantial number of PDs needs to be preserved to get better result in subsequent layers. It is observed that the PDs giving best results may not combine to yield improved performance in subsequent layers. Very often it may happen that if the PD giving better result is combined with PD giving inferior result it may improve the performance in subsequent layer. Therefore, it is always essential to preserve substantial number of PDs with a hope of getting better result in subsequent layers. In turn huge amount of memory and running time are needed for the process of generation of a model for a particular domain. Each PD tries to approximate the input-output relationship of the database. In the proposed model PDs for the first layer have been developed. Along with the output of the first layer, the original features of the database have been considered to approximate input-output relation.[9,27]

Figure 8.4 shows the architecture of the RCPNN model. In this model, m represents the number of features in the database and k represents the number of PDs generated out of m features. One bias has also been included in the network at this level. Then $m + k + 1$ number of weights are optimized to map the input-output relation. Gradient descent as well as PSO techniques have been used to train the model. The general framework for RCPNN algorithm trained with gradient descent technique is given as

(1) *Determine the order of the polynomial forming the PD.*
(2) *Estimate the coefficients of PDs for the first layer.*

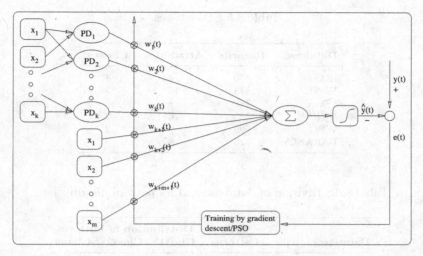

Fig. 8.4.: RCPNN architecture.

(3) *Randomly initialize the weights on the links between first layer PDs, the output layer, and the bias with the output layer.*
(4) *Compute the weighted sum and fed to the output layer.*
(5) *Estimate MSE by comparing the estimated output with the desired output.*
(6) *If stopping criterion is not met, update the weights by gradient descent technique and go to step 4.*

Similarly the general framework for RCPNN algorithm trained with PSO technique is outlined as

(1) *Determine the order of the polynomial forming the PD.*
(2) *Estimate the coefficients of PDs for the first layer.*
(3) *Consider the weights on the links between the first layer PDs, input features and bias with the output layer as the position vector of the particles of the PSO. Randomly initialize the position and velocity vectors of the particles. Estimate the pbest and gbest values.*
(4) *Compute the weighted sum and feed to the output layer.*
(5) *Estimate MSE by comparing the estimated outputs with the desired outputs.*
(6) *If stopping criterion is not met, update the velocity, position, pbest and gbest vectors and go to step 4.*

Table 8.5.: Databases.

Database	Records	Attributes	Classes
IRIS	150	4	3
WINE	178	13	3
PIMA	768	8	2
BUPA	345	6	2
WBC	699	9	2
BALANCE	625	4	3

Table 8.6.: Division of database and its pattern distribution.

Databases		Total Patterns	Distribution of Patterns		
			Class1	Class2	Class3
IRIS	Set1	75	25	25	25
	Set2	75	25	25	25
WINE	Set1	89	29	36	24
	Set2	89	30	35	24
PIMA	Set1	384	134	250	-
	Set2	384	134	250	-
BUPA	Set1	172	72	100	-
	Set2	173	73	100	-
WBC	Set1	350	229	121	-
	Set2	249	229	120	-
BALANCE	Set1	313	144	25	144
	Set2	312	144	24	144

8.4.1. *Experimental Studies with RCPNN*

The IRIS, WINE, PIMA, BUPA, WBC and BALANCE databases are used to evaluate the performance of the RCPNN model.[9] Details of the databases are presented in Table 8.5.

8.4.1.1. *A Two-fold Cross Validation of PNN Model*

For the purpose of comparative analysis of results of different models, a 2-fold cross validation technique is considered. The database is divided into two parts. The division of databases and its class distribution is therefore shown in Table 8.6. The performance of PNN algorithm for both parts of the six databases are evaluated again.

One part is used for building the PNN model and other part is used for testing the model. The layers of PNN models are grown and the residual error between the estimated output and the actual output is calculated at each layer. If the error level is within a tolerable limit then the growth of the model is stopped and the final model is derived taking into account only those PDs that contribute to obtain the best result. Otherwise the next layer is grown. It is observed that the error level decreases rapidly at the first layers of PNN network and is relatively slower near the optimal number of layers, and further increase in the number of layers causes increasing the value of error level because of over-fitting.[131–133] Thus in the simulation the number of layers in the model increases one-by-one till the "stopping rule" i.e., the tolerable error level is met at the layer "r." Subsequently a desired PNN model of nearly optimal complexity from (r-1)th layer is taken. Hence only those PDs that contribute to give better results are preserved. From the simulation study it is seen that the output of best two PDs of previous layer does not necessarily yield the best result in the next layer. Hence PDs that give better result in a layer are preserved for building the next layer.

The PNN model is tested with the test database. The 2-fold cross-validation strategy is used for testing. Each set of the databases is used both for training and testing. From the training set, two third is used to train the model and the remaining one third is used for validation of the model. Then both the train and the test databases are exposed to the model for classification. The percentage of correct classification for each database using PNN model is presented in Table 8.7.

8.4.1.2. *Simulations and Results of RCPNN Model Trained with Gradient Descent Technique*

In this simulation study each set is considered for training while the other set is used for testing. Each set of data while taken for training is simulated for 50 times. Table 8.8 shows the average of results obtained in the different simulation for the 2-fold cross validation.

8.4.1.3. *Simulation Results of RCPNN using PSO*

The databases with the set divisions as at Table 8.6 are again used here for simulation of RCPNN model using PSO technique for optimizing the weights. Again each set is used for training and testing alternatively. Each set is simulated for 50 times for every cross validation procedure. Table 8.9 shows the parameters of the PSO for RCPNN.

Table 8.7.: Classification accuracy of databases simulated in PNN model.

Databases used for Training		Hit Percentage in	
		Training Set	Test Set
IRIS	Set1	78.667	95.555
	Set2	99.111	76.889
	Average	88.889	86.223
WINE	Set1	98.876	73.782
	Set2	77.528	95.880
	Average	88.202	84.831
PIMA	Set1	69.790	72.743
	Set2	74.305	66.146
	Average	72.047	69.444
BUPA	Set1	73.984	64.922
	Set2	71.291	65.667
	Average	72.637	65.294
WBC	Set1	96.667	95.047
	Set2	97.517	96.753
	Average	97.092	95.900
BALANCE	Set1	71.991	66.240
	Set2	72.756	75.855
	Average	72.373	71.048

For each cross validation 50 simulations have been performed and the average of 50 simulations has been obtained. In Table 8.10 the average of the results obtained in the 2-fold cross validation are shown.

8.4.1.4. *Performance Analysis of RCPNN*

The average percentage of correct classification obtained for the test set and training set are taken in Tables 8.11 and 8.12 respectively for the purpose of comparison.

In almost all cases the gradient descent as well as the PSO technique for optimizing the weights of the RCPNN model perform better over the PNN model. Only in case of WINE database the performance of RCPNN with PSO is slightly inferior to that of the RCPNN with gradient descent method.

The processing time in seconds is given for all the models in Table 8.13 and it is observed that the processing time of proposed model is substantially less in comparison to the PNN model.

Further Figs. 8.5 and 8.6 show a comparison among the models such as RCPNN based on gradient decent and RCPNN based on PSO with PNN

Table 8.8.: Classification accuracy of RCPNN model for gradient descent training.

Databases used for Training		Hit Percentage in	
		Training Set	Test Set
IRIS	Set1	99.556	95.111
	Set2	98.223	96.000
	Average	98.889	95.555
WINE	Set1	100.00	93.633
	Set2	97.378	96.629
	Average	98.689	95.131
PIMA	Set1	72.829	73.871
	Set2	80.208	72.829
	Average	76.519	73.350
BUPA	Set1	69.942	69.961
	Set2	76.879	69.186
	Average	73.410	69.573
WBC	Set1	97.714	97.612
	Set2	97.238	96.657
	Average	97.476	97.135
BALANCE	Set1	79.766	70.833
	Set2	77.316	83.333
	Average	78.541	77.083

Table 8.9.: Parameters considered for simulation of RCPNN model using PSO.

Parameters	Values
Population Size	20
Maximum Iterations	50
Inertia Weight	0.729844
Cognitive Parameter	1.49445
Social Parameter	1.49445
Constriction Factor	1.0

using the training set and the test set. The X-axis is labeled with the numbers from 1-6 represents the database IRIS, WINE, PIMA, BUPA, WBC and BALANCE respectively. Similarly Y-axis is labeled with performance of the classifiers.

Table 8.10.: Classification accuracy of RCPNN model for PSO training.

Databases used for Training		Hit Percentage in Training Set	Test Set
IRIS	Set1	97.333	97.333
	Set2	99.111	100.00
	Average	98.222	98.665
WINE	Set1	99.756	90.955
	Set2	99.925	90.936
	Average	99.840	90.945
PIMA	Set1	81.180	75.937
	Set2	80.972	76.145
	Average	81.076	76.041
BUPA	Set1	75.799	70.968
	Set2	78.112	70.775
	Average	76.956	70.872
WBC	Set1	97.714	97.612
	Set2	98.238	97.657
	Average	97.976	97.635
BALANCE	Set1	79.766	76.833
	Set2	78.316	83.333
	Average	79.014	80.083

Table 8.11.: Comparison of average percentage of correct classification of test sets with PNN, RCPNN with gradient descent and RCPNN with PSO.

Databases	PNN	RCPNN Trained with	
		Gradient Descent	PSO
IRIS	86.22	95.55	98.66
WINE	84.83	95.13	90.94
PIMA	69.44	73.35	76.04
BUPA	65.29	69.57	70.87
WBC	95.90	97.13	97.63
BALANCE	71.048	77.08	80.08

8.5. Conclusions

In this chapter, the details of the design procedure of the Polynomial Neural Network (PNN) models for the task of classification is presented. PNN is a flexible neural architecture whose structure is developed through learning. The number of layers of the PNN is not fixed in advance but becomes

Table 8.12.: Comparison of average percentage of correct classification of training sets with PNN, RCPNN with gradient descent and RCPNN with PSO.

Databases	PNN	RCPNN Trained with	
		Gradient Descent	PSO
IRIS	88.88	98.88	98.22
WINE	88.20	98.68	99.84
PIMA	72.04	76.51	81.07
BUPA	72.63	73.41	76.95
WBC	97.09	97.47	97.97
BALANCE	72.37	78.54	79.01

Table 8.13.: Comparison of time requirement for development of model for PNN, RCPNN with gradient descent and RCPNN with PSO in second.

Database		PNN	RCPNN with	
			Gradient Descent	PSO
IRIS	Set1	129	0.219	0.418
	Set2	125	0.209	0.412
WINE	Set1	225	1.107	1.500
	Set2	224	1.008	1.506
PIMA	Set1	783	3.320	3.783
	Set2	793	1.964	3.783
BUPA	Set1	352	0.657	1.215
	Set2	353	0.618	1.153
WBC	Set1	713	2.177	4.059
	Set2	712	2.131	4.039
BALANCE	Set1	493	0.865	1.789
	Set2	499	0.845	1.744

generated on the fly, for which PNN is called a self-organizing network. The essence of the design procedure dwells on the group methods of data handling (GMDH). Each node of the PNN exhibits a high level of flexibility and realizes a polynomial type of mapping between the input and output variables. Depending on the type of polynomials used, it generates Partial Descriptions (PDs) by taking a combination of two to three inputs. In each layer the number of PDs generated grows exponentially, as a result requirement of computation time as well as the requirement of storage space increases.

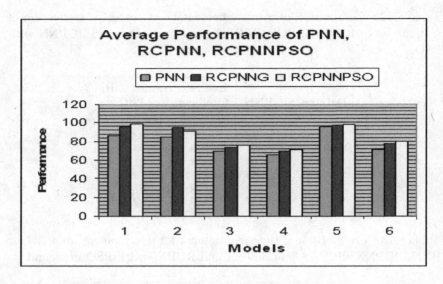

Fig. 8.5.: Comparison of performance of RCPNN models with PNN using test samples.

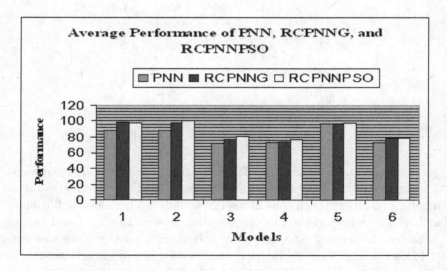

Fig. 8.6.: Comparison of performance of RCPNN models with PNN using training samples.

In this chapter a reduced and comprehensive polynomial neural network (RCPNN), which takes only one layer of the PNN is used for classification task. The outputs of the PDs of this layer along with the input features are fed to a single perceptron model of the Artificial Neural Network (ANN). The weights used for this ANN is adjusted using gradient descent and Particle Swarm Optimization (PSO) techniques. Experimental studies show that the RCPNN is a better candidate than the PNN model in its predictive capability. Further the RCPNN model substantially reduces the time requirements in comparison to the PNN model.

References

1. L. Anastasakis and N. Mort. The development of self-organization techniques in modelling: A review of the group method of data handling (GMDH). Technical report, Department of Automatic Control & Systems Engineering, The University of Sheffield, Mappin St, Sheffield, S1 3JD, United Kingdom (October, 2001).
2. A. J. Muller, F. Lemke, and A. G. Ivakhnenko, Gmdh algorithms for complex systems modelling, *Mathematical and Computer Modelling of Dynamical Systems*. **4**(4), 275–316, (1998).
3. W. Sarle. Neural networks and statistical models. In *the Annual SAS User Group International Conference, Dallas*, pp. 1538–1549, (1994).
4. H. Madala, Comparison of inductive versus deductive learning networks, *Complex Systems*. **5**(2), 239–258, (1991).
5. A. G. Ivakhnenko, The group method of data handling in prediction problems, *Soviet Automatic Control c/c of Avtomatika*. **9**(6), 21–30, (1976).
6. A. G. Ivakhnenko and N. Ivakhnenko, Self-organization of mathematical models for creating an artificial intelligence system, *Soviet Journal of Automation and Information Sciences c/c of Avtomatika*. **19**(2), 24–33, (1986).
7. J. H. Holland, *Adaptation in Natural and Artificial Systems*. (Ann Arbor, University of Michigan Press, MI, 1975).
8. B. B. Misra, S. C. Satapathy, B. N. Biswal, P. K. Dash, and G. Panda. Pattern classification using polynomial neural networks. In *IEEE Int. Conf. on Cybernetics and Intelligent Systems (CIS)*, (2006).
9. B. B. Misra, S. Dehuiri, G. Panda, and P. Dash, A reduced comprehensible polynomial neural network for classification in data mining, *Pattern Recognition Letters*. **29**(12), 1705–1712, (2008).
10. A. G. Ivakhnenko and H. R. Madala, *Inductive learning algorithm for complex systems modelling*. (CRC Inc, 1994).
11. S. J. Farlow, *The GMDH algorithm*, In ed. S. Farlow, *Self-organizating methods in modelling:GMDH type algorithm*, pp. 1–24. Marcel Dekker, (1984).
12. B. Kim, D. W. Kim, and G. T. Park, Prediction of plasma etching using a polynomial neural network, *IEEE Trans. on Plasma Science*. **31**(6(2)), 1330–1336, (2003).

13. B. Kim, D. .-W. Kim, and G. .-T. Park, Prediction of plasma-induced dc bias using polynomial neural network, *Vacuum*. **79**(3-4), 111–118, (2005).

14. W. Woo and L. Khor, Blind restoration of nonlinearly mixed signals using multilayer polynomial neural network, *IEE Proc.-Vis. Image Signal Process.* **151**(1), 51–61, (2004).

15. R. Ghazali, A. J. Hussain, and M. N. Mohd. Salleh. Application of polynomial neural networks to exchange rate forecasting. In *Eighth International Conference on Intelligent Systems Design and Applications*, pp. 90–95, (2008).

16. R. Ghazali, A. J. Hussain, N. M. Nawi, and B. Mohamad, Non-stationary and stationary prediction of financial time series using dynamic ridge polynomial neural network, *Neurocomputing*. **72**(10-12), 2359–2367, (2009).

17. A. J. Hussain, A. Knowles, P. J. G. Lisboa, and W. El-Deredy, Financial time series prediction using polynomial pipelined neural networks, *Expert Systems with Applications*. **35**(3), 1186–1199, (2008).

18. E. Gmez-Ramrez, K. Najim, and E. Ikonen, Forecasting time series with a new architecture for polynomial artificial neural network, *Applied Soft Computing*. **7**(4), 1029–1216, (2007).

19. N. Y. Nikolaev and H. Iba, Polynomial harmonic gmdh learning networks for time series modeling, *Neural Networks*. **16**(10), 1527–1540, (2003).

20. M. R. Mosavi. Recurrent polynomial neural networks for enhancing performance of gps based line fault location. In *Proceedings of ICSP2008*, pp. 1668–1672, (2008).

21. Y. Yamamoto. Identification of nonlinear discrete time systems using trigonometric polynomial neural networks. In *International Conference on Control, Automation and Systems 2008 Oct. 14-17, 2008 in COEX, Seoul, Korea*, pp. 366–370, (2008).

22. A. Patrikar and J. Provence, Nonlinear system identification and adaptive control using polynomial networks, *Mathematical and Computer Modelling*. **23**(1-2), 159–173, (1996).

23. R. V. Rajandran, N. C. Sahoo, and R. Gobbi. Mathematical modeling of flux-linkage characteristics of switched reluctance motors using polynomial neural networks. In *First International Power and Energy Coference PECon 2006 November 28-29, 2006, Putrajaya, Malaysia*, pp. 378–382, (2006).

24. H. .-S. Park, K. .-W. Jang, S.-K. Oh, and T. .-C. Ahn. Evolutionary design of self-organizing fuzzy polynomial neural networks for modeling and prediction of nox emission process. In *SICE-ICASE International Joint Conference 2006 Oct.18-21, 2006 in Bexco, Busan, Korea*, pp. 3796–3799, (2006).

25. B. B. Misra, S. C. Satapathy, P. K. Dash, and G. Panda. Polynomial neural swarm classifier. In *Multimedia University International Symposium on Information and Communication Technology, Malaysia*, pp. 5–8, (2005).

26. N. E. Mitrakis and J. B. Theocharis. A self-organizing fuzzy polynomial neural network - multistage classifier. In *2006 International Symposium on Evolving Fuzzy Systems, September, 2006*, pp. 74–79, (2006).

27. B. B. Misra, B. N. Biswal, P. K. Dash, and G. Panda. Simplified polynomial

neural network for classification task in data mining. In *IEEE Congress on Evolutionary Computation (CEC)*, pp. 721–728, (2007).

28. B. B. Misra, S. Dehuri, P. K. Dash, and G. Panda. Reduced polynomial neural swarm net for classification task in data mining. In *IEEE World Congress on Computational Intelligence*, (2008).

29. S. Dehuri and S. .-B. Cho, Multi-criterion pareto based particle swarm optimized polynomial neural network for classification: A review and state-of-the-art, *Computer Science Review.* **3**(1), 19–40, (2009).

30. B. B. Misra, P. K. Dash, and G. Panda. Optimal polynomial fuzzy swarm net for handling data classification problems. In *IEEE International Advance Computing Conference (IACC'09)*, pp. 1235 – 1240, (2009).

31. C. L. Philip Chen and A. D. McAulay. Robot kinematics learning computations using polynomial neural networks. In *1991 IEEE international Conference on Robotics and Automation Sacramento, California -April 1991*, pp. 2638–2643, (1991).

32. L. M. de Menezes and N. Y. Nikolaev, Forecasting with genetically programmed polynomial neural networks, *International Journal of Forecasting.* **22**(2), 249–265, (2006).

33. M. H. F. Zarandi, I. Trksen, J. Sobhani, and A. A. Ramezanianpour, Fuzzy polynomial neural networks for approximation of the compressive strength of concrete, *Applied Soft Computing.* **8**(1), 488–498, (2008).

34. L. .-L. Huang, A. Shimizu, Y. Hagihara, and H. Kobatake, Face detection from cluttered images using a polynomial neural network, *Neurocomputing.* **51**, 197–211, (2003).

35. L. .-X. Zhao, L. .-L. Shao, and C. .-L. Zhang. Steady-state hybrid modeling of èconomized screw water chillers using polynomial neural network compressor model. International Journal of Refrigeration, in press, (2010).

36. H. Ardalan, A. Eslami, and N. Nariman-Zadeh, Piles shaft capacity from cpt and cptu data by polynomial neural networks and genetic algorithms, *Computers and Geotechnics.* **36**(4), 616–625, (2009).

37. B. .-J. Park, W. Pedrycz, and S. .-K. Oh, An approach to fuzzy granule-based hierarchical polynomial networks for empirical data modeling in software engineering, *Information and Software Technology.* **50**(9-10), 912–923, (2008).

38. K. Atashkari, N. Nariman-Zadeh, M. Glc, A. Khalkhali, and A. Jamali, Modelling and multi-objective optimization of a variable valve-timing spark-ignition engine using polynomial neural networks and evolutionary algorithms, *Energy Conversion and Management.* **48**(3), 1029–1041, (2007).

39. N. Nariman-Zadeh, A. Darvizeh, A. Jamali, and A. Moeini, Evolutionary design of generalized polynomial neural networks for modelling and prediction of explosive forming process, *Journal of Materials Processing Technology.* **164-165**(15), 1561–1571, (2005).

40. A. Jamali, N. Nariman-zadeh, A. Darvizeh, A. Masoumi, and S. Hamrang, Multi-objective evolutionary optimization of polynomial neural networks for modelling and prediction of explosive cutting process, *Engineering Applications of Artificial Intelligence.* **22**(4-5), 676–687, (2009).

41. N. Amanifard, N. Nariman-Zadeh, M. Borji, A. Khalkhali, and A. Habib-doust, Modelling and pareto optimization of heat transfer and flow coefficients in microchannels using gmdh type neural networks and genetic algorithms, *Energy Conversion and Management*. **49**(2), 311–325, (2008).

42. M: .-W. Cho, G. .-H. Kim, T. .-I. Seo, Y. .-C. Hong, and H. H. Cheng, Integrated machining error compensation method using omm data and modified pnn algorithm, *International Journal of Machine Tools and Manufacture*. **46**(12-13), 1417–1427, (2006).

43. S. .-K. Oh, W. Pedrycz, and H. .-S. Park, Multi-layer hybrid fuzzy polynomial neural networks: a design in the framework of computational intelligence, *Neurocomputing*. **64**, 397–431, (2005).

44. S. .-K. Oh and W. Pedrycz, A new approach to self-organizing multi-layer fuzzy polynomial neural networks based on genetic optimization, *Advanced Engineering Informatics*. **18**, 29–39, (2004).

45. S. .-K. Oh and W. Pedrycz, Self-organizing polynomial neural networks based on polynomial and fuzzy polynomial neurons: analysis and design, *Fuzzy Sets and Systems*. **142**, 163–198, (2004).

46. S. .-K. Oh and W. Pedrycz, Multi-layer self-organizing polynomial neural networks and their development with the use of genetic algorithms, *Journal of the Franklin Institute*. **343**, 125–136, (2006).

47. S. .-K. Oh, W. Pedrycz, and S. .-B. Roh, Genetically optimized fuzzy polynomial neural networks with fuzzy set-based polynomial neurons, *Information Sciences*. **176**, 3490–3519, (2006).

48. S. .-K. Oh, S. .-B. Roh, W. Pedrycz, and T. C. Ahn, Ig-based genetically optimized fuzzy polynomial neural networks with fuzzy set-based polynomial neurons, *Neurocomputing*. **70**, 2783–2798, (2007).

49. H. .-S. Park, W. Pedrycz, and S. .-K. Oh, Evolutionary design of hybrid self-organizing fuzzy polynomial neural networks with the aid of information granulation, *Expert Systems with Applications*. **33**, 830–846, (2007).

50. S. .-K. Oh, W. Pedrycz, and H. .-S. Park, A new approach to the development of genetically optimized multilayer fuzzy polynomial neural networks, *IEEE Trans. On Industrial Electronics*. **53**(4), 1309–1321, (2006).

51. H. .-S. Park, W. Pedrycz, and S. .-K. Oh, Fuzzy polynomial neural networks: hybrid architectures of fuzzy modeling, *IEEE Trans. on Fuzzy Systems*. **10** (5), 607–621, (2002).

52. S. .-K. Oh, W. Pedrycz, and H. .-S. Park, Genetically optimized fuzzy polynomial neural networks, *IEEE Trans. on Fuzzy Systems*. **14**(1), 125–144, (2006).

53. S. .-B. Roh, W. Pedrycz, and S. .-K. Oh, Genetic optimization of fuzzy polynomial neural networks, *IEEE Trans. on Industrial Electronics*. **54**(4), 2219–2238, (2007).

54. Y. L. Karnavas and D. P. Papadopoulos, Excitation control of a synchronous machine using polynomial neural networks, *Journal of Electrical Engineering*. **55**(7-8), 169–179, (2004).

55. J. A. Muller and F. Lemke, *Self-organizing data mining extracting knowledge from data*. (Trafford Publishing, Canada British Columbia, 2003).

56. V. Schetinin and J. Schult, The combined technique for detection of artifacts in clinical electroencephalograms of sleeping newborn, *IEEE Trans. on Information Technologies in Biomedicine.* **8**(1), 28–35, (2004).

57. N. L. Nikolaev and H. Iba, *Automated discovery of polynomials by inductive genetic programming*, In eds. J. Zutkow and J. Ranch, *Principles of Data Mining and Knowledge Discovery (PKDD'99)*, pp. 456–462. Springer, Berlin, (1999).

58. P. J. Angeline, *Evolutionary optimization versus particle swarm optimization: philosophy and performance differences*, In eds. V. W. Porto, N. Saravanan, D. Waagen, and A. E. Eiben, *Evolutionary Programming VII*, vol. 1447, *Lecture Notes in Computer Science*, pp. 601–610. Springer, (1998).

59. R. C. Eberhart and Y. Shi, *Comparison between genetic algorithms and particle swarm optimization*, In eds. V. W. Porto, N. Saravanan, D. Waagen, and A. E. Eiben, *Evolutionary Programming VII*, vol. 1447, *Lecture Notes in Computer Science*, pp. 591–600. Springer, (1998).

60. Y. Shi and R. C. Eberhart, *Parameter selection in particle swarm optimization*, In eds. V. W. Porto, N. Saravanan, D. Waagen, and A. E. Eiben, *Evolutionary Programming VII*, vol. 1447, *Lecture Notes in Computer Science*, pp. 611–616. Springer, (1998).

61. Y. Shi and R. Eberhart. A modified particle swarm optimizer. In *IEEE Conference on Evolutionary Computation*, (1998).

62. J. Kennedy, *The behavior of particles*, In eds. V. W. Porto, N. Saravanan, D. Waagen, and A. E. Eiben, *Evolutionary Programming VII*, vol. 1447, *Lecture Notes in Computer Science*, pp. 581–590. Springer, (1998).

63. A. Carlisle and G. Dozier. An off-the-shelf pso. In *Particle Swarm Optimization Workshop*, pp. 1–6, (2001).

64. R. C. Eberhart and J. Kennedy. A new optimizer using particle swarm theory. In *6th Symp. MicroMachine and Human Science, Nagoya, Japan*, pp. 39–43, (1995).

65. C. A. Coello Coello and M. S. Lechuga. Mopso: A proposal for multiple objective particle swarm optimization. In *IEEE Congr. Evolutionary Computation*, pp. 1051–1056, (2002).

66. E. C. Laskari and K. E. Parsopoulos. article swarm optimization for minimax problems. In *article swarm optimization for minimax problems*, pp. 1582–1587, (2002).

67. S. Y. and R. A. Krohling. Co-evolutionary particle swarm optimization to solve min-max problems. In *IEEE Conf. Evolutionary Computation*, pp. 1682–1687, (2002).

68. E. C. Laskari, K. E. Parsopoulos, and M. N. Vrahatis. Particle swarm optimization for integer programming. In *IEEE 2002 Congr. Evolutionary Computation*, pp. 1576–1581, (2002).

69. A. Carlisle. *Applying the particle swarm optimizer to non-stationary environments*. PhD thesis, Auburn Univ., Auburn, AL, (2002).

70. R. C. Eberhart and Y. Shi. Tracking and optimizing dynamic systems with particle swarms. In *IEEE Congr. Evolutionary Computation Seoul, South Korea*, pp. 94–100, (2001).

71. K. E. Parsopoulos and M. N. Vrahatis, *Particle swarm optimizer in noisy and continuously changing environments*, In ed. M. Hamza, *Artificial Intelligence and Soft Computing*, pp. 289–294. Cancun, Mexico: IASTED/ACTA Press, (2001).

72. K. E. Parsopoulos, E. C. Laskari, and M. N. Vrahatis, *Solving norm errors-in-variables problems using particle swarm optimizer*, In ed. M. Hamza, *Artificial Intelligence and Applications*, pp. 185–190. Marbella, Spain:IASTED/ACTA Press, (2001).

73. K. E. Parsopoulos and M. N. Vrahatis. Investigating the existence of function roots using particle swarm optimization. In *IEEE 2003Congr. Evolutionary Computation*, pp. 1448–1455, (2003).

74. A. A. E. Ahmed, L. T. Germano, and Z. C. Antonio, A hybrid particle swarm optimization applied to loss power minimization, *IEEE Trans. Power Syst.* **20**(2), 859–866, (2005).

75. T. O. Ting, M. V. C. Rao, and C. K. Loo, A novel approach for unit commitment problem via an effective hybrid particle swarm optimization, *IEEE Trans. Power Syst.* **21**(1), 411–418, (2006).

76. B. Zhao, C. X. Guo, and Y. J. Cao, A multiagent-based particle swarm optimization approach for optimal reactive power dispatch, *IEEE Trans. Power Syst.* **20**(2), 1070–1078, (2005).

77. J. G. Vlachogiannis and K. Y. Lee, A comparative study on particle swarm optimization for optimal steady-state performance of power systems, *IEEE Trans. Power Syst.* **21**(4), 1718–1728, (2006).

78. C. M. Huang, C. J. Huang, and M. L. Wang, A particle swarm optimization to identifying the armax model for short-term load forecasting, *IEEE Trans. Power Syst.* **20**(2), 1126–1133, (2005).

79. Y. Liu and X. Gu, Skeleton-network reconfiguration based on topological characteristics of scale-free networks and discrete particle swarm optimization, *IEEE Transactions on Power Systems.* **22**(3), 1267 – 1274, (2007).

80. Y. d. Valle, G. K. Venayagamoorthy, S. Mohagheghi, J. C. Hernandez, and R. G. Harley, Particle swarm optimization: Basic concepts, variants and applications in power systems, *IEEE Transactions on Evolutionary Computation.* **12**(2), 171 – 195, (2008).

81. M. M. Jannett and T. C. Jannett. Simulation of a new hybrid particle swarm optimization algorithm. In *36th Southeastern Symp. Syst. Theory*, pp. 150–153, (2004).

82. C. Zhang, H. Shao, and Y. Li. Particle swarm optimisation for evolving artificial neural network. In *IEEE Int. Conf. Syst., Man, Cybern., vol. 4*, pp. 2487–2490, (2000).

83. Y. Song, Z. Chen, and Z. Yuan, New chaotic pso-based neural network predictive control for nonlinear process, *IEEE Trans. Neural Netw.* **18**(2), 595–601, (2007).

84. Y. S. Wang, K. J. Wang, J. S. Qu, and Y. R. Yang. Adaptive inverse control based on particle swarm optimization algorithm. In *IEEE Int. Conf. Mechatronics, Autom.*, pp. 2169–2172, (2005).

85. L. dos Santos Coelho and B. M. Herrera, Fuzzy identification based on a

chaotic particle swarm optimization approach applied to a nonlinear yo-yo motion system, *IEEE Transactions on Industrial Electronics.* **54**(6), 3234 – 3245, (2007).

86. V. Kadirkamanathan, K. Selvarajah, and P. J. Fleming, Stability analysis of the particle dynamics in particle swarm optimizer, *IEEE Transactions on Evolutionary Computation.* **10**(3), 245 – 255, (2006).

87. W. Liu, K. Wang, B. Sun, and K. Shao. A hybrid particle swarm optimization algorithm for predicting the chaotic time series mechatronics and automation. In *IEEE Int. Conf. Mechatronics*, pp. 2454–2458, (2006).

88. C. J. Lin, C. H. Chen, and C. T. Lin, A hybrid of cooperative particle swarm optimization and cultural algorithm for neural fuzzy networks and its prediction applications, *IEEE Transactions on Systems, Man, and Cybernetics, Part C: Applications and Reviews.* **39**(1), 55 – 68, (2009).

89. W. C. Liu, Design of a multiband cpw-fed monopole antenna using a particle swarm optimization approach, *IEEE Transactions on Antennas and Propagation.* **53**(10), 3273 – 3279, (2005).

90. R. Marinke, E. Araujo, L. S. Coelho, and I. Matiko. Particle swarm optimization (pso) applied to fuzzy modeling in a thermal-vacuum system. In *5th Int. Conf. Hybrid Intell. Syst.*, pp. 67–72, (2005).

91. Y. Liu and X. He. Modeling identification of power plant thermal process based on pso algorithm. In *Amer. Control Conf.,vol. 7*, pp. 4484–4489, (2005).

92. R. Saeidi, H. R. S. Mohammadi, T. Ganchev, and R. D. Rodman, Particle swarm optimization for sorted adapted gaussian mixture models, *IEEE Transactions on Audio, Speech, and Language Processing.* **17**(2), 344 – 353, (2009).

93. M. A. Abido, Optimal design of power system stabilizers using particle swarm optimization, *IEEE Trans. Energy Conversion.* **17**, 406–413, (2002).

94. D. K. Agrafiotis and W. Cedeno, Feature selection for structure-activity correlation using binary particle swarms, *J. Medicinal Chem.* **45**(5), 1098–1107, (2002).

95. A. R. Cockshott and B. E. Hartman, Improving the fermentation medium for echinocandin b production part ii: Particle swarm optimization, *Process Biochem.* **36**, 661–669, (2001).

96. P. C. Fourie and A. A. Groenwold, The particle swarm optimization algorithm in size and shape optimization, *Struct. Multidisc. Optim.* **23**, 259–267, (2002).

97. W. Z. Lu, H. Y. Fan, A. Y. T. Leung, and J. C. K. Wong, Analysis of pollutant levels in central hong kong applying neural network method with particle swarm optimization, *Environ. Monitoring Assessment.* **79**, 217–230, (2002).

98. C. O. Ourique, E. C. Biscaia, and J. C. Pinto, The use of particle swarm optimization for dynamical analysis in chemical processes, *Comput.Chem. Eng.* **26**, 1783–1793, (2002).

99. K. E. Parsopoulos, E. I. Papageorgiou, P. P. Groumpos, and M. N. Vrahatis. A first study of fuzzy cognitive maps learning using particle swarm

optimization. In *IEEE 2003 Congr. Evolutionary Computation*, pp. 1440–1447, (2003).

100. T. Ray and K. M. Liew, A swarm metaphor for multiobjective design optimization, *Eng. Opt.* **34**(2), 141–153, (2002).

101. A. Saldam, I. Ahmad, and S. Al-Madani, Particle swarm optimization for task assignment problem, *Microprocess. Microsyst.* **26**, 363–371, (2002).

102. V. Tandon, H. El-Mounayri, and H. Kishawy, Nc end milling optimization using evolutionary computation, *Int. J. Mach. Tools Manuf.* **42**, 595–605, (2002).

103. J. C. Tillett, R. M. Rao, F. Sahin, and T. M. Rao. Particle swarm optimization for the clustering of wireless sensors. In *SPIE, vol. 5100, Orlando, FL*, (2003).

104. S. E. Easter Selvan, S. Subramanian, and S. T. S. Theban Solomon. Novel technique for pid tuning by particle swarm optimization. In *7th Annu. Swarm Users/Researchers Conf. (SwarmFest 2003), Notre Dame, IN*, (2003).

105. J. Kennedy and R. Eberhart, *Swarm Intelligence*. (Morgan Kaufmann Academic Press, 2001).

106. J. Kennedy and R. Mendes. Population structure and particle swarm performance. In *the Congress on Evolutionary Computation, Piscatawat,Honolulu, HI. Piscataway: IEEE.*, pp. 1671–1676, (2002).

107. R. Mendes, P. Cortes, M. Rocha, and J. Neves. Particle swarms for feed-forward neural net training. In *the international joint conference on neural networks, Honolulu, HI. Piscataway: IEEE*, pp. 1895–1899, (2002).

108. V. Miranda and N. Fonseca. New evolutionary particle swarm algorithm (epso) applied to voltage/ var control. In *the 14th power systems computation conference (PSCC), Seville, Spain.*, pp. 1–6, (2002).

109. J. Kennedy and R. C. Eberhart. A discrete binary version of the particle swarm algorithm. In *the conference on systems, man, and cybernetics, Piscataway: IEEE.*, pp. 4104–4109, (1997).

110. C. K. Mohan and B. Al-Kazemi. Discrete particle swarm optimization. In *the workshop on particle swarm optimization, Indianapolis, IN, Purdue School of Engineering and Technology, IUPUI*, (2001).

111. G. Pampara, N. Franken, and A. P. Engelbrecht. Combining particle swarm optimization with angle modulation to solve binary problems. In *the IEEE congress on evolutionary computation (CEC), Piscataway: IEEE*, pp. 225–239, (2005).

112. M. Clerc, *Discrete particle swarm optimization, illustrated by the traveling salesman problem*, In eds. B. V. Babu and G. C. Onwubolu, *New optimization techniques in engineering*, pp. 219–239. Berlin: Springer, (2004).

113. M. Clerc, *Particle swarm optimization*. (London: ISTE, 2006).

114. A. Moraglio, C. Di Chio, and R. Poli, *Geometric particle swarm optimization*, In ed. M. e. a. Ebner, *Lecture notes in computer science: Proceedings of the European conference on genetic programming (EuroGP)*, vol. 4445, pp. 125–136. Berlin: Springer, (2007).

115. J. Kennedy. Bare bones particle swarms. In *the IEEE swarm intelligence symposium (SIS), Indianapolis, IN. Piscataway: IEEE*, pp. 80–87, (2003).
116. T. Richer and T. M. Blackwell. The lvy particle swarm. In *IEEE congress on evolutionary computation, Vancouver. Piscataway: IEEE*, pp. 3150–3157, (2006).
117. M. Loovbjerg, T. K. Rasmussen, and T. Krink. Hybrid particle swarm optimiser with breeding and subpopulations. In *the third genetic and evolutionary computation conference (GECCO), San Francisco: Kaufmann*, pp. 469–476, (2001).
118. C. Wei, Z. He, Y. Zhang, and W. Pei. Swarm directions embedded in fast evolutionary programming. In *the IEEE congress on evolutionary computation (CEC), Honolulu, HI. Piscataway: IEEE*, pp. 1278–1283, (2002).
119. T. Krink and M. Loovbjerg. The lifecycle model: combining particle swarm optimization, genetic algorithms and hillclimbers. In *Lecture notes in computer science. Proceedings of parallel problem solving from nature (PPSN),Granada, Spain. Berlin: Springer*, pp. 621–630, (2002).
120. R. Poli and C. R. Stephens, *Constrained molecular dynamics as a search and optimization tool*, In ed. M. e. a. Keijzer, *Lecture notes in computer science: Proceedings of the 7th European conference on genetic programming (EuroGP)*, vol. 3003, pp. 150–161. Berlin: Springer, (2004).
121. J. S. Vesterstrom, J. Riget, and T. Krink. Division of labor in particle swarm optimization. In *the IEEE congress on evolutionary computation (CEC), Honolulu, HI. Piscataway: IEEE*, pp. 1570–1575, (2002).
122. N. Holden and A. A. Freitas. A hybrid particle swarm/ant colony algorithm for the classification of hierarchical biological data. In *IEEE swarm intelligence symposium (SIS), Piscataway: IEEE*, pp. 100–107, (2005).
123. T. Hendtlass, *A combined swarm differential evolution algorithm for optimization problems*, In eds. L. Monostori, J. Vncza, and M. Ali, *Lecture notes in computer science: Proceedings of the 14th international conference on industrial and engineering applications of artificial intelligence and expert systems (IEA/AIE)*, vol. 2070, pp. 11–18. Berlin: Springer, (2001).
124. W.-J. Zhang and X.-F. Xie. Depso: hybrid particle swarm with differential evolution operator. In *the IEEE International conference on systems, man and cybernetics (SMCC), Washington, DC. Piscataway: IEEE*, pp. 3816–3821, (2003).
125. R. Poli, W. B. Langdon, and O. Holland, *Extending particle swarm optimization via genetic programming*, In ed. M. K. et al., *Lecture notes in computer science: Proceedings of the 8th Eu*, vol. 3447. (2005).
126. R. Poli, C. Di Chio, and W. B. Langdon, *Exploring extended particle swarms: a genetic programming approach*, In ed. e. a. H.-G. Beyer, *the conference on genetic and evolutionary computation, Washington, DC. New York*, pp. 169–176. ACM, (2005).
127. M. Lovbjerg and T. Krink. Extending particle swarms with self-organized criticality. In *the IEEE congress on evolutionary computation (CEC-2002), Piscataway: IEEE*, pp. 1588–1593, (2002).

128. T. Blackwell and P. J. Bentley. Don't push me! collision-avoiding swarms. In *the IEEE congress on evolutionary computation (CEC), Honolulu, HI. Piscataway: IEEE*, pp. 1691–1696, (2002).

129. T. Krink, J. S. Vesterstrom, and J. Riget. Particle swarm optimization with spatial particle extension. In *the IEEE congress on evolutionary computation (CEC-2002),Piscataway: IEEE*, pp. 1474–1479, (2002).

130. X. Xie, W. Zhang, and Z. Yang. Dissipative particle swarm optimization. In *the IEEE congress on evolutionary computation (CEC), Honolulu, HI. Piscataway: IEEE*, pp. 1456–1461, (2002).

131. T. M. Mitchel, *Machine Learning.* (McGraw Hill, 1997).

132. A. C. Tsoi and A. R. Pearson, Comparison of three classification techniques, cart, c4.5, and multiplayer perceptrons, *Advances in Neural Information Processing Systems.* **3**, 963–969, (1991).

133. J. R. Quinlan. Generating production rules from decision trees. In *International Joint Conference on Artificial Intelligence, San Francisco: CA: Morgan Kaufmann*, pp. 304–307, (1987).

Chapter 9

Efficient Prediction of Retail Sales Using Differential Evolution Based Adaptive Model

Ritanjali Majhi[*], Babita Majhi[†] and Ganapati Panda[‡]

School of Management, National Institute of Technology, Warangal,
Dept. of Information Technology,
Institute of Technical Education and Research, Bhubaneswar, India
School of Electrical Sciences, Indian Institute of Technology,
Bhubaneswar, India
[*] *ritanjalimajhi@gmail.com*
[†] *babita.majhi@gmail.com*
[‡] *ganapati.panda@gmail.com*

The chapter develops an efficient adaptive prediction model using recently developed differential evolution (DE) technique. The forecasting model employs an adaptive linear combiner architecture and DE based learning rule to predict seasonally adjusted (SA) and non seasonally adjusted (NSA) sales data for short and long ranges. The prediction performance of proposed model is assessed through simulation study and using real life data. For comparison purpose the corresponding results are also obtained using genetic algorithm (GA), bacterial foraging optimization (BFO) and particle swarm optimization (PSO) based forecasting models. It is, in general, observed that the new DE forecasting model offers fastest training, best sales prediction and least mean square error after training compared to other three evolutionary computing based models.

9.1. Introduction

Sales forecasting is the prediction of future sales based on past historical data. It plays a vital role in business strategy. Manufactures, organizations and business houses require accurate and reliable forecast of sales data. Future sales plan facilitates effective control of inventory, scheduling, optimal utilization of facility and conveyance. These in turn result in decrease in production cost and enhancement of clients' satisfaction. In recent past various techniques on sales forecasting have been reported.[1] Numerous statis-

---

[*]Corresponding author: Dr. Ritanjali Majhi, Cell: +91-9959204274

tical methods such as regression, auto-regressive moving average (ARMA), auto-regressive integrated moving average (ARIMA) have been employed for developing sales forecasting models.[2] But it is a complex problem and is influenced by internal and external environments.[3] The statistical methods have two major drawbacks. Firstly an individual model is required to be chosen for each problem which makes some assumptions about understanding trends. Secondly, these methods fail for multidimensional time series with mutual nonlinear dependencies. The statistical forecasting models do not provide promising performance if domain models are not well defined. Under such situations the artificial neural networks can be conveniently applied as they do not require the specification of functional model of the problem. In recent past artificial neural network (ANN) has been successfully applied for prediction of stock market[4] and exchange rate.[5] In the area of sales forecasting of various commodities, the ANN has also been introduced as the forecasting models.[6-12] With an objective to improve the prediction performance fuzzy-neural networks have been applied to sales forecasting.[13-16] The radial basis function neural network (RBFN)[17,18] has been shown to be a potential model for prediction of sales data. In recent years hybrid methods of forecasting have been developed of sales data[19-21] by suitably combining good features of two methods. Support vector machine based sales forecasting models[22,23] have also been proposed to improve prediction performance.

The literature survey reveals that very few work has been reported on the use of swarm intelligence technique in effectively training the forecasting models.[24,25] In recent past the differential evolution (DE) has emerged[34] as a powerful alternative of GA[35] for many applications. The choice of DE over GA is mainly due to lesser involvement of computational complexity, better optimization performance and faster training. Keeping these advantages of DE in mind, we are motivated to use this algorithm to train the parameters of the forecasting model. The particle swarm optimization (PSO) and bacterial foraging optimization (BFO) are effective evolutionary computing tools for optimization and have been proven to provide superior performance in many applications.[26-30] Further we observed that to the best of our knowledge for sales forecasting, the PSO[31] and BFO[33] based models have not been reported. Therefore we are curious to develop and evaluate the performance of these models and compare with that obtained by the new model.

The organization of the chapter proceeds as follows. The basic principle of adaptive linear combiner (ALC) is outlined in Sec. 9.2. The DE and GA are briefly discussed in Sec. 9.3. In Sec. 9.4 the particle swarm optimization and bacterial foraging optimization algorithms are then dealt in brief. Four types of evolutionary computing based learning rules for training the ALC weights are presented in Sec. 9.5. The simulation study of the proposed model and as well as GA, PSO and BFO based models is carried out and various performance measures are evaluated and compared in Sec. 9.6. Finally the conclusion of the chapter is included in Sec. 9.7.

9.2. Adaptive Linear Combiner (ALC)

The adaptive linear combiner model using DE based learning is shown in Fig. 9.1. It is an adaptive finite impulse response (FIR) filter having number of inputs equal to the number of features in the input pattern derived from the retail sales data. The weights of the combiner are considered as the members of the initial population and their values are set to random numbers. A population of such random individual is chosen to represent a set of initial solutions.

For every pattern of input the linear combiner generates the error. Basically the development of the ALC model corresponds to iterative minimization of squared error cost function by changing its weights. This optimization task is accomplished by using either DE, PSO, BFO or GA technique.

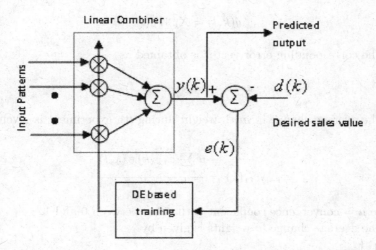

Fig. 9.1.: Retail sales forecasting model using adaptive linear combiner with DE technique.

The basic operation of ALC using epoch based least mean square (LMS) weight update[36] algorithm is outlines as follows

Let the number of sales data available be J (one value for each month). Corresponding to rth month first r data are used to calculate M features such as mean, variance etc. Then by sliding one sample down and using the new data vector, the features of next month is computed. This process is repeated until the features corresponding all available data are evaluated. In this way $S = (J - r + 1)$ feature patterns are generated from J available sales data. A major portion of the input patterns (say K) is used for training the ALC based forecasting model and the remaining $(S - K)$ patterns are used for validation or testing of the model. Let $W_k(l)$ represents the weights vector of the model during application of kth $(1 \leq k \leq K)$ input pattern and for lth $(1 \leq l \leq L)$ experiment and is denoted as.

$$W_k(l) = [W_{l,1}(k), W_{l,2}(k), \dots\dots\dots, W_{l,M}(k)]^T \qquad (9.1)$$

Let the kth input pattern of the model be represented as

$$X_k = [x_1(k), x_2(k), \dots\dots\dots, x_M(k)]^T \qquad (9.2)$$

The output of the model due to application of the kth pattern at lth experiment is given by

$$y(k, l) = X_k^T W_k(l) \qquad (9.3)$$

The corresponding error vector is obtained as

$$e(k, l) = d(k) - y(k, l) \qquad (9.4)$$

The average change in m th weight during lth experiment is given by

$$\Delta w_{l,m}(k) = \frac{\mu \sum\limits_{k=1}^{K} x_m(k)\, e(k, l)}{K} \qquad (9.5)$$

where μ = convergence coefficient with value between 0 and 1.

The average change in weights is given by

$$\Delta W(l) = [\Delta w_{l,1}, \Delta w_{l,2}, \dots\dots \Delta w_{l,M}]^T \qquad (9.6)$$

Applications of K successive patterns constitutes one experiment. After every experiment the weight vector is updated as

$$W(l+1) = W(l) + \Delta W(l) \tag{9.7}$$

When the mean square error, $\frac{1}{k}\sum_{k=1}^{K} e^2(k,l)$, attains the lowest possible value, the learning process is stopped and the final weight vector thus obtained represents the weight vector of the model.

9.3. Basics of GA and DE Algorithms

9.3.1. *Genetic Algorithm (GA)*

The Genetic algorithm (GA) was first introduced by John Holland and was developed by Goldberg and De Jong.[35] It is a global optimization technique and has been successfully applied to many multimodal optimization problems. It is based on the mechanics of natural selection and genetics to mimic the evolutionary behavior of biological system. Figure 9.2 shows the basic operation of GA. The different operations involved in basic GA cycle are encoding and population creation, selection, crossover, mutation, fitness evaluation and terminating condition. These operations are briefly described in sequel.

Operators of GA

9.3.1.1. *Population*

In GA, a string of numeric, binary, or a user defined type data represents a solution. The string structure is called chromosome. The population consists of a group of chromosomes from which candidates are selected for the solution of the problem. In case of binary coding, each parameter is coded using a string of L bits having random binary value 0 or 1. A linear mapping procedure is used to decode any unsigned integer from zero to a specific interval. For multi parameter optimization, the coded parameter values are joined to form a large string which then forms one chromosome of the population.

9.3.1.2. *Crossover*

Crossover is a recombination operator which combines subparts of two parent chromosomes to produce offsprings which employ some genetic material

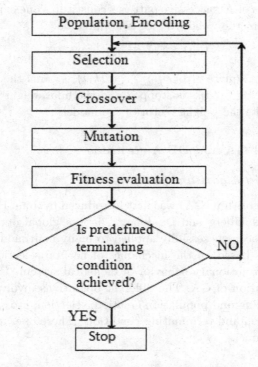

Fig. 9.2.: Block diagram of Genetic Algorithm cycle.

of both parents. A probability term, P_c, is used to determine the operation rate which is usually very close to unity.

9.3.1.3. *Mutation*

Mutation operation introduces global/local variations into the chromosomes with a relatively smaller probability P_m, but randomly alters the value of a string position. Each bit of a chromosome is replaced by a randomly generated bit if a probability test is passed. Within a specific probability, certain bits are altered from either 0 to 1 or 1 to 0 in binary coding.

9.3.1.4. *Selection*

Finally the fitness or cost function of original parents, offspring and mutated offspring are evaluated and the best ones are selected from the entire pools.

These are then treated as the parents of the next generation. The entire process continues till the global optimum situation or terminating condition is reached.

9.3.2. *Differential Evolution (DE)*

In this section the basic DE algorithm[34] is explained. Each individual of initial target population has an J-dimensions with parameter values selected randomly and uniformly between x_{ij}^{min} and x_{ij}^{max} as

$$x_{ij}^0 = x_{ij}^{\min} + \left(x_{ij}^{\max} - x_{ij}^{\min}\right) \times r \tag{9.8}$$

where x_{ij}^t represents the ith target individual with respect to Jth dimension at tth generation and r is a random number between 0 and 1. The vectors form the target population are perturbed by adding the weighted difference between two randomly selected target population members to a third member in the target population to produce the mutant vector. It is obtained as

$$v_{ij}^t = x_{mj}^{t-1} + F \times \left(x_{nj}^{t-1} - x_{pj}^{t-1}\right) \tag{9.9}$$

where m, n and p are three randomly chosen individuals from the target population such that $m \neq n \neq p \neq i$; $1 < i \leq NP$ and $(j = 1, 2, \cdots J)$. F is positive and represents a mutation scale factor which influences the differential variation between two individuals. After mutation the recombination of mutant individual with its corresponding target individual is performed. To achieve this, a crossover operator is applied to obtain the trial individual as

$$u_{ij}^t = \begin{cases} v_{ij}^t & if\ r_{ij}^t \leq CR\ or\ j = k \\ x_{ij}^t & otherwise \end{cases} \tag{9.10}$$

where the index k refers to a randomly chosen dimension $(j = 1, 2, \cdots J)$. It is used to ensure that at least one parameter of each trial individual u_i^t differs from its previous counterpart u_i^{t-1}. CR represents crossover constant in the range [0,1] and r_{ij}^t is a uniform random number between 0 and 1.

During the reproduction phase, it is possible to extend the search outside of the initial range of the search space. For this reason, parameter value which violates the search range is restricted according to:

$$u_{ij}^t = x_{ij}^{\min} + \left(x_{ij}^{\max} - x_{ij}^{\min}\right) \times r_1 \quad j = 1, 2, ...J \tag{9.11}$$

where r_1 is a uniformly distributed random number between 0 and 1.

The fitness functions of the trial individual u_i^t and its counterpart target individual x_i^{t-1} at the previous generation are evaluated and the better fitted one is chosen to be a member of the target population for the next generation. That is

$$x_i^t = \begin{cases} u_i^t & if \ f\left(u_i^t\right) \leq f\left(x_i^{t-1}\right) \\ x_i^{t-1} & otherwise \end{cases} \tag{9.12}$$

where $f(.)$ denotes the fitness function of an individual. The pseudo code of a simple DE algorithm is outlined

Initialize parameters
Initialize target population
Evaluate target population
Do
{
Obtain mutant population
Obtain trial population
Evaluate trial population
Make selection
}
While (Not Termination)

9.4. Basics of PSO and BFO Algorithm

9.4.1. *The Particle Swarm Optimization (PSO)*

The PSO algorithm[31] has evolved from the simulation of social behavior. In evolutionary computational algorithms, evolutionary operators are used to manipulate the individuals. But in PSO these individuals are evolved through generations by cooperation and competition among the individuals. Each individual in PSO flies in the search space with a velocity which is dynamically adjusted according to its own as well as its companions' flying experiences. Each individual is treated as a volume-less particle in a D-dimensional space. The ith particle is represented as $X_i = [x_{i1}, x_{i2}, \cdots x_{iD}]^T$. The position giving the best fitness value of ith particle is called its personal best and is represented as $P_i = [p_{i1}, p_{i2}, \cdots p_{iD}]^T$. The velocity of ith particle is represented as $V_i = [v_{i1}, v_{i2}, \cdots v_{iD}]^T$. The overall best location (global best) obtained so far by the particles is also tracked for velocity update and is represented by $P_g = [p_{g1}, p_{g2}, \cdots p_{gD}]^T$. Using the information of kth search, the velocity and position of dth element of ith particle at $(k + 1)$th search are updated according to (9.13) and (9.14) respectively.

$$v_{id}(k + 1) = H(k)*v_{id}(k)+c_1*r_1*(p_{id}(k) - x_{id}(k))+c_1*r_1*(p_{gd}(k) - x_{id}(k)) \tag{9.13}$$

$$x_{id}(k + 1) = x_{id}(k) + v_{id}(k + 1) \tag{9.14}$$

The acceleration constants c_1 and c_2 in (9.13) represent the weighting of the stochastic acceleration terms that pull each particle towards their best and global best positions. Low values allow particles to roam far from target regions while high values results in abrupt movement towards or past target regions. The acceleration constants are usually taken to be 2.0 for almost all applications. The selection of population size is problem dependent but the most common population size selected is 20 to 50. The inertia weight, H has characteristics that are similar to the temperature parameter in the simulated annealing. A large inertia weight facilitates a global search while a small inertia weight facilitates a local search. By linearly decreasing the inertia weight from a large value (close to unity) to a small value through the course of PSO run, the PSO tends to have more global search ability at the beginning of the run while possessing more local search ability towards the end of the run. The inertia weight is chosen according to the relation[32]

$$H(k) = H_0 - \frac{(H_0 - H_1) * k}{I} \tag{9.15}$$

where
k=search number
I=max. no of iterations
$H_0 = 0.9$
$H_1 = 0.4$

The steps in implementing the PSO algorithms are:

(a) Initialize a population of particles with random positions and velocities on D-dimensions in the problem space (the problem consists of D variables)
(b) For each particle, evaluate the fitness function in D variables.
(c) Compare the fitness value of each particle with that obtained from the personal best value. If the current position, X_i provides better fitness value than provided by P_i, then it is made equal to the current value and the previous location is updated to the current location in D-dimensional space.

(d) Compare each fitness value with that given by overall best value, P_g of the population. If the current value is better than that offered by P_g then reset P_g to the current best position vector.

(e) Update the velocity and position of each particle according to (9.13) and (9.14) respectively.

(f) Loop to step (b) until a pre-specified criterion is met, for example, the model achieves the minimum mean square error (MMSE).

9.4.2. The Bacterial Foraging Optimization (BFO)

Bacterial Foraging Optiomization (BFO) is a new evolutionary computation technique proposed by Passino.[33] It is inspired by the pattern exhibited by bacterial foraging bahaviour. Bacteria have the tendency to gather to the nutrient-rich areas by an activity called chemotaxis. The bacteria swim by rotating whip like flagella driven by a reversible motor embedded in the cell wall. E. coli has 8-10 flagella placed randomly on a cell body. When all flagella rotate counterclockwise, they move forward, which is called run but for clockwise rotation they pull on the bacterium in different directions and causes the bacteria to tumble. The bacterial foraging system primarily consists of four sequential mechanisms namely chemotaxis, swarming, reproduction and elimination-dispersal. A brief outline of each of these processes is given in this section.

9.4.2.1. Chemotaxis

An E. coli bacterium can move in two different ways : it can run (swim for a period of time) or it can tumble, and alternate between these two modes of operation in the entire lifetime. In the BFO, a unit walk with random direction represnts a tumble and a unit walk in the same direction indicates a run. In computational chemotaxis, the movement of the ith bacterium after one step is represented as

$$\theta^i(j+1,k,l) = \theta^i(j,k,l) + C(i)\phi(j) \qquad (9.16)$$

where $\theta^i(j,k,l)$ denotes the location of ith bacterium at jth chemotactic, kth reproductive and lth elimination and dispersal step. $C(i)$ is the length of unit walk, which is a constant in basic BFO and $\phi(j)$ is the direction angle of the jth step. When its activity is run, $\phi(j)$ is same as $\phi(j-1)$, otherwise, $\phi(j)$ is a random angle directed within a range of $[0, 2\pi]$.

If the cost at $\theta^i(j+1,k,l)$ is better than the cost at $\theta^i(j,k,l)$ then the bacterium takes another step of size $C(i)$ in that direction otherwise it is allowed to tumble. This process is continued until the number of steps taken is greater than the number of chemotactic loop, N_c.

9.4.2.2. *Swarming*

The bacteria in times of stresses release attractants to signal bacteria to swarm together. Each bacterium also releases a repellent to signal others to be at a minimum distance from it. Thus all of them will have a cell to cell attraction via attractant and cell to cell repulsion via repellent. The cell to cell signaling in E. coli swarm may be mathematically represented as

$$
J_{cc}(\theta, P(j,k,l)) = \sum_{i=1}^{S} J_{cc}(\theta, \theta^i(j,k,l)) = \sum_{i=1}^{S} [-d_a \exp(-w_a \sum_{m=1}^{P} (\theta_m - \theta_m^i)^2)]
$$
$$
+ \sum_{i=1}^{S} h_r \exp(-w_r \sum_{m=1}^{P} (\theta_m - \theta_m^i)^2)
$$
(9.17)

where $J_{cc}(\theta, P(j,k,l))$ represents the objective function value to be added to the actual objective function, S is the total number of bacteria, P is the number of variables to be optimized and $\theta = [\theta_1, \theta_2 \cdots \theta_p]^T$ is a point in the p-dimensional search domain. The symbols d_a, w_a, h_a and w_r are coefficients to be chosen properly to achieve best performance.

9.4.2.3. *Reproduction*

After all N_c chemotactic steps have been covered, a reproduction step takes place. The fitness values of the bacteria are sorted in ascending order. The lower half of the bacteria having higher fitness die and the remaining $S_r = S/2$ bacteria are allowed to spilt into two identical ones. Thus the population size after reproduction is maintained constant.

9.4.2.4. *Elimination and Dispersal*

Since bacteria may stuck around the initial or local optima positions, it is required to diversify the bacteria either gradually or suddenly so that the possibility of being trapped into local minima is eliminated. The dispersion operation takes place after a certain number of reproduction process. A bacterium is chosen, according to a preset probability P_{ed}, to be dispersed and moved to another position within the environment. These events help to prevent the local minima trapping and provide better solution.

9.5. New Adaptive Models Using DE, GA, PSO and BFO Based Learning

9.5.1. *Steps Involved in DE Based Training of the Model*

(a) The coefficients or weights of the model are initially chosen from a target population of I individuals. Each member of the population constitutes D number of parameters and each parameter represents a weight of the linear combiner.

(b) K number of patterns each containing three features (mean, variance and actual data of a month) are obtained from the past retail sales time series.

(c) Each of the K patterns, x_n is passed through the linear combiner, multiplied with the corresponding weight and the partial sums are added together to give y_i, where

$$y_i = \sum_{n=1}^{N} w_n x_n \tag{9.18}$$

(d) The output of the linear combiner, y_i is then compared with corresponding normalized desired value, d_i to produce the error, e_i. After the application of all patterns, K errors are produced. The mean square error (MSE) corresponding to ith particle is determined by using the relation.

$$MSE\,(i) = \frac{\sum\limits_{k=1}^{K} e^2\,(k)}{K} \tag{9.19}$$

This is repeated for I times.

(e) The perturbs vector is then calculated using (9.9) and the mutant population is obtained.

(f) the crossover operation is carried out using (9.10) and the trial population is created.

(g) The selection of members of the target population for the next generation is performed using (9.12) and the whole process is repeated for some generations.

(h) The learning process is stopped when the MMSE reaches the minimum possible floor level.

(i) The minimum MSE (MMSE), the generation at which it is achieved and the number of squared errors per generation are obtained.
(j) The sales prediction of the models are found and compared.

9.5.2. *Steps Involved in GA Based Training of the Model*

(a) The coefficients or weights of the model are initially chosen from a population of I chromosomes. Each chromosome constitutes D number of parameters and each parameter represents a weight of the linear combiner.
(b) K number of patterns each containing three features (mean, variance and actual data of a month) are obtained from the past retail sales time series.
(c) Each of the K patterns, x_n is passed through the linear combiner, multiplied with the corresponding weight and the partial sums are added together to give y_i using (9.18).
(d) The output of the linear combiner, y_i is then compared with corresponding normalized desired value, d_i to produce the error, e_i. After the application of all patterns, K errors are produced. The mean square error (MSE) corresponding to ith particle is determined by using the relation (9.19). This is repeated for I times.
(e) The crossover, mutation and selection operator are sequentially carried out following the steps as given in Sec. 3 and the whole process is repeated for some generations.
(f) The learning process is stopped when the MMSE reaches the minimum possible floor level.
(g) The minimum MSE (MMSE), the generation at which it is achieved and the number of squared errors per generation are obtained.
(h) The sales prediction of the models are computed for comparison.

9.5.3. *Steps Involved in PSO Based Training of the Model*

The following steps are involved to develop a PSO based prediction model

(a) Using the retail sale data of first day of a month and similar previous values D number of features are extracted. One set of features of a day is termed as a pattern. In this way M patterns are generated from known sales data of which N patterns are used for training and the remaining $(M - N)$ patterns are used for testing the model. An nth test pattern is represented as

$$\underline{X_n} = [X_{n1}, X_{n2}, \ldots\ldots\ldots X_{nD}]^T \tag{9.20}$$

(b) The basic structure of the model is an adaptive linear combiner with the number of inputs equal to the number of features (D) in a pattern.

(c) The (D) weights of the model are trained using a PSO based learning algorithm.

(d) A particle (agent) consists of (D) weights of a linear combiner. The position of each particle is considered as a set of (D) random numbers each lying between 0 to 1. In total, (I) sets are initialized to represent a swarm moving around in the search space looking for the best location.

(e) For the ith particle, the position vector (which represents in this model as the weight vector) is given by

$$\underline{W_i} = [w_{i1}, w_{i2}, \ldots\ldots\ldots w_{id} \ldots\ldots w_{iD}]^T \tag{9.21}$$

where w_{id} is the dth weights of ith particle. Similarly the velocity of the ith particle is expressed as

$$\underline{V_i} = [v_{i1}, v_{i2}, \ldots\ldots\ldots v_{id} \ldots\ldots v_{iD}]^T \tag{9.22}$$

Initially the personal best or pbest, the ith particle has achieved is same as the initial i)th weight vector $\underline{W_i}$ and is represented as

$$\underline{W_{il}} = \underline{W_i} = [w_{il1}, w_{il2}, \ldots\ldots\ldots w_{ild} \ldots\ldots w_{ilD}]^T \tag{9.23}$$

(f) Each pattern $(1 \leq n \leq N)$ is applied to the ith particle and the output of the linear combiner is computed using the relation

$$y_i(n) = \underline{X_n^T} \underline{W_i} \tag{9.24}$$

This output is then compared with the corresponding desired value, $d(n)$ to produce the error signal

$$e_i(n) = d(n) - y_i(n) \tag{9.25}$$

The desired value $d(n)$ represents n) months ahead sales value in a normalized form.

(g) After application of all patterns N errors are generated from which the fitness function known as the MSE for the ith particle is computed as

$$MSE\,(i) = \frac{\sum\limits_{n=1}^{N} e_i^2\,(n)}{N} \qquad (9.26)$$

(h) In this way the fitness values of all I particles are evaluated following steps 6 and 7. The best fitness value that is, the MMSE is obtained and its corresponding D weights are identified and is termed as the global best or gbest. It is denoted as

$$\underline{W}_g = [w_{g1},\, w_{g2}, \ldots\ldots\ldots w_{gd}\ldots\ldots w_{gD}]^T \qquad (9.27)$$

(i) The velocity and the position of dth weight of each ith particle for the next search are obtained by using (9.13) and (9.14). Steps 6 to 9 constitute one search operation K. Using the new position and velocity of each particle, these steps are repeated in the next search. The search process continues until all the particles in the swarm (the weight vectors) have converged to the global best position or the cost function (the MSE) attains the least value.

9.5.4. *Steps Involved in BFO Based Training of the Model*

The development of the forecasting model using an adaptive linear combiner and BFO based training proceeds as follows

Initialization of parameters

S = No. of bacteria used for optimization
N = Number of input patterns
P = Number of parameters to be optimized = no. of weights of the model
N_s = Swimming length after which tumbling of bacteria is undertaken in a chemotactic loop.
N_c = Number of iterations to be carried out in a chemotactic loop. Always $N_c > N_s$.
N_{re} = Number of reproduction loops
N_{ed} = Maximum number of elimination and dispersal events to be imposed over the bacteria.

P_{ed} = Probability with which the elimination and dispersal continues.
θ^i = Location of ith $(i = 1, 2 \cdots, S)$ bacterium which is initially specified by random numbers between 0 to 1.
$C(i)$ = Step size of ith bacterium in the random direction.

Iterative Algorithm for optimization

This section models the bacterial population, chemotaxis, reproduction, elimination and dispersal processes. The swarming operation of bacteria is not considered in the training rule to keep the algorithm computationally simple and without much sacrificing the accuracy of prediction.

Initially $j = k = l = 0$

Elimination dispersal loop $l = l + 1$
Reproduction loop $k = k + 1$
Chemotaxis loop $j = j + 1$

(a) For $i = 1, 2 \cdots, S$

    i. N training samples are passed through the model (three different features computed from the retail sales data are used at a time as input) and the weighted sum is computed to obtain the output.

    ii. The output is then compared with the corresponding desired value to calculate the error.

    iii. The cost function, $J(i, j, k, l)$ for each ith bacterium which in the present case is the mean squared error is calculated using (9.28)

$$J(i, j, k, l) = \frac{1}{N} \sum_{n=1}^{N} e^{i^2}(n) = \frac{1}{N} \sum_{n=1}^{N} \left[d(n) - y^i(n) \right]^2 \quad (9.28)$$

The output of the linear combiner due to ith bacterium when n th pattern applied is given by

$$y^i(n) = \sum X^T(n) \underline{\theta}^i \quad (9.29)$$

$d(n) = n$th desired pattern (normalized sales value)
$e^i(n)$=error value for th pattern due to th bacterium

    iv. End of For loop.

(b) For $i = 1, 2 \cdots, S$, the tumbling/swimming decision is taken.
Tumble A random vector $\Delta(i)$ with each element$\Delta_m(i), m = 1, 2, \cdots, p$ a random number in the range $[-1, 1]$ is generated.
Move The location is updated as
Let

$$\theta^i(j+1, k, l) = \theta^i(j+1, k, l) + C(i) \times \frac{\Delta(i)}{\sqrt{\Delta^T(i) \Delta(i)}} \quad (9.30)$$

The second part of (9.6) is an adaptable step size in the direction of tumble for bacterium i.

The new cost function $J(i, j + 1, k, l)$ is computed corresponding to new location of bacteria.

Swim (i) Let $c = 0$; (counter for swim length)

(ii) While $c < N_s$ (the bacteria have not climbed down too long)

Let $c = c + 1$

If $J^i(j+1, k, l) < J_{last}$ (if doing better), let $J_{last} = J^i(j+1, k, l).\theta^i(j+1, k, l)$ is again computed by (9.6) and then used to compute new $J^i(j + 1, k, l)$

ELSE let $c = N_s$. This is end of the WHILE statement.

(c) Go to next bacterium $(i + 1)$ if $i \neq S$ to process the next bacterium.

(d) If $min(J)$ minimum value of J among all the bacteria is less than the tolerance limit then break all the loops.

If $j < N_c$, go to (iii) i.e., continue chemotaxis loop since the life of the bacteria is not over.

Reproduction

(a) For the given k and l and for each $(i = 1, 2, \cdots, S)$ let J^i be the health of ith bacterium. Sort bacteria in ascending order of cost J (higher cost means lower health)

(b) The $S_r = S/2$ bacteria with highest J value die and other S_r bacteria with the best value split and the copies that are made are placed at the same location as their parents.

If $K = N_{re}$ goto 2.

Elimination-Dispersal

For$(i = 1, 2 \cdots, S)$ with probability P_{ed}, eliminate and disperse each bacterium(this keeps the number of bacteria in the population constant). To achieve this, a bacterium is eliminated by simply dispersing it to a random location on the optimization domain.

9.6. Simulation Study

9.6.1. *Experimental Data for Training and Testing*

For carrying out simulation study two sets of data for the retail sales have been collected from www.forecasts.org. Total number of data for the retail sales available are 652 starting from 1st January 1947 to 1st January 2001. One set of data is seasonally adjusted (SA) and other one is not seasonally adjusted (NSA). Each set of data are normalized by dividing each value by the maximum value of each set such that each normalized value is less than or equal to unity. The normalized rate on the first day of a month, the mean and variance value computed up to this month are considered as the inputs to model. To obtain the mean and variance value of a given month twelve data (present one and eleven preceding values) are used. The available data is divided into training and testing sets. The training set consists of 600 samples and the rest is used for testing purpose.

9.6.2. *Training and Testing of the Forecasting Model*

Training of the ALC model is carried out using DE, GA, PSO and BFO algorithms given in Secs. 9.3 and 9.4 and the optimum weights are obtained. Then using the trained model, the forecasting performance is validated using test patterns for one, three, six, nine and twelve months in advance. The Root Mean Absolute Percentage Error (RMAPE) defined in (9.31) is computed to compare the performance of various models.

$$RMAPE = sqrt\left(\frac{1}{k}\sum_{k=1}^{K}\left|\frac{d_k - y_k}{d_k}\right| \times 100\right) \qquad (9.31)$$

where is the number of test patterns.

Table 9.1.: Performance comparison of different algorithms.

Method	No. of generations required to converge	No. of squared error evaluation ($\times 10^5$)	MSE floor obtained ($\times 10^{-4}$)
DE	05	0.90	0.560
PSO	150	27	0.629
BFO	470	225.60	0.999
GA	500	999.84	11.0

Table 9.1 shows the vital simulation results. It is observed that DE takes minimum generations to converge, minimum number of squared error per generation and minimum MSE level. Hence DE offers the best and fast prediction performance. The actual and predicted NSA and SA sales values obtained from DE model for twelve months are plotted in Figs. 9.3(a) and 9.3(b) respectively and those for 6 months are shown in Figs. 9.4(a) and 9.4(b). Excellent agreement in the actual and predicted values are observed in all cases. For SA sales prediction it is observed that with increase in month the RMAPE increases gradually but the range lies between 0.75% to 1.62% for DE, PSO and BFO based prediction models. However the

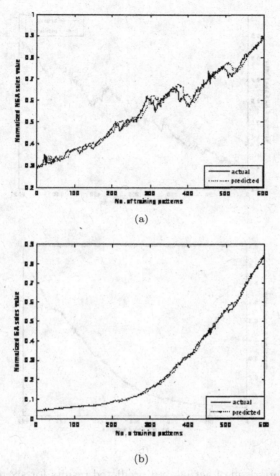

(a)

(b)

Fig. 9.3.: Comparison of actual and predicted results for twelve months ahead prediction using DE. (a) NSA sales value. (b) SA sales value.

GA model offers higher RMAPE for all months which is about 3.96%. These results are displayed in Table 9.2. However for NSA sales data, the same trend is observed in all cases (Table 9.4) but the RMAPE obtained is slightly higher. The maximum MAPE in this case is about 1.89%. In case of GA model the RMAPE varies between 2.1% to 2.7% which is less than the corresponding RMAPE of SA sales. From Tables 9.3 and 9.5 the CPU time during training for SA and NSA sales prediction respectively is

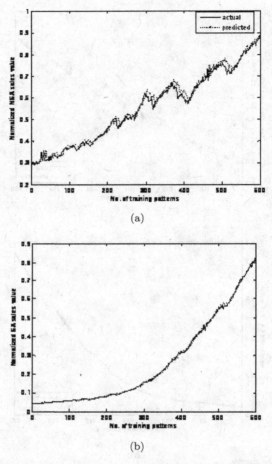

Fig. 9.4.: Comparison of actual and predicted results for six months ahead prediction using DE. (a) NSA sales value. (b) SA sales value.

observed to be the least in case of DE and the highest in case of BFO model. Based on the CPU time the performance of various model may be graded in the order DE, PSO, GA and BFO. The comparison of actual/predicted NSA and SA sales volume for one and twelve months ahead prediction is shown in Tables 9.6 and 9.7 respectively. The prediction performance of PSO, BFO and DE based models is observed to be similar but better than that of GA.

Table 9.2.: Comparison of square root of mean average percentage error (seasonally adjusted).

	DE	GA	PSO	BFO
1 month	0.7507	4.0045	0.7418	0.7689
3 month	1.0083	3.9909	1.0060	1.0192
6 month	1.1954	3.8413	1.2396	1.2467
9 month	1.4331	3.9567	1.4351	1.4736
12 month	1.6271	3.9529	1.6044	1.6015

Table 9.3.: Comparison of CPU Time (seasonally adjusted).

	DE	GA	PSO	BFO
1 month	0.2572	3.0931	0.3714	7.1953
3 month	0.2283	3.0867	0.3687	7.1968
6 month	0.2287	3.0837	0.3794	7.2149
9 month	0.2345	3.0053	0.3730	7.2056
12 month	0.2252	3.0311	0.3775	7.1950

Table 9.4.: Comparison of square root of mean average percentage error (not seasonally adjusted).

	DE	GA	PSO	BFO
1 month	0.7404	2.1303	0.7113	0.9418
3 month	1.0169	2.1150	1.0313	1.0634
6 month	1.3372	2.1619	1.3354	1.2120
9 month	1.5956	2.7258	1.5966	2.6791
12 month	1.8336	2.5452	1.8351	1.8908

Table 9.5.: Comparison of CPU Time (not seasonally adjusted).

	DE	GA	PSO	BFO
1 month	0.2388	3.3621	0.4361	7.8315
3 month	0.2384	2.9425	0.3909	7.6967
6 month	0.2378	3.0470	0.3948	7.6613
9 month	0.2380	3.0782	0.4014	7.2928
12 month	0.2334	3.0786	0.3895	7.3951

9.7. Conclusion

This chapter proposes an efficient sales forecasting model using an ALC and DE based training of its parameters. The performance of the new model is obtained in terms of CPU time, rate of convergence, accuracy of short and long range prediction and computational complexity. To facilitate compar-

Table 9.6.: Comparison of actual/predicted value (NSA) during testing.

Actual value	Genetic algorithm	Particle swarm optimization	Bacterial foraging optimization	Differential evolution
1 month ahead				
1.5620	1.6975	1.5680	1.5580	1.5627
1.5589	1.6719	1.5670	1.5600	1.5620
1.5680	1.6328	1.5633	1.5602	1.5588
1.5666	1.6050	1.5710	1.5671	1.5680
1.5709	1.5860	1.5694	1.5680	1.5665
1.5669	1.5581	1.5727	1.5720	1.5709
1.5553	1.5175	1.5681	1.5712	1.5669
1.5541	1.4943	1.5570	1.5656	1.5552
1.5656	1.4938	1.5559	1.5650	1.5541
1.5587	1.4976	1.5664	1.5714	1.5655
1.5522	1.4941	1.5600	1.5674	1.5586
1.5603	1.4879	1.5538	1.5629	1.5521
12 months ahead				
1.5620	1.6257	1.5145	1.5055	1.5196
1.5589	1.6692	1.5173	1.5092	1.5238
1.5680	1.6752	1.5242	1.5165	1.5307
1.5666	1.6708	1.5412	1.5320	1.5471
1.5709	1.6439	1.5398	1.5328	1.5445
1.5669	1.6258	1.5428	1.5369	1.5467
1.5553	1.6304	1.5599	1.5521	1.5635
1.5541	1.6657	1.5809	1.5706	1.5852
1.5656	1.6718	1.5816	1.5731	1.5860
1.5587	1.7134	1.5952	1.5857	1.6007
1.5522	1.7377	1.6029	1.5937	1.6090
1.5603	1.6965	1.5897	1.5850	1.5943

ison of performance other evolutionary computing models such as BFO, PSO and GA are also simulated and corresponding results are obtained. It is observed that in all performance counts, the DE based prediction model outperforms the other three models.

Table 9.7.: Comparison of actual/predicted value (SA) during testing.

Actual value	Genetic algorithm	Particle swarm optimization	Bacterial foraging optimization	Differential evolution
1 month ahead				
2.6742	3.7788	2.6275	2.7152	2.6808
2.6844	3.6639	2.6374	2.7176	2.6854
2.7064	3.4752	2.6589	2.7233	2.6969
2.7055	3.3260	2.6945	2.7386	2.7183
2.7271	3.2015	2.6990	2.7379	2.7201
2.7249	3.0597	2.7089	2.7527	2.7409
2.7093	2.8277	2.7427	2.7476	2.7417
2.7135	2.6288	2.7902	2.7333	2.7306
2.7491	2.5568	2.7995	2.7359	2.7354
2.7433	2.5412	2.8392	2.7629	2.7667
2.7334	2.5405	2.8723	2.7606	2.7630
2.7555	2.5450	2.8463	2.7547	2.7553
12 months ahead				
2.6742	3.2279	2.7043	2.6411	2.6155
2.6844	3.3770	2.7072	2.6589	2.6231
2.7064	3.4146	2.7146	2.6806	2.6442
2.7055	3.4282	2.7318	2.7118	2.6800
2.7271	3.3747	2.7317	2.7183	2.6857
2.7249	3.3468	2.7484	2.7291	2.6964
2.7093	3.3800	2.7450	2.7593	2.7300
2.7135	3.5448	2.7313	2.8054	2.7753
2.7491	3.6145	2.7346	2.8207	2.7837
2.7433	3.8032	2.7632	2.8624	2.8206
2.7334	3.9624	2.7604	2.8988	2.8514
2.7555	3.8520	2.7537	2.8816	2.8273

References

1. C. W. Chu and G. P. Zhang, A comparative study of linear and nonlinear models for aggregate retails sales forecasting, *Int. Journal Production Economics*, **86**, 217–231, 2003.
2. A. S. Weigend and N. A. Gershenfeld, Time series prediction : Forecasting the future and understanding the past, *Addison-Wesley*, 1994.
3. R. J. Kuo and K. C. Xue, A decision support system for sales forecasting through fuzzy neural networks with asymmetric fuzzy weights, *Decision Support System*, **24**, 105–126, 1998.

4. R. Majhi, G. Panda and G. Sahoo, Development and performance evaluation of FLANN based model for forecasting of stock markets, *Expert Systems with Applications*, **36**, (3), 6800–6808, April 2009.

5. R. Majhi, G. Panda and G. Sahoo, Efficient prediction of exchange rates with low complexity artificial neural network models, *Expert Systems with Applications*, **36**,(1), 181–189, January 2009.

6. Min Qi and G. Peter Zhang, Trend time series modeling and forecasting with neural networks, *IEEE Trans. on Neural Networks*, **19**, (5), 808–816, May 2008.

7. Ilona Jagielska and Ashok Jacob, A neural network model for sales forecasting, *First New Zealand International Two Stream Conference on Artificial neural networks and expert systems*, 24-26, 284–287,Nov. 1993.

8. J H L Kong and G P M D Martin, A backpropagation neural network for sales forecasting, *IEEE International Conference on Neural Networks*, **2**, 1007–1011,Dec. 1995.

9. Abdul Sahli Fakharudin, Mohd Azwan Mohamad and Mohd Usaid Johan, Newspaper vendor sales prediction using artificial neural networks, *IEEE International conference on education technology and computer*, 339–343,2009.

10. Frank M. Thiesing, Ulrich Middelberg, Oliver Vornberger, Short term prediction of sales in supermarkets, *IEEE International conference on Neural networks*, **2**, 1028–1031,1995.

11. Devil H. F. Yip, E. L. Hines and William W. H. Yu, Application of artificial neural networks in sales forecasting, *IEEE International conference on Neural networks*, **4**, 2121–2124,1997.

12. Frank M. Thiesing and Oliver Vornberger, Sales forecasting using neural networks, *IEEE International conference on Neural networks*, **4**, 2125–2128,1997.

13. Pei Chann Chang and Yen Wen Wang,Fuzzy Delphi and back propagation model for sales forecasting in PCB industry, *Expert systems with applications*, **30**, 715–726, 2006.

14. R. J. Kuo and K. C. Xue, A decision support system for sales forecasting through fuzzy neural networks with asymmetric fuzzy weights, *Decision support systems*, **24**, 105–126, 1998.

15. Pei Chann Chang, Chen Hao Liu, Chin Yuan Fan, Hsiao Ching Chang, Data clustering and fuzzy neural network for sales forecasting in printed circuit board industry, *IEEE Symposium on Computational Intelligence and Data Mining*, 107–113,2007.

16. Yihang Liu and Lieli Liu, Sales forecasting through fuzzy neural networks, *IEEE Int. Conference on Electronic Computer technology*, 511–515,2009.

17. R. J. Kuo, Tung Lai Hu and Zhen Yao Chen,Application of radial basis function neural networks for sales forecasting, *Proc. of Int. asia Conference on Informatics in control, automation and robotics*, 325–328, 2009.

18. Liu Hong, Cui Wenhua and Zhang Qingling, Nonlinear combination forecasting model and application based on radial basis function neural networks, *IITA International Conference on Control, Automation and Systems Engineering*, 387–390,2009.

19. Xiao Jiwei, Wu Yaohua, Wang Qian, Liao Li and Hu Hongchun, A hybrid sales forecasting method based on stable seasonal pattern models and BPNN, *Proc. of IEEE Int. Conf. On automation and logistics*, 2868–2872, 2007.

20. Philip Doganis, Alex Alexandridis, Panagiotis Patrinos and Haralambos Sarimveis, Time series sales forecasting for short-life food products based on artificial neural networks and evolutionary computing, *Journal of Food Engineering*, **75**, 196–204, 2006.

21. Julie Marcoux and Sid-Ahmed Selouani, A hybrid subspace-connectionist data mining approach for sales forecasting in the video game industry, *World Congress on Computer Science and Information Engineering*, **5**, 666–670,2009.

22. Qi Wu, Hong Sen Yan and Hong Bing Yang, A forecasting model based support vector machine and particle swarm optimization, *Proc. of IEEE workshop on Power electronics and intelligent transportation system*, 218–222, 2008.

23. Min Zhou and Qiwan Wang, The online electronic commerce forecast based on least square support vector machine, *IEEE Second International Conference on Information and Computing Science*, 75–78,2009.

24. R. Majhi, G. Panda, G. Sahoo and A. Panda,On the development of Improved Adaptive Models for Efficient Prediction of Stock Indices using Clonal-PSO (CPSO) and PSO Techniques, *International Journal of Business Forecasting and Market Intelligence*, **1**, 50–67, 2008.

25. R. Majhi, G. Panda, B. Majhi and G. Sahoo, Efficient prediction of stock market indices using adaptive bacterial foraging optimization (ABFO) and BFO based techniques, *Expert Systems with Applications, Elsevier*, **36**,(6),10097–10104, August 2009.

26. M. Tripathy and S. Mishra, Bacteria foraging based to optimize both real power loss and voltage stability limit, *IEEE Trans. on Power Systems*, **22**, (1), 240–248, 2007.

27. R. Majhi, G. Panda, G. Sahoo and D.P. Das, Stock market prediction of S&P 500 and DJIA using Bacterial Foraging Optimization technique, *IEEE Congress on Evolutionary Computation (CEC 2007)*, 2569–2575,2008.

28. G. Panda, D. Mohanty, Babita Majhi and G. Sahoo, Identification of Nonlinear Systems using Particle Swarm Optimization Technique, *Proc. of IEEE International Congress on Evolutionary Computation*, 3253–3257, 2007.

29. L. Messerschmidt and A. P. Engelbrecht, Learning to play games using a PSO-based competitive learning approach, *IEEE Trans. Evolutionary Computation*, **8**, 280–288, June 2004.

30. M. P. Wachowiak, R. Smolikova, Y. F. Zheng, J. M. Zurada and A. S. Elmaghraby, An approach to multimodal biomedical image registration utilizing particle swarm optimization, *IEEE Trans. on Evolutionary Computation*,**8**, 289–301, June 2004.

31. Y. Shi and R. C. Eberhart, A modified particle swarm optimizer, *in Proc. IEEE Congress on Evolutionary Computation*, 69–73,1998.

32. J. J. Liang, A. K. Qin., P. N. Suganthan and S. Baskar, Comprehensive learning particle swarm optimizer for global optimization of multimodal functions, *IEEE Trans. on Evolutionary Computation*, **10**, (3), 281–295, June 2006.

33. K. M. Passino, Biomimicry of Bacterial Foraging for distributed optimization and control, *IEEE control system magazine*, **22**, (3), 52–67, June 2002.

34. A.K.Qin, V. L. Huang and P.N. Suganthan, Differential evolution algorithm with strategy adaptation for global numerical optimization, *IEEE Trans. on Evolutionary Computation*, **13**, (2), 398–417, April 2009.

35. D.H.Goldberg, Genetic algorithms in search, optimization and machine learning,*Addition-Wesley*,1989.

36. B. Widrow and S. D. Sterns, Adaptive Signal Processing, *Pearson Education*, 22, Inc. 1985.

Chapter 10

Some Studies on Particle Swarm Optimization for Single and Multi-Objective Problems

Madhabananda Das* and Satchidananda Dehuri[†]

School of Computer Engineering,
KIIT University,
Bhubaneswar, 751024, India
* mndas12@gmail.com

Department of Information and Communication Technology,
Fakir Mohan University,
Vyasa Vihar, Balasore 756019, India
† satchi.lapa@gmail.com

This chapter deals with particle swarm optimization (PSO)(i.e., one of the components of swarm intelligence) for single and multi-objective problems. In PSO, an initial swarm propagates in the design space towards the optimal solution in cooperation of members of the swarm by updating velocity and position of each member. A considerable number of algorithms have been and are being proposed for single and multi-objective problems based on either tuning or introducing various parameters of PSO. These algorithms have there own merits and demerits. Further, since these algorithms are scattered in various journals, proceedings and book chapters, therefore it is a herculean task for a reader to assemble in one place and learn. Hence, this study can reduce the effort and time of the reader to digest the various approaches of PSO. Additionally, we identify some application areas, where PSO has given a clear edge over other meta heuristic approaches for solving single and multi-objective optimization problems. This can motivate and lead the readers for extensive application of PSO in new domains of interest. The last part of the chapter concludes with a discussion and future research scope followed by an extensive set of relevant references.

---

*Professor and Associate Dean
†Reader and Head

10.1. Introduction

When an optimization problem, modeling a physical system, involves only one objective function, the task of finding the lone optimal solution is called single-objective optimization and the problem is called single-objective optimization problem (SOP). A multi-objective optimization problem (MOP) deals with more than one objective functions. In most practical decision-making problems, multiple objectives are evident. In such problems, the objectives to be optimized are normally in conflict with respect to each other, which means that there is no single optimum solution. Instead, we aim to find some acceptable "trade-off" solutions that represent the best possible compromise among the objectives. Particle Swarm Optimization (PSO)[1] is a heuristic search method that simulates the movements of a flock of birds which aim to find food. The relative simplicity of PSO and the population-based technique and the information sharing mechanism associated with this method have made it a natural candidate to be extended for single as well as multi-objective optimization.

After the early attempt by Moore *et al.*,[2] a great interest to extend PSO to MOPSO generated among researchers. Nevertheless, there are currently a large number of different proposals of MOPSO reported in the specialized literature. This chapter provides a study on single and multi-objective PSO with a greater emphasis on various implicit and explicit parameters.

10.1.1. *Definitions of Single and Multi-Objective Problem*

A single objective optimization problem can be stated in the general form (10.1):

$$
\begin{aligned}
&Minimize/Maximize \qquad f(x) \\
&\quad subject\ to \\
&\quad g_j(x) \geq 0 \qquad j = 1, 2, ..., J; \\
&\quad h_k(x) = 0 \qquad k = 1, 2, .., K; \\
&\quad x_i^{(L)} \leq x_i \leq x_i^{(U)} \qquad i = 1, 2, ..., n.
\end{aligned} \tag{10.1}
$$

Similarly, a multi-objective optimization problem can be stated in the general form (10.2):

$$Minimize/Maximize \ f_m(x), m = 1, 2, ..., M$$
$$subject \ to$$
$$g_j(x) \geq 0 \qquad j = 1, 2, ..., J;$$
$$h_k(x) = 0 \qquad k = 1, 2, .., K; \qquad (10.2)$$
$$x_i^{(L)} \leq x_i \leq x_i^{(U)} \qquad i = 1, 2, ..., n.$$

A solution x is a vector of n decision variables: $x = (x_1, x_2, ..., x_n)^T$. The last set of constraints are called variable bounds, restricting each decision variable x_i to take a value within a lower $x_i^{(L)}$ and an upper $x_i^{(U)}$ bounds. These bounds constitute a decision variable space D, or simply the decision space. Associated with the problem are J inequality and K equality constraints and the terms $g_j(x)$ and $h_k(x)$ are called constraint functions. Although the inequality constraints are treated as $\geq$ types, the $\leq$ constraints can also be considered in the above formulation by converting those to $\geq$ types simply by multiplying the constraint function by -1.[3]

In single objective optimization, there is a single goal - the search for an optimum solution (i.e., global one). Although the search may have a number of local optimal solutions, the goal is always to find the global optimal solution. However, most single-objective optimization algorithms aim at finding one optimum solution, even when there exist a number of optimal solutions. In a single-objective optimization algorithm, as long as a new solution has a better objective function value than an old solution, the new solution can be accepted. In single objective optimization, there is only one search space i.e., the decision variable space. An algorithm works in this space by accepting and rejecting solutions based on their function values.

In multi-objective optimization, the M objective functions $f(x) = (f_1(x), f_2(x), ..., f_M(x))^T$ can be either minimized or maximized or both. Many optimization algorithms are developed to solve only one type of optimization problems, such as e.g., minimization problems. When an objective is required to be maximized by using such an algorithm, the duality principle[3-5] can be used to transform the original objective for maximization into an objective for minimization by multiplying objective function by -1. It is to be noted that for each solution x in the decision variable space, there exists a point in the objective space, denoted by $f(x) = z = (z_1, z_2, ..., z_M)$. There are two goals in a multi-objective optimization: firstly, to find a set of solutions as close as possible to the Pareto-optimal front; secondly, to find a set of solutions as diverse as possible. Multi-objective optimization involves two search spaces i.e., the decision variable space and the objective space. Although these two spaces are related by an unique mapping between them, often the mapping is non-linear and the properties of the two search spaces are not similar. In any optimization algorithm, the search is performed in the decision variable space. However, the proceedings of an algorithm in the decision variable space can be traced in the objective space. In some algorithms, the resulting proceedings in the objective space are used to steer the search in the decision variable space. When this happens, the proceedings in both spaces must be coordinated in such a way that the creation of new solutions in the decision variable space is complementary to the diversity needed in the objective space.

A multi-objective problem can be handled as a single objective optimization problem by using various classical methods such as weighted sum approach,[6-8] ϵ-constraint method,[9] weighted metric method,[8] value-function method,[8-10] and goal programming method.[11-17] In the weighted sum approach, multiple objectives are weighted and summed together to create a composite objective function. Optimization of this composite objective results in the optimization of individual objective functions. The outcome of such an optimization strategy depends on the chosen weights. The ϵ-constraint method chooses optimizing one of the objective functions and treats the rest of the objectives as constraints by limiting each of them within certain pre-defined limits. This fix-up converts the multi-objective optimization problem into a single objective optimization problem. Here too the outcome of the single-objective constrained optimization results in a solution which depends on the chosen constraint values. Weighted metric method suggests minimizing an L_p -metric constructed from all objectives. The value function method suggests maximizing an overall value function (or utility function) relating all objectives. Goal programming methods suggest minimizing a weighted sum of deviations of objectives from user-specified targets. These conversion methods result in a single-objective optimization problem, which must be solved by using a single-objective optimization algorithm. These classical multi-objective optimization algorithms are having some difficulties particularly if the user wants to find multiple Pareto-optimal solutions. First, only one Pareto-optimal solution can be expected to be found in one simulation run of a classical algorithm. Second, not all Pareto-optimal solutions can be found by some algorithms in non-convex MOPs. Third, all algorithms require some problem knowledge, such as suitable weights or ϵ or target values. Multi-objective optimization for finding multiple Pareto-optimal solutions eliminates all such fix-ups and can in principle find a set of optimal solutions corresponding to different weights and ϵ vectors. Although only one solution is needed for implementation, the knowledge of such multiple optimal solutions may help a designer to compare and choose a compromised optimal solution. A multi-objective optimization is, in general, more complex than a single-objective optimization, but the avoidance of multiple simulation runs, no artificial fix-ups, availability of efficient population-based optimization algorithms, and above all, the concept of dominance help to overcome some of the difficulties and give a user the practical means to handle multiple objectives.

Most multi-objective optimization algorithms use the concept of domination. In these algorithms, two solutions are compared on the basis of whether one dominates the other solution or not. The concept of domination is described in the following definitions(assuming, without loss of generality, the objective functions to be minimized).

Definition 1. Given two decision or solution vectors x and y, we say that decision vector x weakly dominates (or simply dominates) the decision vector y (denoted by $x \preceq y$) if and only if $f_i(x) \preceq f_i(y) \forall\ i = 1, ..., M$ (i.e., the solution x is no worse than y in all objectives) and $f_i(x) \prec f_i(y)$ for at least one $i \in 1, 2, ..., M$ (i.e., the solution x is strictly better than y in at least one objective).

Definition 2. A solution x strongly dominates a solution y (denoted by $x \prec y$), if solution x is strictly better than solution y in all M objectives.

Figure 10.1 illustrates a particular case of the dominance relation in the presence of two objective functions. However, if a solution x strongly dominates a solution y, the solution x also weakly dominates solution y, but not vice versa.

Definition 3. The decision vector $x \in P$ (where P is the set of solution or decision vectors) is non-dominated with respect to set P, if there does not exit another $x' \in P$ such that $f'(x) \preceq f(x)$.

Definition 4. Among a set of solution or decision vectors P, the non-dominated set of solution or decision vectors P' are those that are not dominated by any member of the set P.

Definition 5. A decision variable vector $x \in P$ where P is the entire feasible region or simply the search space, is Pareto-Optimal if it is non-dominated with respect to P.

Definition 6. When the set P is the entire search space, the resulting non-dominated set P' is called the Pareto-Optimal set. In other words, $P' = \{x \in P \mid x\ is\ Pareto - Optimal\}$. The non-dominated set P' of the entire feasible search space P is the globally Pareto-Optimal set.

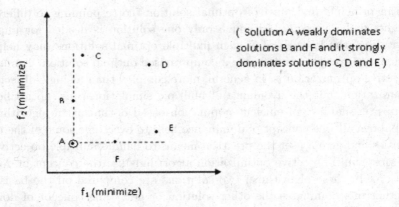

Fig. 10.1.: Illustration of dominance relation in bi-objective functions.

Definition 7. All Pareto-Optimal solutions in a search space can be joined with a curve (in two-objective space) or with a surface (in more than two-objective space). This curve or surface is termed as Pareto optimal front or simply Pareto front. In other words, $PF = \{f(x) \mid x \in P'\}$

Figure 10.2 illustrates a particular case of the Pareto front in the presence of two objective functions. We thus wish to determine the Pareto optimal set form the set P of all the feasible decision variable vectors that satisfy (10.2). It is to be noted that in practice, the complete Pareto Optimal set is

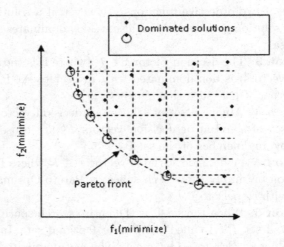

Fig. 10.2.: Pareto front of a set of solutions in a bi-objective problem.

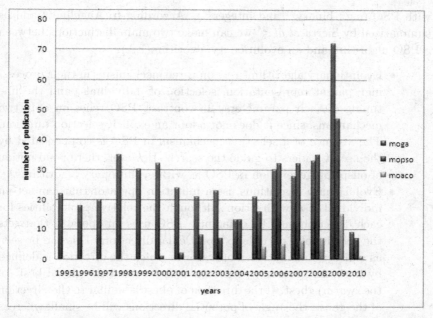

Fig. 10.3.: Growth of multi-objective problem solving approaches using GA, PSO, and ACO.

not normally desirable (e.g., it may not be desirable to have different solutions that map to the same values in objective function space) or achievable. Thus a preferred set of Pareto optimal solutions should be obtained from practical point of view. In Fig. 10.3, an overall growth of multi-objective optimization problem solving approaches using various meta-heuristic approaches like GA, PSO, and ant colony optimization (ACO) from the period 1995 to 2010 is given. However, the year 2010 is incomplete.

10.1.2. *Particle Swarm Optimization (PSO)*

Kennedy and Eberhart originally proposed the PSO algorithm for single objective optimization.[1] PSO is a population-based search algorithm based on the simulation of the social behavior of birds within a flock. Although originally adopted for neural network training and non-linear function optimization,[18,19] PSO soon became a very popular global optimizer, mainly in problems in which the decision variables are real numbers.[19,20] It is worth noting that there have been proposals to use alternative encodings with PSO (e.g., binary[21] and integer[22]). According to Angeline[23] and as summarized by Sierra *et al.*,[24] we can make two main distinctions between a PSO algorithm and an evolutionary algorithm(EA):

- Evolutionary algorithms rely on three mechanisms in their processing: parent representation, selection of individuals, and the fine tuning of their parameters. In contrast, PSO relies on only two mechanisms, since it does not adopt an explicit selection function. The absence of a selection mechanism in PSO is compensated by the use of leaders to guide the search. However, there is no notion of offspring generation in PSO as with evolutionary algorithms.
- Evolutionary algorithms use a mutation operator that can set an individual in any direction (although the relative probabilities for each direction may be different). PSO uses an operator that sets the velocity of a particle to a particular direction. This can be seen as a directional mutation operator in which the direction is defined by both the particle's personal best pbest and the global best (of the swarm) gbest. If the direction of pbest is similar to the direction of the gbest, the angle of potential directions will be small, whereas a larger angle will provide a larger range of exploration. In fact, the limitation exhibited by the directional mutation of PSO has led to the use of mutation operators similar to those adopted in evolutionary algorithms.

PSO has become so popular because its main algorithm is relatively simple and easier to implement. It is also straightforward and has been found to be very effective in a wide variety of applications with very good results at a very low computational cost.[19,25]

As a basic principle, in PSO, a set of randomly generated particles in the initial swarm are flown (have their parameters adjusted) through the hyper-dimensional search space (problem space) according to their previous flying experience. Changes to the position of the particles within the search space are based on the social-psychological tendency of individuals to emulate the success of other individuals. Each particle represents a potential solution to the problem being solved. The position of a particle is determined by the solution it currently represents. The position of each particle is changed according to its own experience and that of its neighbors. These particles propagate towards the optimal solution over a number of generations (moves) based on large amount of information about the problem space that is assimilated and shared by all members of the swarm. PSO algorithm finds the global best solution by simply adjusting the trajectory of each individual toward its own best location (pbest) and towards the best particle of the entire swarm (gbest) at each time step (generation). In this algorithm, the trajectory of each individual in the search space is adjusted by dynamically altering the velocity of each particle according to its own flying experience and the flying experience of the other particles in the search space.

The position vector and the velocity vector of the ith particle in the d-dimensional search space can be expressed as $x_i = (x_{i1}, x_{i2}, ..., x_{id})$ and $v_i = (v_{i1}, v_{i2}, ..., v_{id})$ respectively. According to a user defined fitness function, the best position of each particle (which corresponds to the best fitness value obtained by that particle at time t) is $p_i = (p_{i1}, p_{i2}, ..., p_{id})$, denoted as pbest and the fittest particle found so far in the entire swarm at time t is $p_g = (p_{g1}, p_{g2}, ..., p_{gd})$, denoted as gbest. Then the new velocities and the new positions of the particles for the next fitness evaluation are calculated at time t+1 using the following two self-updating equations:

$$v_{id}(t+1) = wv_{id}(t) + c_1 rand_1()(p_{id}(t) - x_{id}(t)) + c_2 rand_2()(p_{gd}(t) - x_{id}(t)) \tag{10.3}$$

$$x_{id}(t+1) = x_{id}(t) + v_{id}(t) \tag{10.4}$$

where $rand_1()$ & $rand_2()$ are two separately generated uniformly distributed random values in the range [0,1], w is inertia weight (or inertia factor) which is employed to control the impact of the previous history of velocities on the current velocity of a given particle, c_1 & c_2 are constants known as acceleration coefficients (or learning factors); c_1 is the cognitive learning factor (or self confidence factor) which represents the attraction that a particle has toward its own success and c_2 is the social learning factor (or swarm confidence factor) which represents the attraction that a particle has toward the success of its neighbors. Hassan *et al.*[26] have presented an idea about the ranges of c_1 to be from 1.5 to 2, c_2 to be from 2 to 2.5 and w to be from 0.4 to 1.4. From (1.3), it is observed that it has three components which are incorporated via a summation approach and effect the new search direction. The first component is known as current motion influence component which depends on previous velocity and provides the necessary momentum for particles to roam across the search space. The second component is known as cognitive (or particle own memory influence) component which represents the personal thinking of each particle and encourages the particles to move toward their own best positions found so far. The third

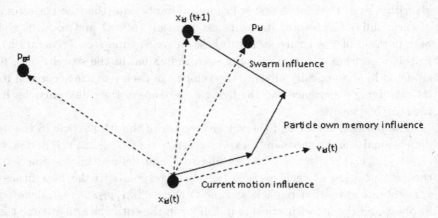

Fig. 10.4.: Depiction of velocity and position updates in PSO.[26]

component is known as social (or swarm influence) component which represents the collaborative effect of the particles, in finding the global optimal solution. This component always pulls the particles toward the global best particle found so far.

Figure 10.4 depicts the velocity and position updates of a particle as per Eqs. (10.3) and (10.4) in PSO. PSO algorithms are quite promising in the applications to single objective optimization problems as compared to evolutionary algorithm (EA) techniques.[19,26,27] These are very popular due to their simplicity in their implementations (a few parameters are needed to be tuned). A PSO algorithm is computationally cheap in the updating of the individuals per iteration, as the core updating mechanism in the algorithm relies only on two simple self-updating Eqs. (10.3) and (10.4) as compared to using mutation and crossover operations in typical EA which requires a substantial computational cost.

PSO uses an operator that sets the velocity of a particle to a particular direction. This can be seen as a directional mutation operator in which the direction is defined by both the particle's personal best and the global best (of the swarm). If the direction of the personal best is similar to the direction of the global best, the angle of potential directions will be small, whereas a larger angle will provide a larger range of exploration. In contrast, evolutionary algorithms use a mutation operator that can set an individual in any direction (although the relative probabilities for each direction may be different). In fact, the limitations exhibited by the directional mutation of PSO has led to the use of mutation operators (sometimes called turbulence operators) similar to those adopted in evolutionary algorithms. The pseudo code for a basic PSO algorithm is illustrated in Algorithm 1.

Algorithm 1. General Single-Objective Particle Swarm Optimization

(1) BEGIN
(2) Parameter settings and initialization of swarm.
(3) Evaluate fitness and locate the leader (i.e., initialize *pbest* and *gbest*).
(4) I = 0 /* I = Iteration count */
(5) WHILE (the stopping criterion is not met, say, I ¡ I_{max})
(6) DO
(7) FOR each particle
(8) Update position & velocity (flight) as per Eqs. (10.3)& (10.4)
(9) Evaluate fitness
(10) Update *pbest*
(11) END FOR
(12) Update leader (i.e., *gbest*)
(13) I++
(14) END WHILE
(15) END

First, the swarm is initialized. This initialization includes both positions and velocities. The corresponding pbest of each particle is initialized and the leader is located (the gbest solution is selected as the leader). Then, for a maximum number of iterations, each particle flies through the search space updating its position (using (10.3) and (10.4)) and its pbest and, finally, the leader is updated too.

10.2. Topological Structure of PSO

In the original PSO,[1] two different kinds of neighborhoods were defined for PSO;one is gbest swarm and other is lbest swarm.

In the gbest swarm, all the particles are neighbors of each other; thus, the position of the best overall particle in the swarm is used in the social term of the velocity update equation. It is assumed that gbest swarms converge fast, as all the particles are attracted simultaneously to the best part of the search space. However, if the global optimum is not close to the

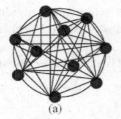
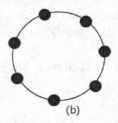

Fig. 10.5.: Graphical representation of gbest and lbest model.

best particle, it may be impossible to the swarm to explore other areas; this means that the swarm can be trapped in local optima. In the lbest swarm, only a specific number of particles (neighbor count) can affect the velocity of a given particle. The swarm will converge slower but can locate the global optimum with a greater chance. Figure 10.5 illustrate the graphical representation of the topological structure of both models.

Both gbest and lbest can be seen as "social" neighborhoods, as the relations among particles does not depend on their positions in the search space, but on "external" relationships that are not dependent on the problem that is being solved.

In Ref. 28 the effects of neighborhood topologies was investigated. Different neighborhoods can be characterized in terms of two factors:

- The degree of connectivity, k, that measures the number of neighbors of a particle.
- The amount of clustering C, that measures the number of neighbors of a particle that are also neighbors of each other.

The following additional topologies were tested[29]: i) random, ii) Von Neumann, a two dimensional grid with neighbors to the North, East, West and South; iii) Pyramid, a three-dimensional triangular grid; iv) Star, all the particles connected to a central particle; and iv) Heterogeneous, particles are grouped in several cliques. Figure 10.6 shows the graphical representation of these models.

In Ref. 1, the concept of a "dynamic" neighborhood is explored. In this work, neighborhood of a particle changes over time in two senses:

The number of neighbors of a particle starts as in the lbest swarm, with neighbor count 1; that is, the particle itself is its only neighbor. This neighbor count increases over time, finally becoming a gbest swarm. The

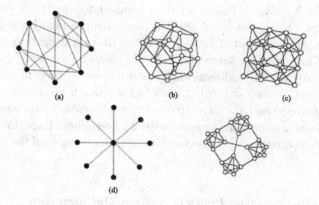

Fig. 10.6.: Graphical representation of random, Von Neumann, pyramid, star, and heterogeneous topologies.

selection of neighbors for a particle is calculated on each iteration, based on the hamming distance between the potential neighbors and the particle. The advantage of this topology, however, is not clearly stated in the mentioned paper. The drawbacks are the extra computational cost required to calculate the neighborhood for each particle, and the separation from the original social metaphor. In Ref. 5, a dynamic neighborhood topology is used for multimodal function optimization. Although many topological structures have so far been developed under the umbrella of PSO but a very rare has been seen in MOPSO.

Veenhuis[30] proposes an extended Meta-PSO approach to optimize the PSO parameters as well as the neighborhood topology for a given problem by PSO itself. The Meta-PSO concept proposed here intents to optimize the PSO configuration. For this, a particle encodes all important parameters of a PSO as well as its kind of neighborhood topology with its corresponding parameter. During evaluation the encoded parameters are normalized and mapped to intervals for the corresponding PSO parameters. The discrete neighborhood topology (global PSO, local PSO, etc.) is encoded by a point in a real-valued vector space. This way it becomes possible to provide the arithmetical operations needed to establish the dynamics of a PSO.

Liu and Xiao[31] have applied LPSO which is a local version of the PSO algorithm to design a long period grating (LPG) filters with 3dB flatten loss spectrum within the bandwidth about 100nm. To make the algorithm convergence to a global optimal or a better sub-optimal, a special topol-

ogy was used by them. Based on their studied results, they concluded
that the LPSO algorithm is an effective inveiglement algorithm for optimally designing complicated LPG and other fiber grating filters. Wang
and Xiang[32] have proposed a dynamically changing ring topology, in which
particles are connected unidirectionally with respect to their personal best
fitness. Meanwhile, two strategies, namely the "Learn From Far and Better Ones" strategy and the "Centroid of Mass" strategy are used to enable certain particle to communicate with its neighbors. Experimental results on six benchmarks functions validate the effectiveness of the proposed
algorithm.

10.2.1. *Multi-Objective Particle Swarm Optimization (MOPSO)*

Even though PSO is good and fast search algorithm, it has the limitations when solving real world problems. The two PSO equations (10.3) and
(10.4), which are in the mathematical format, restrict additional heuristics related to the real-world problem to be incorporated in the algorithm.
Thus, PSO in its basic form will not perform well in its search in complex multi-objective solution spaces, which are the case for many complex
real world scenarios. However, since PSO and EA algorithms have structural similarities (such as presence of population searching for optima and
information sharing between population members) and since EAs have already been successfully applied to multi-objective optimization problems, a
transfer of PSO to the multi-objective domain can be a natural progression
with some intelligent modifications in the basic PSO algorithm. Changing
a PSO to a MOPSO requires a redefinition of what a guide is, in order to
obtain a front of optimal solutions (Pareto front). In MOPSO, the Pareto-
optimal solutions are used to determine the guide for each particle. A
number of different studies have been published on Pareto approach based
multi-objective PSO (MOPSO).[2,27,33-38]

Each of these studies implements MOPSO in a different fashion. However, the PSO heuristic puts a number of constraints on MOPSO. In PSO
itself the swarm population is fixed in size, and its members can not be
replaced, only adjusted by their pbest and gbest, which are by themselves
easy to define. However, in order to facilitate an multi-objective approach
to PSO, a set of non-dominated solutions (the best individuals found so far
using the search process) must replace the single global best individual in
the standard single-objective PSO case. Besides, there may be no single

previous best individual for each member of the swarm. Choosing which gbest and pbest to direct a swarm member's flight is therefore important in MOPSO. Main focus of various MOPSO algorithms is how to select gbest and pbest with a separate divergence on whether an elite archive is maintained.

In order to apply the PSO strategy for solving multi-objective optimization problems, the original scheme has to be modified. The algorithm needs to search a set of different solutions (the so-called Pareto front) instead of a single solution (as in single objective optimization). We need to apply Multi-Objective Particle Swarm Optimization (MOPSO) to search towards the true Pareto front (non-dominated solutions). Unlike the single objective particle swarm optimization, the algorithm must have a solution pool to store non-dominated solutions found by searching upto stopping criterion (say, upto iteration I_{max}). Any of the solutions in the pool can be used as the global best (gbest) particle to guide other particles in the swarm during the iterated process. The plot of the objective functions whose non-dominated solutions are in the solution pool would make up for the Pareto front. The pseudo code for a general MOPSO is illustrated in Algorithm 2.

Algorithm 2. Multi-Objective Particle Swarm Optimization Algorithm

(1) BEGIN
(2) Parameter Settings and initialize Swarm
(3) Evaluate Fitness and initialize leaders in a leader pool or external archive
(4) Archive the top best leader from the external archive through evaluation of some sort of quality measure for all leaders.
(5) I = 0   /* I = Iteration count */
(6) WHILE (the stopping criterion is not met, say, I ¡ I_{max})
(7) Do
(8) FOR each particle
(9) Select leader in the external archive
(10) Update velocity
(11) Update position
(12) Mutate periodically /*optional */
(13) Evaluate Fitness
(14) Update pbest

(15) END FOR
(16) Crowding of the leaders
(17) Update the top best into external archive
(18) I=I+1.
(19) END WHILE
(20) Report results in the external archive
(21) END

In the above general MOPSO algorithm, first the swarm is initialized. Then, a set of leaders is also initialized with the non-dominated particles from the swarm. This set of leaders are stored in an external archive. Later on, some sort of quality measure is calculated for all the leaders in order to select usually one leader for each particle of the swarm. At each generation, for each particle, a leader is selected and the flight is performed. Most of the existing MOPSOs apply some sort of mutation operator after performing the flight. Then, the particle is evaluated and its corresponding pbest is updated. A new particle replaces its pbest particle usually when this particle is dominated or if both are incomparable (i.e., they are both non-dominated with respect to each other). After all the particles have been updated, the set of leaders is updated, too. Finally, the quality measure of the set of leaders is re-calculated. This process is repeated for a certain fixed number of iterations.

In the case of multi-objective optimization problems, each particle might have a set of different leaders from which just one can be selected in order to update its position. Such set of leaders is usually stored in a different place from the swarm, that is called external archive. This is a repository in which the non-dominated solutions formed so far are stored. The solutions contained in the external archive are used as leaders when the positions of the particles of the swarm have to be updated. Furthermore, the contents of the external archive is also usually reported as the final output of the algorithm.

10.3. Comprehensive Review of PSO for Single Objective Optimization Problems

The framework for Particle Swarm approaches was laid by Kennedy and Eberhart[1] for optimization of nonlinear functions. It is discovered through simulation of the social behavior of the birds. The initial PSO had two cognitive aspects, individual learning and learning from its social group.

Whenever a particle finds itself in the problem space, it uses its own experience and that of its flock to reach to the optimal solution. Equations (10.5) and (10.6) were the initial PSO equations as per Ref. 21.

$$v_{i+1} = v_i + \varphi_1.\beta_1.(p_i - x_i) + \varphi_2.\beta_2.(p_g - x_i) \qquad (10.5)$$

$$x_{i+1} = x_i + v_{i+1} \qquad (10.6)$$

Where φ_1 and φ_2 determine the balance between the influence of the individual's knowledge and the group knowledge respectively; β_1 and β_2 are uniformly distributed random numbers generated by some upper limit, β_{max}. p_i and p_g are previous best position of an individual and groups, best position respectively, x_i is the current position of the particle in the dimension considered.

Further study by Eberhart et al.,[18] revealed that this suffers from instability caused by particles accelerating out of the problem space. So he proposed for clamping the velocity of each particle in the range of $[-v_{max}, v_{max}]$, where v_{max} being somewhere between 0.1 and 1.0 times the maximum position of the particle. However, to better control the search scope, Shi and Eberhart[39] revealed in their study that without a velocity memory the swarm would contract to the global best solution with the initial boundary. Hence to provide a balance between exploration and exploitation, PSO was modified, incorporating a inertia weight w, and thus:

$$v_{i+1} = w.v_i + \varphi_1.\beta_1.(p_i - x_i) + \varphi_2.\beta_2.(p_g - x_i) \qquad (10.7)$$

$$x_{i+1} = x_i + v_{i+1} \qquad (10.8)$$

Ozacan and Mohan[40] have extended Kennedy and Eberhart[1,21] works and proposed a generalization to obtain a closed form equations for trajectories of particles in a multi-dimensional search space. They have shown that a particle does not "fly" in the search space, but rather "surfs" it on sine waves. A particle seeking an optimal location attempts to "catch" another wave randomly, manipulating its frequency and amplitude. Suganthan[41] has introduced a variable neighborhood operator to enhance the performance of particle swarm optimization. During the initial stages of optimization, the PSO algorithms neighborhood will be an individual particle itself. As the number of generations increases, the neighborhood will be gradually extended to include all particles. In other words, the variable gbest in the PSO algorithm is replaced by lbest where a local neighborhood size is gradually increased. In addition, he has suggested the magnitudes of the random walk and inertia weight in the PSO are gradually adjusted in order to perform a fine grain search during the final stages of optimization.

So far the algorithms developed under the umbrella of PSO is categorized into various groups. Let us discuss the members of each group.

10.3.1. *Discrete/Binary PSO*

Kennedy and Eberhart[21] immediately after proposing the framework for particle swarm optimization gave an alternative algorithm to operate on discrete binary variables. In this binary version, the trajectories are changes in the probability that a coordinate will take on a zero or one value. The binary PSO can handle the discrete optimization problem.

Yang et al.[42] has proposed a new discrete particle swarm optimization algorithm based on quantum individual. They defined a particle $Q(t)$ based on the quantum bit (qubit). A random observation is used to replace the sigmoid function, and the best chromosome's guidance is also used to draw close to the optimum step by step. They have shown its application in CDMA.

Liao et al.[43] have applied discrete particle swarm optimization to solve the flow shop scheduling problem. In this proposed algorithm, the particle and the velocity are redefined, and an efficient approach is developed to move a particle to the new sequence. They have also incorporated a local search scheme into the proposed algorithm, called PSO-LS. Although PSO-LS has performed better for some problems, the results are obtained by taking more computation times.

Correa et al.[44] have proposed a Discrete Particle Swarm Optimization (DPSO) algorithm designed for attribute selection. The proposed algorithm deals with discrete variables, and its population of candidate solutions contains particles of different sizes. It differs from other traditional PSO algorithms because its particles do not represent points inside an n-dimensional Euclidean space or lattice as in the standard PSO algorithms. Instead, they represent a combination of selected attributes. DPSO has shown efficiency in finding the smallest subset of attributes and maximizing the predictive accuracy.

Wang[45] has suggested a novel discrete particle swarm optimization algorithm based on estimation of distribution and has proposed this algorithm for combinatorial optimization problems. The proposed algorithm combines the global statistical information collected from local best solution information of all particles and the global best solution information found so far in the whole swarm. The algorithm uses univariate marginal distribution (UMD) model to estimate the distribution of good regions over the search space based on the selected local best solutions. Mutation is also incorporated into the algorithm to keep the diversity of the algorithm. The results, on simulating the knapsack problem, have shown that the proposed algorithm has superior performance to other discrete particle swarm algorithms as well as having less parameters. Sarfaraz and Al-Awami[46] have applied particle swarm optimization with selected and weighted moments invariants for identification of Arabic character. The proposed scheme works in such a way that the features are selected as well as weighted using a swarm-based optimization technique. PSO has been used to assign an optimal set of weights for these features so as to maximize the recognition rate using the minimum number features. They have achieved an recognition of 82% successfully. Jarboui *et al.*[47] have proposed a combinatorial PSO (CPSO) algorithm that is used in order to solve a multi-mode resource-constrained project scheduling problem (MRCPSP). MRCPSP consists of two different sub-problems: the assignment of modes to tasks and then the scheduling of these tasks in order to minimize the makespan of the project. The combinatorial PSO proposed here deals with the first problem to generate an assignment of modes to activities which is called particle. A local search will optimize the sequences when a new assignment is made.

Zhen *et al.*[48] have proposed the probability based binary particle swarm optimization (PBPSO) algorithm and has applied for wet flues gas desulphurization (WFGD) technique which directly determines the desulphurization performance and the quality of product. PBPSO algorithm uses the position updating formula and the velocity updating formula of the PSO. But the position x is not the final solution any more in PBPSO; it is defined as the pseudo-probability which is used to determine the value of solution. The simulation results demonstrate that PBPSO has better optimization ability than the traditional discrete binary PSO. By constructing a proper fitness function, PBPSO can search and find the optimal proportional integral derivative (PID) parameters and achieve the expected control performance.

Tasgetiren *et al.*[49] applied discrete particle swarm optimization to solve the generalized travelling sales person problem. The discrete particle swarm optimization algorithm exploits the basic features of its continuous counterpart. It is also hybridized with a local search, variable neighborhood descend algorithm, to further improve the solution quality.

Lee *et al.*[50] applied discrete particle swarm optimization to review course composition system to quickly pick up suitable materials and can be customized in accordance with learners' intention. As a result, such a composition system satisfies the majority of learners with the customized review courses based on their needs. DPSO is adopted to select better and more subject-related materials. By DPSO, these materials which are conformed to the learner's intention will be appropriately selected.

10.3.2. *Adaptive PSO*

Shi and Eberhart[51] have performed an empirical study of particle swarm optimization and explored the advantages and disadvantages of PSO. Four different benchmark functions with asymmetric initial range settings are selected as test functions. Under all the testing cases, the PSO always converges very quickly towards the optimal positions but may slow its convergence speed when it is near minimum. The PSO may fail to find the required optima in cases when the problem to be solved is too complicated and complex. They have suggested a self-adapting strategy for adjusting the inertia weight to overcome the problem.

Charlsie and Dozier[52] have worked on Shi and Eberhart[53] proposal for adaptive approach and proposed a method for adapting the particle swarm optimizer for dynamic environments. The process consists of causing each particle to reset its record of its best position as the environment changes, to avoid making direction and velocity decisions on the basis of outdated information. In this method, periodical resetting the particle memories to the current positions allows the swarm to track a changing goal with minimum overhead.

Hu and Eberhart[54] have devised an adaptive particle swarm optimization to track the changes in dynamic environments. They have focused on location varying in the problem space where optimum values occur. They have varied the location simultaneously and equally in each dimension. The modified PSO has introduced two new features, namely, environment and response. They have also introduced two environment detection methods: changed gbest value method and fixed gbest method. Both these methods can successfully detect the various dynamic changes. Xie *et al.*[55] have also presented an adaptive particle swarm optimization (PSO) on individual level. By analyzing the social model of PSO, a replacing criterion based on the diversity of fitness between current particle and the best historical experience is introduced to maintain the social attribution of swarm adaptively by taking off inactive particles. The proposed algorithm adapted the swarm at individual level, which is realized by replacing the inactive particle with a fresh one in order to maintain the social attribution of swarm, according to the analyzing for the model of PSO.

Zhang and Liu[56] have presented an adaptive particle swarm optimization(APSO) which adjusts the parameters automatically in the optimization process. In the APSO, all three parameters, swarm size N, coefficient φ_i and neighborhood size h_i of each particle, are adapted based on the fitness values of particles during optimization process. They have applied this approach to reactive power and voltage control in power systems. The result has shown that the APSO can adjust parameters automatically in searching process and yield better solution comparing with PSO.

Zhen-su *et al.*[57] presents a new adaptive mutation particle swarm optimizer, which is based on the variance of the population's fitness. During the running time, the mutation probability for the current best particle is determined by two factors: the variance of the population's fitness and the current optimal solution. The ability of particle swarm optimization (PSO) algorithm to break away from the local optimum is greatly improved by the mutation. The algorithm has not only the powerful ability to search the global optimal solutions, but also the ability to effectively avoid the premature convergence that exists in PSO and GA. Although its computation is slightly increased compared with the PSO algorithm, it is much fewer than that of GA. Zhen *et al.*[57] have proposed modified PSO (MPSO) which is based on an adaptive strategy, the particle should stop the inertia movement to enhance the learning from its experiences and its neighbors when it is found to be in wrong searching direction, and stop the learning process to fly straight when it is found to be the nearest to the destination in the swarm. They also proposed four different models of MPSO. They have shown the effectiveness of the modified approach by comparing the results with the standard PSO on some unconstrained and constrained global optimization functions.

Chunxia and Youhong[58] have argued that swarm optimization algorithm with constriction factor (CFPSO) has some demerits, such as relapsing into local extremum, slow convergence velocity and low convergence precision in the late evolutionary. Thus they have proposed an adaptive simple particle swarm optimization with constriction factor (AsCFPSO) which is when combined with chaotic optimization, a new CFPSO is developed, i.e., a chaotic optimization-based adaptive simple particle swarm optimization equation with constriction factor (CAsCFPSO). Distribution vector of particles is defined as constriction factor in optimization process. Furthermore, piecewise linear chaotic map is employed to perform chaotic optimization due to its ergodicity and stochasticity. Consequently, the particles are accelerated to overstep the local extremum in AsCFPSO algorithm.

Zhan et al.[59] have proposed an adaptive particle swarm optimization (APSO) that features better search efficiency than classical PSO. It can perform a global search over the entire search space with faster convergence speed. The APSO consists of two main steps: first, by evaluating the population distribution and particle fitness, a real-time evolutionary state estimation procedure is performed to identify one of the four defined evolutionary states including exploration, exploitation, convergence and jumping out of each generation. It enables automatic control of inertia weights, acceleration coefficients and other algorithmic parameters at run time to improve the search efficiency and convergence speed. Then second, an elitist learning strategy is performed when the evolutionary state is classified as convergence state. The strategy acts on the globally best particle to jump out of the likely local optima. The APSO substantially enhances the performance of the PSO paradigm in terms of convergence speed, global optimality, solution accuracy and algorithm reliability.

Hongwu[60] has proposed an adaptive chaotic PSO (ACPSO) which applies a short term chaotic search to the best particle in the iteration and an adaptive mechanism is used to control the scale of chaotic which can avoid trapping in local optima and improve the searching performance of chaotic PSO. The adaptive mechanism used is based on following principles: if the particle enters into a range of local optima, normally, the chaotic turbulence added to the particle will generate several better positions. If the number of better positions in the current PSO's iteration is larger than a threshold, adaptive weight (β) is scaled-up, and the swarm enters into the next PSO's iteration; If the better position found is less than the threshold, β is set as the constant 1; If no better position is found in the current PSO's iteration, as the best particle is applied to chaotic turbulence, β is scaled-down and the swarm enters into the next PSO's iteration. The benchmark test of ACPSO algorithm shows that the algorithm has super ability in balancing the exploration and exploitation.

10.3.3. Multi-Swarm PSO

Bergh and Engelbrecht,[61] have proposed a variation on the traditional PSO algorithm, called the cooperative particle swarm optimizer, or CPSO, employing cooperative behavior to significantly improve the performance of the original algorithm. This is achieved by using multiple swarms to optimize different components of the solution vector cooperatively. They have applied potter's technique to the PSO. Since the CPSO algorithms decompose the larger search space into several smaller spaces, the rate at which each of these subswarms converge onto solutions contained in their subspaces is significantly faster than the rate of convergence of the standard PSO on the original, n-dimensional search space.

Ahmadi *et al.*,[62] have proposed a clustering technique by use of multiple swarms. The proposed method considers multiple cooperating swarms to find centers of clusters. By assigning a portion of the solution space to each swarm, the exploration ability to find the solution is enhanced. Moreover, the cooperation among swarms increases the between-class distance. The technique is capable of considering several objective functions simultaneously. It assigns a portion of the solution space to each swarm. This strategy boosts its exploration ability as each swarm deals with a part of solution space. Each swarm explores its own region while cooperating with other swarms. It knows the global best of other swarms and attempts to find a point whose accumulative distance from the other clusters' centers is maximum. Each swarm also tends to decrease within-class distance. The proposed method outperforms k-means clustering as well as conventional PSO-based clustering techniques.

Inthachot and Supratid[63] have proposed a a multi-subpopulation particle swarm optimization which combines the coarse-grained model of evolutionary algorithms with particle swarm optimization. The coarse-grained model of evolutionary algorithm supports the idea of exploration while the PSO serves the notion of elitism. This study utilizes two performance measurements: the correctness and the number of iterations required for finding the optimal solution. According to both types of performance measurement, the multi-subpopulation particle swarm optimization shows distinctly superior performance over the particle swarm optimization does.

Jian *et al.*[64] have provided a dual PSO (Dual-PSO) algorithm which embeds the characteristics of genetic particle swarm optimization (GPSO) and original particle swarm optimization (OPSO) to solve constrained optimization problem. To deal with the constraints, the stochastic ranking algorithm is employed. Based on which Dual-PSO is introduced, where at each generation GPSO and OPSO generate a new position for the particle synchronously and respectively, with the original position of the particle, and the better one is accepted as the new position. Dual-PSO provides better or comparable solutions for most problems with proper stochastic ranking parameters, which have shown the robust and consistent effectiveness of Dual-PSO for constrained optimization problems.

Li and Xiao[31] propose a novel Multi-Swarm and Multi-Best particle swarm optimization (MSBPSO) algorithm. The algorithm divides initialized particles into several populations randomly. After calculating certain generations respectively, every population is combined into one population and continues to calculate until the stop condition is satisfied. At the same time, the algorithm updates particles' velocities and positions by following multi-gbest and multi-pbest instead of single gbest and single pbest. The algorithm is not only a generalization of the basic particle swarm optimization, but can improve the searching efficiency, help the algorithm fly out of local optimum and increase the possibility of finding the real global best solution greatly. The performance test result with Griewank function shows that the MSBPSO convergences faster and the searching efficiency is higher than basic PSO.

10.3.4. *Hybrid PSO*

Ratnaweera *et al.*[65] have introduced a novel parameter automation strategy for the particle swarm algorithm and two further extensions to improve its performance after a predefined number of generations. Initially, to efficiently control the local search and convergence to the global optimum solution, time-varying acceleration coefficients (TVAC) are introduced in addition to the time-varying inertia weight factor in particle swarm optimization (PSO). From the basis of TVAC, two new strategies are discussed to improve the performance of the PSO. First, the concept of "mutation" is introduced to the particle swarm optimization along with TVAC (MPSO-TVAC), by adding a small perturbation to a randomly selected modulus of the velocity vector of a random particle by predefined probability. Second, they introduce a novel particle swarm concept "self-organizing hierarchical particle swarm optimizer with TVAC (HPSO-TVAC)." Under this method, only the "social" part and the "cognitive" part of the particle swarm strategy are considered to estimate the new velocity of each particle and particles are reinitialized whenever they are stagnated in the search space. In addition, to overcome the difficulties of selecting an appropriate mutation step size for different problems, a time-varying mutation step size is introduced by them.

Ling *et al.*[66] have developed a hybrid particle swarm optimization algorithm by hybridizing simulated annealing with PSO to solve capacitated vehicle routing problem which is an NP-hard problem and is difficult to achieve optimal results with traditional optimization techniques. In the hybrid algorithm, discrete particle swarm optimization (DPSO) combines global search and local search to search for the optimal results and simulated annealing (SA) uses certain probability to avoid being trapped in a local optimum. The computational study shows that the proposed algorithm is a feasible and effective approach for capacitated vehicle routing problem, especially for large scale problems. Blackwell[67] has proposed a diversity measure and examined its time development for charged and neutral swarms. These results facilitate predictions for optima tracking given knowledge of the amount of dynamism. He has defined particle diversity as maximum component separation of the particle positions in the swarm. A number of experiments test these predictions and demonstrate the efficacy of charged particle swarms in a simple dynamic environment. Voss[68] introduces the Principal Component Particle Swarm Optimization (PCPSO) procedure which flies the particles in two separate spaces at the same time; the traditional n-dimensional x space and a rotated m-dimensional z space where $m < n$. The PCPSO algorithm is grouping certain ideas together based on past experience. This grouping of ideas can be seen as a form of linkage-learning or dynamic probabilistic building block discovery. Grundy and Stacey[69] have proposed a particle swarm optimization with mutation and hill climbing. In the proposed algorithm, at each generation, a small number of particles are mutated and are allowed to hill climb. The mutation has the effect of randomly bouncing particles towards other parts of the search space, while the effect of hill-climbing is to greatly increase the effective size of the "target" region of interest around the global optimum. They have identified an undesirable feature of the particle swarm algorithm, in that, in some circumstances, particles fly over regions of significance and the swarm converges to a local but not global optimum. So, to address this problem, they have allowed for certain particles to be mutated and then behave as hill climbers for a fixed number of steps between swarm generations.

Pan *et al.*[70] have introduced two improvements on particle swarm optimization namely swarm-core evolutionary particle swarm optimization (SCEPSO) and PSO with Simulated Annealing (PSOwSA). In PSOwSA, the particle x_i would not move to the next position x'_i directly if the next position is worse than the current position. It just moves with probability P_T. The probability P_T can be controlled by the temperature Meissner *et al.*,[71] present a method for parameter meta-optimization based on PSO and its application to neural network training. The concept of the Optimized Particle Swarm Optimization (OPSO) is to optimize the free parameters of the PSO by having swarms within a swarm. They have applied the OPSO method to neural network training with the aim to build a quantitative model for predicting blood-brain barrier permeation of small organic molecules. On the average, training time decreases by a factor of four and two in comparison to the other PSO methods respectively. By applying the OPSO method, they have obtained a prediction model showing good correlation with training, test and validation data.

Mohemmed and Sahoo[72] have proposed a novel hybrid algorithm based on particle swarm optimization (PSO) and noising metaheuristics for solving the single-source shortest-path problem (SPP). This hybrid search process combines PSO for iteratively finding a population of better solutions and noising method for diversifying the search scheme to solve this problem. A new encoding/decoding scheme based on heuristics has been devised for representing the SPP parameters as a particle in PSO. Noising-method-based metaheuristics (noisy local search) have been incorporated in order to enhance the overall search efficiency. In particular, an iteration of the proposed hybrid algorithm consists of a standard PSO iteration and few trials of noising scheme applied to each better/improved particle for local search, where the neighborhood of each such particle is noisily explored with an elementary transformation of the particle so as to escape possible local minima and to diversify the search. They have devised a new cost-priority-based particle encoding/decoding scheme so as to incorporate the network-specific heuristic information in the path construction process.

Tian and Li[73] have presented a novel fuzzy particle swarm optimization (NFPSO), in which inertia weight as well as the learning coefficient can be adaptively adjusted according to the control information translated from the fuzzy logic controller (FLC) during the search process. The algorithm introduces two input-two output FLC into the canonical PSO. The two inputs for the FLC are the increment of global optimum and deviation of the particle fitness. The two input variables are defined with Gaussian membership function. The NFPSO algorithm proposed here helps in dynamically adapting the performance of the system. Lung and Dumitrescu[74] have presented an evolutionary swarm cooperative algorithm (ESCA) based on the collaboration between a particle swarm optimization algorithm and an evolutionary algorithm. ESCA is designed to deal with moving optima of optimization problems in dynamic environments. ESCA uses three populations of individuals: two EA populations and one Particle Swarm population. The EA populations evolve by the rules of an evolutionary multimodal optimization algorithm being used to maintain the diversity of the search. The particle swarm confers precision to the search process. The performance of ESCA has been evaluated using some numerical experiments on the MPB (Moving Peaks Benchmark). They have shown that ESCA is able to cope better than any other method considered with severe changes in the fitness landscape.

Yeh et al.[75] have developed a hybrid data mining approach to separate from a population of patients who have and who do not have breast cancer. The proposed data mining approach consists of two phases. In first phase, the statistical method will be used to pre-process the data which can eliminate the insignificant features. It can reduce the computational complexity and speed up the data mining process. In second phase, a new data mining methodology is proposed which is based on the fundamental concept of the standard particle swarm optimization (PSO) namely discrete PSO (DPSO). This phase aims at creating a novel PSO in which each particle is coded in positive integer numbers and has a feasible system structure. Authors have argued that when compared with the previous research, the proposed hybrid approach shows the improvement in both accuracy and robustness. Natarajan et al.[76] applied particle swarm optimization in artificial neural network to predict the life of a tool. The network was trained with particle swarm optimization algorithm instead of backpropogation algorithm.

10.3.5. *Other PSO*

Xie et al.[77] have developed a dissipative particle swarm optimization according to the self-organization of dissipative structure. The negative entropy is introduced to construct an opening dissipative system that is far-from-equilibrium so as to drive the irreversible evolution process with better fitness. With the internal nonlinear interactions among particles, the self-organization of dissipative structure comes into being with the dissipative processes for the introduced negative entropy, which drives the irreversible evolution process toward higher fitness by the selection of keeping best experience.

Parsopoulos and Vrahatis[78] have introduced the method of initializing PSO with the nonlinear simplex method which helps in exploring the search space more effectively and producing better solution.

Kennedy[79] has revisited the constricted PSO and examined whether the added components are necessary, and whether any further components can be removed. Various experiments are performed with a view to paring the process for further efficiency gains. To achieve this, a Gaussian PSO is developed. In this implementation the entire velocity vector is replaced by a random number generated around the mean $(p_{id} + p_{gd})/2$ with a Gaussian distribution of $|p_{id} - p_{gd}|$ in each dimension (d). This effectively means that the particles no longer 'fly' but are 'teleported'. Kennedy justifies this departure on the grounds that it is the social aspect of the swarm that is more important to its effectiveness. This is empirically backed up with the Gaussian influenced swarms performing competitively. DePuy et al.[80] have analyzed the effects of maximum velocity, societal factor and individual factor on the swarm's ability to find the optimum point in the solution space.

Yang and Simon[81] have proposed a new particle swarm optimization method (NPSO). In NPSO, each particle adjusts its position according to its own previous worst solution and its group's previous worst based on similar formulae of the regular PSO. The NPSO exhibits the social behavior where the particle learns not only from its previous best but also from its and other individuals' mistakes.

Koh et al.[82] introduce a parallel asynchronous PSO (PAPSO) algorithm to enhance computational efficiency. They have evaluated a parallel asynchronous particle swarm optimization (PAPSO) algorithm that dynamically adjusts the workload assigned to each processor, thereby making efficient use of all available processors in a heterogeneous cluster. The performance of the PAPSO algorithm is compared to that of a parallel synchronous PSO (PSPSO) algorithm in homogeneous and heterogeneous computing environments for small- to medium-scale analytical test problems and a medium-scale biomechanical test problem. Authors concluded that for all problems, the robustness and convergence rate of PAPSO are comparable to those of PSPSO. However, they observe that the parallel performance of PAPSO is significantly better than that of PSPSO for heterogeneous computing environments or heterogeneous computational tasks. Khosla et al.,[83] propose a systematic based on Taguchi method reasoning scheme for rapidly identifying the strategy parameters for the PSO algorithm. The proposed method has the ability to quickly identify the strategy parameters for the PSO algorithm. Chandramouli and Izquierdo[84] have applied chaotic particle swarm optimization (CPSO) for image classification. Chaotic parameter wind speed is introduced along with the wind direction in modeling PSO. Their simulation result shows a considerable increase in the image classification accuracy with the proposed CPSO algorithm. The increased accuracy is due to the better exploration of the problem space in search of the optimal solution rather than converging at local minima. Kadirkamanathan *et al.*[85] have made the stability analysis of the particle dynamics in PSO using Lyapunov stability analysis and the concept of passive systems. The analysis is made feasible by representing the particle dynamics as a nonlinear feedback controlled system. The passivity theorem and Lyapunov stability methods are applied to the particle dynamics to determine sufficient conditions for asymptotic stability and, hence, convergence to the equilibrium point. Since the results are based on the Lyapunov function approach, they are conservative, and, hence, violation of these conditions does not imply instability. Nevertheless, the results can be used to infer qualitative design guidelines.

Wang *et al.*[45] introduces a PSO based algorithm for classification of rule mining. The objective was to investigate the capability of the PSO algorithm to discover classification rule with higher predictive accuracy and a much smaller rule list. Tan and Xiao[86] have proposed a novel particle swarm optimization algorithm based on immunity clonal strategies, called as clonal particle swarm optimization (CPSO). Clonal expansion in natural immune system (NIS) helps to guide or direct the standard PSO (SPSO) escaping from local optima whilst searching for the global optima efficiently. By cloning the best individual of ten succeeding generations, CPSO has better optimization solving capability and faster convergence performance than the conventional standard particle swarm optimization (SPSO) based on a number of simulations. Xiang *et al.*,[87] introduce the concept of time-delay into PSO to control the process of information diffusion and keep the particle diversity. Time-delay is applied to the velocity updating of PSO, i.e., once a new gbest is discovered, it spreads over all other particles with certain time-delay. On this occasion, the gbest used to update the particle velocity may not be the latest one, and different particles may use different gbest in the same iteration. Particle diversity thus increases. They have also proposed four time-delay schemes for particle swarm optimization. On simulation, the results reveal that time-delay PSO (TPSO) schemes with appropriate time-delay have their great superiority in optimizing multimodal functions.

Arumugam and Rao[88] present an efficient method for solving the optimal control of single-stage hybrid manufacturing systems which are composed with two different categories: continuous dynamics and discrete dynamics. Three different inertia weights, a constant inertia weight (CIW), time-varying inertia weight (TVIW), and global-local best inertia weight

(GLbestIW), are considered with the particle swarm optimization (PSO) algorithm to analyze the impact of inertia weight on the performance of PSO algorithm. The PSO algorithm is simulated individually with the three inertia weights separately to compute the optimal control of the single-stage hybrid manufacturing system, and they have observed that the PSO with the proposed inertia weight yields better result in terms of both optimal solution and faster convergence. They have concluded that PSO performs significantly well on high-dimensional optimal problems with great speed, reliability, and accuracy than Real Coded Genetic Algorithms (RCGAs). Also, they have clearly showed the impact of the inertia weight in improving the performance of the PSO towards obtaining the optimal solution for the optimal control of the single-stage hybrid system. Liu et al.[89] have modified the basic PSO algorithm by proposing an improvement in initial solution and search precision. They have suggested that the initial solution should be homogenous in nature that will help in maintaining the diversity of the population. Also they have suggested that it is not necessary to update the speed and location of the particles after getting the current individual optimal solutions at each iteration. First, they intercept a certain incremental or decreasing dimension from every dimension of each individual optimal solution is made, then get the fitness values of the right solutions, finally gain the minimum solution by comparing the right fitness values with the fitness value of solutions obtained from iteration and record them. With this method, it can expand the scope of the searching in the early part of the optimization, and do meticulous searching around the optimal solution in the latter part of optimization. Poli,[90] presents a method that allows one to exactly determine all the characteristics of a PSO's sampling distribution and explain how it changes over time during stagnation (i.e., while particles are in search for a better personal best) for a large class of PSO's. Panigrahi et al.[91] present a novel heuristic optimization approach to constrained economic load dispatch (ELD) problems using the adaptive variable population- PSO technique. The proposed methodology APSO easily takes care of different constraints like transmission losses, dynamic operation constraints (ramp rate limits) and prohibited operating zones and also accounts for non-smoothness of cost functions arising due to the use of multiple fuels. On simulation, the findings affirmed the robustness, fast convergence and proficiency of the proposed methodology over other existing techniques such as GA, etc. Arumugam et al.[88] have used extrapolation technique with PSO for solving optimization problem. The current particle position is updated by extrapolating the global best position and the current position in the search space. The position equation is formulated by extrapolating the gbest, pbest and the current position of the particle.

Zhang *et al.*[92] have proposed a model for particle swarm optimization with diverse curiosity (PSO/DC) to obtain superior search performance. The mechanism of diverse curiosity in PSO can prevent premature convergence and ensure exploration. Diverse curiosity signifies general tendency to seek novelty to take risks and to search adventure. Here an internal indicator is defined to indicate the state of the particle whether it is searching its surrounding for cognition or it has lost its interest. The indicator detects premature convergence and stagnation in particle. The PSO/DC algorithm maintains the exploration and exploitation trade-off well to explore the optimal solution corresponding to the given problem over search space. Hsieh *et al.*[93] have proposed an efficient population utilization strategy for PSO (EPUS-PSO), adopting a population manager to significantly improve the efficiency of PSO. This is achieved by using variable particles in swarms to enhance the searching ability and drive particles more efficiently. Moreover, sharing principles are constructed to stop particles from falling into the local minimum and make the global optimal solution easier found by particles. The population size proposed in this algorithm is variable and controlled by the population manager. Population manager adds more particles to the population to keep the gbest updated and to find better solutions in the current generation. The proposed population manager and sharing principles can significantly improve particles' searching ability to move easily to find the global optimal solution. Zhu *et al.*[94] have proposed a modification of standard PSO named Euclidean PSO (EPSO) by introducing the concept that if the global best fitness has not been updated for k times, velocities of particles will get an inference factor to make most of the particles fly out of the local optimum but the best one is kept continuing to local search. This modification of PSO prevents the particle from getting stuck to the local minima. On comparisons with GPSO (Gaussian PSO) and SPSO (Standard PSO) based on benchmark functions, the EPSO has shown better convergence efficiency and precision. Hong[95] has proposed a new algorithm that has both properties of the original PSO and immune mechanism and can improve the ability of seeking the global optimum and evolution speed. For adding one most important parameter of PSO (V_{max}), the theory of immune mechanism is used to perform parameter learning. Analogous to the biological immune system, the proposed algorithm has the capability of seeking feasible solution while maintaining diversity. The immune mechanism based PSO (IMPSO) algorithm utilizes the global exploration capability of PSO and the properties of clonal selection mechanism and idiotypic immune network theory to improve the global convergence performance. Winner et al.[96] have proposed a method for controlling particle swarm optimization with non-explicit control parameters: parameters that describe self-organizing systems at an abstract level. Effectively, this process converts intuitive control parameter values into explicit configurations that particle swarm optimization can directly apply. Basically these mapping translates control parameters that are intitutive for a human user into configurations that PSO understands.

Arfia et al.[97] have proposed a particle swarm optimization algorithm to adjust the parameters of the nonlinear filters and to make this type of filters more powerful for elimination of the Gaussian noise and also the impulse noise. They have compared the PSNR, MSE & SNR to another adaptive nonlinear filter adopted by LMS and non-adaptive rational filter for both types of noise: Gaussian and impulse noise. They have observed that the PSO rational filter is more effective for impulse noise than other used filters.

de Pina et al.[98] have applied particle swarm optimization algorithm for design of off-shore oil production risers. They have applied the PSO for designing of steel catenary risers in a lazy wave configuration.

10.4. Comprehensive Review of Pareto-Based MOPSO Approaches

These approaches use leader selection techniques based on Pareto dominance. The basic idea of all the approaches considered here is to select as leaders to the particles that are non-dominated with respect to the swarm. Note, however, that several variations of the leader selection scheme are possible since most authors adopt additional information to select leaders (e.g., information provided by a density estimator) in order to avoid a random selection of a leader from the current set of nondominated solutions.

The PSO heuristic was first proposed by Kennedy and Eberhart[1] for the optimization of continuous non-linear function. The computational efficiency of PSO has been shown to be significantly better than GA by Rania Hassan et al.[26] They have come to this conclusion by carrying out the performance comparison between GA and PSO using a set of benchmark test problems as well as two space systems design optimization problems. As per Kennedy and Eberhart[19] basic PSO algorithm has been successfully used to solve a number of both continuous nonlinear and discrete binary single-objective optimization problems. According to them, PSO provides high speed of convergence by the use of flying potential solutions through hyperspace.

Moore *et al.*[2] have modified the basic single-objective PSO algorithm in the line of Pareto preference by modifying its position vector which contains the locations of all the non-dominated solutions instead of a single best solution as is the case with basic single-objective PSO without modification. They have tested their modified PSO by using two multi-objective functions which Lis and Eiben[99] have already used to test their multi-objective algorithm based on traditional Evolutionary Algorithm (EA). According to Moore *et al.*,[2] the results obtained by them are superior to those obtained by Lis and Eiben.[99] However, Coello and Lechuga[27] have pointed out that algorithm of Moore *et al.*[2] does neither use a secondary population nor is compared against other evolutionary multi-objective techniques using standard test functions and metrics. Coello and Lechuga[27] have extended PSO to deal with multi-objective optimization problems using the similar approach of Pareto dominance to determine the flight direction of a particle and their MOPSO algorithm maintains previously found non-dominated vectors in a global repository (secondary memory) that is later used by other particles to guide their own flight. Their approach is population based as well as geographically based to maintain diversity. They have validated their MOPSO algorithm using some standard test functions and metrics such as average running time of the algorithm using same number of fitness function evaluations and average distance to the Pareto optimal set as reported in the specialized literatures by Deb[100] and by Schaffer.[101] As per Coello and Lechuga,[27] their MOPSO algorithm performs reasonably well and it requires lower computational times as compared to two highly competitive evolutionary multi-objective (EMO) algorithms such as Pareto Archived Evolution Strategy (PAES) as proposed by Knowles and Corne[102] and Non-Dominated Sorting Genetic Algorithm-II (NSGA-II) as proposed by Deb *et al.*[103]

Fieldsend *et al.*[33] utilizes the dominated tree data structure to enable the selection of an appropriate Pareto archive member to act as the global "best" for any given particle and also maintains a local set of "best" solutions for each swarm member. They have demonstrated that this approach is significantly better than the method used by Coello and Lechuga[27] and also PAES derived from the unified model proposed by Laumanns *et al.*[104] They have demonstrated that by including a stochastic turbulence variable within MOPSO, its performance has been significantly increased. Even though Coello and Lechuga[27] maintains an archive of global best solutions, Fieldsend et al.[33] have pointed out that there is a better way to select from this archive than by simple density based selection. Thus, they have included a new data structure called dominated tree, as this data structure facilitates rapid selection of an appropriate archive member for their new MOPSO method.

Hu and Eberhart[34] attempt to optimize MOPSO having two objectives through the a priori knowledge of the test function properties. Instead of a single gbest, a local lbest is found for each swarm member selected from the "closest" two swarm members. The concept of closeness is calculated in terms of only one of the evaluated objective dimensions, with the selection of the local optima from the two based upon the other objective. Selecting which objective to fix first and which to optimize second is based on the knowledge of the test function design - the relatively simple objective function being fixed. Fieldsend et al.[33] have pointed out that although the model of Hu and Eberhart[34] has been used on a number of test functions from the literature, however, no comparison was made with any other models, or the true Pareto fronts for the problems.

Parsopoulos et al.[35] introduce two methods to optimize MOP having two objectives. One uses a weighted aggregate approach and another is loosely based on Schaffer's MOEA[101] i.e., Vector Evaluated PSO (VEPSO) method.

They compare these two methods on a number of two dimensional (i.e., having two objectives) problems. The weighted aggregate algorithms need to be run K times to produce K estimated Pareto optimal points (meaning each run had a single global best). Although Parsopoulos et al.[35] states that this approach has a low computational cost, however according to Fieldsend et al.,[33] the need for a separate run for each solution found does not necessarily support this. The second method - the Vector Evaluated Particle Swarm Optimizer (VEPSO) of Parsopoulos et al.[35] uses one swarm for each objective. According to them, the best particle of the second swarm is used to determine the velocities of the first swarm (act as its global best), and vice versa. Fieldsend et al.[33] point out that comparison of VEPSO algorithm applied on a number of two dimensional problems, made by Parsopaulus et al.,[35] is qualitative (based on visual inspection of the found fronts), and no comparison is made to competitive methods in the MOEA domain. In addition, this VEPSO model is also only designed for two dimensional problems.

Mostaghim and Teich[36] introduces a new method called sigma method for finding best local guides for each particle of the population from a set of Pareto optimal solutions. According to them such a technique has a great impact on the convergence and diversity of solutions, especially when optimizing problems with a high number of objectives. The sigma method which uses clustering techniques for fixing the archive size has a better computational time, diversity and covergence than the dominated tree method of Fieldsend *et al.*[33] and Strength Pareto Evolutionary Algorithm (SPEA) method of Zitzler *et al.*[105] In order to select a leader for each particle of the swarm, a sigma value is assigned to each particle of the swarm and of the external archive. Each particle of the swarm selects as its leader the particle of the external archive with the closest sigma value. The use of the sigma values makes the selection pressure of PSO even higher, which may cause premature convergence in some cases. Authors use a turbulence operator, which is applied on a decision variable space. This approach has been successfully applied to the molecular force field parameterization problem as shown by Mostaghim *et al.*[106]

Again same authors Mostaghim and Teich[107] demonstrate that in the sigma method by using ϵ-dominance technique instead of clustering technique, the results for computational time, diversity and convergence are still far better than those obtained by these authors in Ref. 36. The ϵ-dominance bounds the number of solutions in the archive and decreases the computational time. However, they observe that with higher number of objectives (more than two), the results are not good and also the results are not good, if the number of generations is less.

Coello *et al.*[37] proposes an improved version of the algorithm reported by Coello and Lechuga in Ref. 27 in which they have added a constraint handling mechanism and a mutation operator that considerably improves the exploratory capabilities of their original algorithm. Their MOPSO is validated using several standard test functions reported in the specialized literature. They have compared this improved MOPSO algorithm against three highly competitive evolutionary multi-objective (EMO) algorithms, NSGA - II of Deb *et al.*,[103] PAES of Knowles & Corne[102] and micro GA of Coello & Pulido.[108] Results indicate that the approach is highly competitive and that can be considered as a viable alternative to solve MOPs.

Fieldsend[109] compares a number of selection regimes for the choosing of global best (gbest) and personal best (pbest) for swarm members in MOPSO. He has shown two distinct gbest selection techniques, one that does not restrict the selection of archive members and the other with distance based gbest selection techniques. According to him, these two methods promote two types of search. He has also described the potential problem of particle clumping in MOPSO. He has compared various popular MOPSO techniques[27,33,34,78,110] for pbest and gbest selections. He has viewed the effect of turbulence in terms of the additional search it promotes.

Ray and Liew[38] propose an algorithm which uses Pareto dominance and combines concepts of evolutionary techniques with the particle swarm. According to them, the approach uses a nearest neighbor density estimator to promote diversity (by means of a roulette selection scheme of leaders based on this value), a multilevel sieve to handle constraints (for this the authors adopt the constraint and objective matrices) and to generate a set of leaders, a probabilistic crowding radius based strategy for leader selection and a simple generalized operator for information transfer. The authors apply this algorithm to two complex test problems and three well-studied engineering design unconstrained & constrained optimization problems. They observe that each case it generates an extended Pareto front consisting of well spread Pareto points with a significantly less number of function evaluations when compared to the NSGA-II of Deb *et al.*[103]

Pulido and Coello[111] use the concept of Pareto dominance to determine the flight direction of a particle. The authors adopt clustering techniques to divide the population of particles into several swarms. This aims to provide a better distribution of solutions in decision variable space. Each sub-swarm has its own set of leaders (non-dominated particles). In each sub-swarm, a PSO algorithm is executed (leaders are randomly chosen) and, at some point, the different sub-swarms exchange information: the leaders of each swarm are migrated to different swarm in order to variate the selection pressure. This approach does not use an external archive as in Coello *et al.*[27,37] since elitism in this case is an emergent process derived from the migration of leaders.

Srinivasan and Seow[112] propose this approach called Particle Swarm Inspired Evolutionary Algorithm (PSEA) which is a hybrid between PSO and an Evolutionary algorithm. The main aim is to use EA operator (mutation, for example) to emulate the workings of PSO mechanisms. They have illustrated the feasibility of PSEA as a multi-objective search algorithm by applying to solve a classic multi-objective problem called Fonseca two-objective minimization problem. A Dynamic Inheritance Probability Adjuster (DIPA) is incorporated in Self-Updating Mechanism (SUM) to dynamically adjust the inheritance probabilities in Probability Inheritance

Tree (PIT) which in turn does the updating operation of each individual in the population. Since the authors mention that the final swarm constitutes the final solution (Pareto front), it is concluded that a plus selection is performed in each iteration of the algorithm. Also authors use a niche count and a Pareto ranking approach in order to assign a fitness value to the particles of the swarm. However, authors have not described the selection technique in this chapter.

Also, based on the idea that the initial external archive from which the particles have to select a leader has influence on the diversity of solutions, the authors propose the use of successive improvements adopting a previous external archive of solutions. In this way, Mostaghim and Teich[113] propose a new method called covering MOPSO (cvMOPSO) which retakes this idea. This method works in two phases. In phase 1, a MOPSO algorithm is run with an external archive with restricted size and the goal is to obtain a good approximation of the Pareto front. In the phase 2, the non-dominated solutions obtained from the phase 1 are considered as the input external archive of the cvMOPSO. The particles in the swarm of the cvMOPSO are divided into sub-swarms around each non-dominated solution after the first generation. The task of the sub-swarms is to cover the gaps between the non-dominated solutions obtained from the phase 1. No restrictions on the archive size are imposed in the phase 2.

Bartz et al.[114] propose an approach which starts from the idea of introducing elitism (through the use of an external archive) into PSO. Different methods for selecting and deleting particles (leaders) from the archive are analyzed to generate a satisfactory approximation of the Pareto front. The deletion methods analyzed are based on the contribution of each particle to the diversity of the Pareto front. Selecting methods are either inversely related to the fitness value or based on the previous success of each particle. The authors provide some statistical analysis in order to access the impact of each of the parameters used by their approach.

Li[115] proposes an approach which incorporates the main mechanisms of the NSGA - II of Deb et al.[103] to the PSO algorithm. In this approach, once a particle has updated its position, instead of comparing the new position only against the pbest position of the particle, all the pbest positions of the swarm and all the new positions recently obtained are combined in just one set (given a total of 2N solutions, where N is the size of the swarm). Then, the approach selects the best solutions among them to conform the next swarm (by means of a non-dominated sorting). The author does not specify which values are assigned to the velocity of pbest positions, in order to consider them as particles. This approach also selects the leaders randomly from the leaders set (stored in an external archive) among the best of them, based on two different mechanisms: a niche count and a nearest neighbor density estimator. This approach uses a mutation operator that is applied at each iteration step only to the particle with the smallest density estimator value (or the largest niche count).

Sierra and Coello[116] propose an approach which is based on Pareto dominance and the use of a nearest neighbor density estimator for the selection of leaders (by means of a binary tournament). This proposal uses two external archives: one for storing the leaders currently used for performing the flight and another for storing the final solutions. The density estimator factor is used to filter out the list of leaders whenever the maximum limit imposed on such list is exceeded. Only the leaders with the best density estimator values are retained. On the other hand, the concept of ϵ-dominance is used to select particles that will remain in the archive of final solutions. Additionally, the authors propose a scheme in which they subdivide the population (or swarm) into three different subsets. A different mutation operator is applied to each subset. For all other purposes, a single swarm is considered (e.g., for selecting leaders). In Ref. 37, MOPSO implementations have all used metrics in objective space (either explicitly or implicitly) in the selection of guides - thus making them susceptible to different scaling in objective space. However, Alvarez - Benitez et al.[117] propose methods which are based exclusively on Pareto dominance for selecting leaders from an unconstrained non-dominated (external) archive. Three different selection techniques are presented: one technique that explicitly promotes diversity (called ROUNDS by the authors), one technique that explicitly promotes convergence (called RANDOM) and finally one technique that is weighted probabilistic method (called PROB) and forms a compromise between RANDOM & ROUNDS. Also, the authors propose and evaluate four mechanisms for confining particles to the feasible region, that is, constraint - handling methods. The authors show that probabilistic selection favoring archival particles that dominate few particles provides good convergence towards the Pareto front while properly covering it at the same time. Also, they conclude that allowing particles to explore regions close to the constraint boundaries is important to ensure convergence to the Pareto front. This approach uses a turbulence factor that is added to the position of the particles with certain probability.

Ho et al.[118] propose a novel formula for updating velocity and position of particles, based on three main modifications to the known flight formula.

The authors introduce a "craziness" operator in order to promote diversity within the swarm. The "craziness" operator is applied (with certain probability) to the velocity vector before updating the position of a particle. Finally, the authors introduce one external archive for each particle and one global external archive for the whole swarm. The archive of each particle stores the latest Pareto solutions found by the particle and the global archive stores the current Pareto optimal set. Every time a particle updates its position, it selects its personal best from its own archive and the global best from the global archive. In both cases, the authors use a roulette selection mechanism based on the fitness values of the particles and on an "age" variable that the authors introduce and that is increased at each generation.

Villalobos-Anias *et al.*[119] propose a new mechanism to promote diversity in multi-objective optimization problems. Although the approach is independent of the search engine adopted, they incorporate it into the MOPSO proposed in Ref. 37. The new approach is based on the use of stripes that are applied on the objective function space. Based on an analysis for a two objective problem, the main idea of the approach is that the Pareto front of the problem is similar to the line determined by the minimal points of the objective functions. In this way, several points (that the authors call stripe centers) are distributed uniformly along such line, and the particles of the swarm are assigned to the nearest stripe center. When using this approach for solving multi-objective problems with PSO, one leader is used in each stripe. Such leader is selected minimizing a weighted sum of the minimal points of the objective functions. The authors show that their approach overcomes the drawbacks on the popular mechanisms such as ϵ-dominance proposed in Ref. 107 and the sigma method proposed in Refs. 36, 120.

Salazar - Lechuga and Rowe[121] propose an approach whose main idea is to use PSO to guide the search with the help of niche counts (applied on objective function space)[122] to spread the particles along the Pareto front. The approach uses an external archive to store the best particles (non dominated particles) found by the algorithm. Since this external archive helps to guide the search, the niche count is calculated for each of the particles in the archive and the leaders are chosen from this set by means of a stochastic sampling method (roulette wheel). Also, the niche count is used as a criterion to update the external archive. Each time the archive is full and a new particle wants to set in, its niche count is compared with the niche count of the worst solution of the archive. If the new particle is better than the worst particle, then the new particle enters into the archive and the worst particle is deleted. Niche counts are updated when inserting or deleting a particle from the archive.

Raquel and Naval[123] propose an approach which as in Ref. 116 incorporates the concept of nearest neighbor density estimator for selecting the global best particle and also for deleting particles from the external archive of non-dominated solutions. When selecting a leader, the archive of non-dominated solutions is sorted in descending order with respect to the density estimator, and a particle is randomly chosen from the top part of the list. On the other hand when the external archive is full, it is again sorted in descending order with respect to the density estimator value and a particle is randomly chosen to be deleted from the bottom part of the list. This approach uses the mutation operator proposed in Ref. 37 in such a way that it is applied only during a certain number of generations at the beginning of the process. Finally, the authors adopt the constraint - handling technique from the NSGA - II.[103]

Zhao and Cao[124] propose an approach which is similar to the proposal of Coello and Lechuga.[27] However, the authors indicate that they maintain two external archives; but one of them is actually a list that keeps the pbest particle for each member of the swarm. Another external archive stores that non-dominated solutions found along the evolutionary process. This truncated archive is similar to the adaptive grid of PAES of Knowles and Corne.[102] The authors apply their approach to solve the economic load dispatch problem. With this aim, they employ a fuzzy - based mechanism to extract the best compromise solution, in which they incorporate the preferences of the decision maker. The approach adopts a linear membership function to represent the goals of each objective function. This membership function is adopted to modify the ranking of the non-dominated solutions so as to focus the search on the single solution that attains the maximum membership in the fuzzy set. Authors have demonstrated the superiority of their approach as compared to multi-objective evolutionary algorithm (MOEA).

Janson and Merkle[110] propose a hybrid particle swarm optimization algorithm for multi-objective optimization, called ClustMPSO. ClustMPSO combines the PSO algorithm with clustering techniques to divide all particles into several sub-swarms. For this aim, the authors use the K-means algorithm. Each sub-swarm has its own non-dominated front and the total non-dominated front is obtained from the union of the front of all the sub-swarms. Each particle randomly selects its neighborhood best (lbest) particle from the non-dominated front of the swarm to which it belongs. Also, a particle only selects a new lbest particle when the current is no longer a non-dominated solution. On the other hand, the personal best (pbest) of each particle is updated based on dominance relations. Finally, the authors define that a sub-swarm is dominated when none of its particles belong to the total non-dominated front. In this way, when a sub-swarm is dominated for a certain number of consecutive generations, the sub-swarm is relocated. The proposed algorithm is tested on an artificial multi-objective optimization function and on a real - world problem from biochemistry, called the molecular docking problem. The authors reformulate the molecular docking problem as a multi-objective optimization problem and, in this case, the updating of the pbest particle is also based on the weighted sum of the objectives of the problem. ClustMPSO outperforms a Simulated Annealing (SA) docking and a well known Lamarckian GA docking that has been previously adopted to solve such problem.

Sierra and Coello[116] presents a comprehensive review of the various MOPSOs reported in specialized literatures. As a part of this review, they have included the classification of approaches (Aggregating approaches, Lexicographic ordering, Sub-population approaches, Pareto - based approaches, Combined approaches and other approaches). They have indentified main features of each proposal and also listed some of the promising areas of future research.

Santana-Quintero et al.[125] present a new algorithm that approximates real function evaluations using supervised learning with a surrogate method called support vector machine (SVM). They have performed a comparative study among different leader selection schemes in a Multi-Objective Particle Swarm Optimizer (MOPSO), in order to determine the most appropriate approach to be adopted for solving the problems of interest. Since the resulting hybrid presents a poor spread of solutions, in the second phase to their algorithm, they have adopted an approach called rough sets in order to improve the spread of solutions along the Pareto front. Rough sets are used as a local search engine, which is able to generate solutions in the neighborhood of the non-dominated solutions previously generated by the surrogate-based algorithm. The resulting approach is able to generate reasonably good approximations of the Pareto front of problems of up to 30 decision variables with only 2,000 fitness function evaluations. Their results are compared with respect to the NSGA-II, which is a multi-objective evolutionary algorithm representative of the state-of-the-art in the area.

Lewis[126] has proposed a novel MOPSO algorithm called LoCost algorithm through some modification in Eq. (10.3) of the conventional PSO algorithm based on an extension of the concepts of spatial social networks using a model of the behavior of particular types of swarms known as crickets and locusts. He observes that the proposed algorithm has performed quite comparably to a conventional MOPSO algorithm in terms of convergence, and has achieved appreciably greater coverage of the approximation to the Pareto-front.

Leong and Yen[127] present the improvement of two design components (swarm growing strategy and objective space compression and expansion strategies) from the existing multiple-swarm MOPSO, namely DSMOPSO (Dynamic Swarm in Multi-Objective Particle Swarm Optimization). The multiple-swarm concept has been incorporated into PSO to yield more efficient and effective designs, especially in enhancing the population diversity, and to counter PSO's tendency in undesirable premature convergence. Multiple swarm PSO uses a heuristically chosen number of swarms with a fixed swarm size throughout the search process. But the issue is how to choose appropriate number of swarms with a fixed swarm size without incurring unnecessary computational cost and without degrading the quality of the optimal Pareto set. DSMOPSO addresses this issue by employing a dynamic population framework. Authors have conducted sensitivity analysis to study the impact of the five tuning parameters on its performance through two performance metrics. They have shown that improved design is robust with respect to the tuning parameters.

Lewis et al.[128] have proposed an approach of hybridizing a multi-objective optimization method and subsequent single-objective search has been proposed as a means to automate the process of solution selection from the set of Pareto-optimal solutions typically delivered. A method is used to provide surrogate approximations to reduce the computational overhead of the interpolated "second-phase" search. In this way, the search for a single, preferred solution can proceed without requiring any additional function evaluations, except for a single, final evaluation to confirm the feasibility of the chosen solution. Authors have demonstrated using the conventional Multi-Objective Particle Swarm Optimization (MOPSO) algorithm and the Simplex method of Nelder and Mead, and applied to a number of standard test problems. The approach shows an ability to deliver close approximations to user-preferred solutions without manual intervention. The proposed approach is also demonstrated to yield superior results to an alternative, a priori aggregation of weighted objectives and use of a single objective optimization method.

Cagnina *et al.*[129] have proposed a hybrid particle swarm approach called Simple Multi-Objective Particle Swarm Optimizer (SMOPSO) which incorporates Pareto dominance, an elitist policy, and two techniques to maintain diversity: a mutation operator and a grid which is used as a geographical location over objective function space. Laura *et al.* have proposed a hybrid technique that combines a Genetic Algorithm (GA) and a PSO algorithm. Each GA chromosome is an array encoding a meaning for updating the particles of the PSO algorithm. The evolved PSO algorithm is compared to a human-designed PSO algorithm by using ten artificially constructed functions and one real-world problem. The model proposed in this paper is divided into two levels: a macro level and a micro level. The macro level is a GA algorithm that evolves the structure of a PSO algorithm. For this purpose, a particular function is used as a training problem. The micro level is a PSO algorithm used for computing the quality of a GA chromosome from the macro level. The array of integers encoded into a GA chromosome represents the order of update for particles used by a PSO algorithm that solves a particular problem.

Goldbarg[130] presents a particle swarm optimization algorithm for the multi-criteria constrained minimum spanning tree problem. The operators for the particle's velocity are based upon local search and path-relinking approaches. In path-relinking approach, a velocity operator is developed and utilized when a particle goes toward the position of another particle. For the iterations where a particle follows its own way, a local search procedure is used. Ho *et al.*[131] propose a novel intelligent multi-objective particle swarm optimization (IMOPSO) to solve multi-objective optimization problems. High performance of IMOPSO mainly arises from two parts: in first part, by using a generalized Pareto-based scale-independent fitness function (GPSISF), each candidate solution is efficiently given a score, and then the level of each candidate solution is decided. In other part, the conventional particle move process of PSO is replaced with an intelligent move mechanism·(IMM) based on orthogonal experimental design to enhance the search ability. IMM can evenly sample and analyze from the best experience of an individual particle and group particles by using a systematic reasoning method, and then efficiently generate a good candidate solution for the next move of the particle. Ireland *et al.*[132] present quantitative comparison of the performance of different methods for selecting the guide particle for multi-objective particle swarm optimization

(MOPSO). Two principal methods are compared: the recently described Sigma method and a new Centroid method. Drawing on the different dominant behaviors exhibited by the different selection methods, a variety of hybridizations of these is proposed to develop a more robust optimization algorithm.

Koppen et al.[133] introduce a new approach to multi-objective particle swarm optimization. The approach is based on the recently proposed Fuzzy-Pareto-Dominance (FPD) relation. FPD is a generic ranking scheme, where ranking values are mapped to element vectors of a set. These ranking values are directly computed from the element vectors of the set and can be used to perform rank operations (e.g., selecting the "largest") with the vectors within the given set. FPD can be seen as a paradigm or meta-heuristic to formally expand single-objective optimization algorithms to multi-objective optimization algorithms, as long as such vector-sets can be defined.

Chiu et al.[134] present a local guide assignment strategy for MOPSO called cross-searching strategy (CSS) which will distribute suitable local guides for particles to lead them toward the Pareto front and also keeping diversity of solutions. A disturbance operation is also introduced to enhance particle's searching ability to avoid local search.

Leong et al.[135] propose the integration of a dynamic population strategy within the multiple-swarm MOPSO. The proposed algorithm is named as dynamic population multiple-swarm MOPSO. An additional feature, adaptive local archive, is designed to improve the diversity within each swarm. Design aspects that are incorporated in the proposed MOPSO include: i) strategy to facilitate access on the status of the particles when the swarm population size varies; ii) strategy to dynamically adjust the swarm population in order to provide the needs of computational resources at different stages and, at the same time, to promote competition among the swarms so that the convergence towards the optimal solutions and the diversity characteristics are preserved; and iii) adaptive local archive procedure to promote diversity within each swarm.

Peng et al.[136] have studied the application of PSO techniques to multi-objective optimization using decomposition methods. A new decomposition-based multi-objective PSO algorithm is proposed, called MOPSOID. It integrates PSO into a multi-obejective evolutionary algorithm based on decomposition (MOEAID). Like MOEAlD, each particle in MOPSOID carries one unique weight vector. Therefore, each particle has an unique search direction defined by its weight vector.

Padhye et al.[137] review the several proposals for guide selection in multi-objective particle swarm optimization (MOPSO) and compare them with each other in terms of convergence, diversity and computational times. The new proposals made for guide selection, both pbest and gbest, are found to be extremely effective and perform well compared to the already existing methods. The combination of various selection methods is also studied and it turns out that there exist certain combinations which yield an overall superior performance outperforming the others. Furthermore, two new proposals namely velocity trigger (as a substitute for turbulence operator) and a new scheme for boundary handling have been made.

Jiang et al.[138] introduce the particle angle division method as a new method for finding the gbest for each particle of the population. Particle angle division is used to store non-dominated solutions into the archive and provide the gbest from archive for each particle. When each member of archive is generated, its particle angle division of objective space is found. A map of the particle angle division is also maintained, indicating for each division how many and which particles in the archive currently reside there. When a candidate particle to join a full archive, it randomly replaces one of the archived particles which resides in the particle angle division location with the highest population of particles. When each particle of the population selects one of the Pareto-optimal as its gbest from the archive, one random particle residing in the particle angle division location with the lowest population is selected.

Tsou et al.[139] incorporates the local search and clustering mechanism into the multi-objective particle swarm optimization (MOPSO). The local search mechanism prevents premature convergence, hence enhances the convergence of optimizer to true Pareto-optimal front. The clustering mechanism reduces the non-dominated solutions to a handful number such that search can be speeded up and the diversity of the non-dominated solutions can be maintained. Venter et al. introduce an approach for dealing with constraints when using particle swarm optimization. The constrained single objective optimization problem is converted into an unconstrained bi-objective optimization problem that is solved using a multi-objective implementation of the particle swarm optimization algorithm. A specialized bi-objective particle swarm optimization algorithm is presented and an engineering example problem is used to illustrate the performance of the algorithm.

Cabrera et al.[140] present a multi-objective particle swarm optimizer (MOPSO) which is characterized for using a very small population size so that it requires a very low number of objective function evaluations (only 3000 per run) to produce reasonably good approximations of the Pareto front of problems of moderate dimensionality. The proposed approach first selects the leader and then selects the neighborhood for integrating the swarm. The leader selection scheme adopted is based on Pareto dominance and uses a neighbors' density estimator. Additionally, the proposed approach performs a re-initialization process for preserving diversity and uses two external archives: one for storing the solutions that the algorithm finds during the search process and another for storing the final solutions obtained. Furthermore, a mutation operator is incorporated to improve the exploratory capabilities of the algorithm. Wang *et al.* extends the NSGA-II-MOPSO algorithm, which is based on the combination of NSGA-II and multi-objective particle swarm optimizer (MOPSO) for unconstrained multi-objective optimization problems, to accommodate constraints and mixed variables. In order to utilize the valuable information from the objective function values of infeasible solutions, a method called M+1 non-dominated sorting is proposed to check the non-domination levels of all infeasible solutions. Integer and discrete variables are dealt with using a method called stochastic approximation.

Goh *et al.*[141] propose a competitive and cooperative co-evolutionary approach to be adapted for multi-objective particle swarm optimization algorithm design. It appears to have considerable potential for solving complex optimization problems by explicitly modeling the co-evolution of competing and cooperating species. The competitive and cooperative co-evolution model helps to produce the reasonable problem decompositions by exploiting any correlation and interdependency among the components of the problem. Each sub-swarm is assigned a probability of representing a particular variable and only two sub-swarms, the current sub-swarm and competing sub-swarm compete for the right to represent any variable at any one time.

Tsai *et al.*[142] proposes an improved multi-objective particle swarm optimizer with proportional distribution and jump improved operation, named PDJI-MOPSO, for dealing with multi-objective problems. PDJI-MOPSO maintains diversity of new found non-dominated solutions via proportional distribution, and combines advantages of wide-ranged exploration and extensive exploitation of PSO in the external repository with the jump improved operation to enhance the solution searching abilities of particles. Introduction of cluster and disturbance allows the proposed method to shift through representative non-dominated solutions from the external repository and prevent solutions from falling into local optimum. The proposed proportional distribution distributes greater or fewer numbers of particles to the archive member whose solution distance is longer or shorter, respectively. This process can significantly improve the searching abilities of particles and more easily find solutions with lower diversity and also on/near the Pareto front. The disturbance operation makes PSO more robust, and prevents particles from falling into the local optimum. The proposed method defines new pbest and gbest from the original PSO algorithm. Thus the proposed method can help particles explore more areas in the solution space.

Zheng *et al.*[143] propose a hybrid vertical mutation and self-adaptation based MOPSO (VMAPSO) to overcome the disadvantages of existing MOPSOs. Firstly, a hybrid vertical mutation operator can escape local optima and conduct a local search by uniform distribution mutation and Gaussian distribution mutation, respectively. Secondly, the adaptation ratio models of two mutations are fully analyzed and compared. Thirdly, the velocity update equations are improved to reduce the randomness of MOPSOs, and non-dominance based archive strategy is adopted in the proposed algorithm. Yang *et al.* propose a new algorithm for choosing gbest for each particle of the swarm from a Pareto-optimal solutions set. The proposed algorithm finds the gbest of the particles in a swarm from a set of non-dominated solutions in order to quickly converge towards a Pareto-optimal front of high diversity. This algorithm can compromise global and local searching based on the process of evolution. By using the proposed algorithm, a good convergence and diversity of solutions have been achieved.

Wang *et al.*[144] use a new optimality criterion based on preference order (PO) scheme to identify the best compromise in multi-objective particle swarm optimization (MOPSO). Preference order is a generalization of Pareto optimality. It provides a way to designate some Pareto solutions superior to others when the size of the non-dominated solutions set is very large. To find the "best compromise", the non-dominated solutions are ranked according to PO. The ranking procedure can be summarized in three steps: (i) identify the combinations of all subsets to m objectives; (ii) assign the order to all non-dominated solutions for each combination of all subsets based on PO; and (iii) identify the "best compromise" in all non-dominated solutions according to their order. The proposed algorithm is quite effective in maintaining the diversity of the solutions.

10.5. Applications with MOPSO Approaches

Particle Swarm Optimization (PSO) is increasingly being applied to optimization of multi-objective problems in engineering design and scientific investigation.

Bazi & Melgani[145] have presented a novel MOPSO approach for Support Vector Machine (SVM) regression with limited training samples. This approach has been applied to estimate the biophysical parameters from remote sensing images. They have proposed this approach for achieving both stability and accuracy improvement. This approach has been compared with earlier approaches such as PSO and SVM and seems to be more effective.

Chauhan et al.[146] have proposed a MOPSO approach in order to get optimum trade-off between matching of resonant frequency and minimizing power reflection for each of the two designed RF windows, namely double-disc RF window and pillbox type RF window which are used in high power microwave / millimeter wave sources.

Falcon et al.[147] have applied MOPSO approaches to the framework of collaborative fuzzy clustering. They have emphasized on determining the collaboration matrix between data repositories. By considering several fitness functions expressing the quality of the collaboration realized, it is possible to know in advance how strongly the findings coming from other repositories will impact local data. They have provided a more effective way of reconciling the findings between the participating data sites and have applied this methodology to marketing research.

Carvalho et al.[148] have introduced a Multi-Objective Particle Swarm Optimization (MOPSO) algorithm for fault prediction. It allows the creation of classifiers composed by rules with specific properties by exploring Pareto dominance concepts. These rules are more intuitive and easier to understand because they can be interpreted independently one of each other. Furthermore, an experiment using the approach is presented and the results are compared to the other techniques explored in the area.

Martin et al.[149] have explored the response of Multiple Objective Particle Swarm Optimization (MOPSO) for designing Ultra Wide Band (UWB) planar antennas. They have focused on those UWB antennas heuristically designed for taking into account the inclusion of a notch for the 5GHz to 6GHz band. Applying these optimization schemes over a conformal FDTD code, a set of UWB circular monopole antennas has been achieved.

Pang and Chen[150] have proposed a MOPSO method based on Strength Pareto Evolutionary Algorithm (SPEA) to optimize the both evaluation indices, the average delay and stop frequency synchronously for evaluating the level of service (LOS) for signalized intersections in intelligent transportation system (ITS). A well-distributed set of Pareto optimal solutions has been obtained.

Hazra and Sinha[151] propose an effective method of congestion management in power systems. Congestions or overloads in transmission network are alleviated by generation rescheduling and/or load shedding of participating generators and loads. The two conflicting objectives 1) alleviation of overload and 2) minimization of cost of operation are optimized to provide Pareto-optimal solutions. A multi-objective particle swarm optimization (MOPSO) method is used to solve this complex nonlinear optimization problem. A realistic frequency and voltage dependent load flow method which considers the voltage and frequency dependence of loads and generator regulation characteristics is used to solve this problem. The proposed algorithm is tested on IEEE 30-bus system, IEEE 118-bus system, and Northern Region Electricity Board, India (NREB) 390-bus system with smooth as well as non-smooth cost functions due to valve point loading effect.

Qasem *et al.*[152] propose an adaptive evolutionary radial basis function (RBF) network algorithm to evolve accuracy and connections (centers and weights) of RBF networks simultaneously. The problem of hybrid learning of RBF network is discussed with the multi-objective optimization methods to improve classification accuracy for medical disease diagnosis. The authors introduce a time variant multi-objective particle swarm optimization (TVMOPSO) of radial basis function(RBF)network for diagnosing the medical diseases.

Pindoriya *et al.*[153] propose an approach for generation of portfolio allocation based on mean-variance-skewness (MVS) model which is an extension of the classical mean-variance (MV) portfolio theory, to deal with assets whose return distribution is non-normal. The MVS model allocates portfolios optimally by considering the maximization of both the expected return and skewness of portfolio return while simultaneously minimizing the risk. Since, it is a competing and conflicting non-smooth multi-objective optimization problem, the authors employ a multi-objective particle swarm optimization (MOPSO) based meta-heuristic technique to provide Pareto-optimal solutions in a single simulation run.

Sha *et al.*[154] have proposed a modified particle swarm optimization due to the discrete nature of the job scheduling problem. They have proposed a modified particle position representation, particle movement, and particle velocity. Wang *et al.* first formulate the stochastic model for combined heat and power (CHP) dispatch and then develop an improved particle swarm optimization (PSO) method to deal with the economic CHP dispatch by simultaneously considering multiple conflicting objectives. Based on the proposed optimization method, the impact of different problem formulations including stochastic and deterministic models on power dispatch results is investigated and analyzed.

Montavlo *et al.*[155] propose a multi-objective variant of the particle swarm optimization (PSO) algorithm for applying to a water distribution system optimization problem. To improve its ability to find the Pareto front in such a multi-objective optimization problem, some changes have been addressed to:i) the way a particle decides what is a better position; ii) the selection of the leader (which now takes into account the existence of different objectives); and iii) the particles abilities to clone themselves for increasing density when needed in the Pareto front. In addition, authors have introduced real-time human interaction which allows better development of certain zones of the Pareto front.

Liu[156] has proposed a new non-dominated sorting particle swarm optimization (NSPSO) for the calibration of a rainfall-runoff model. The proposed NSPSO combines the operations (fast ranking of non-dominated solutions, crowding distance ranking and elitist strategy of combining parent population and offspring population together) of a known MOGA NSGA-II and the other advanced operations (selection and mutation operations) with the particle swarm optimization.

Zhang *et al.*[157] present a new formulation of multi-objective reactive power and voltage control for power system. The objectives are active power loss, voltage deviation and the voltage stability index of the system. The load constraints and operational constraints are also taken into consideration. The multi-objective formulation of the problem requires a global performance index of the problem. A pseudo-goal function derived on the basis of the fuzzy set theory gives a unique expression for the global objective function by eliminating the use of weighing coefficients or penalty terms. Both objective functions and constraints are evaluated by membership functions. The inequality constraints are embedded into the fitness function by pseudo-goal function which guarantees that the searched optimal solution is feasible. Moreover, a new type of particle swarm optimization (PSO) has been adopted and improved for this problem. To improve the performance of PSO, a fuzzy adaptive PSO (FAPSO) is proposed. A fuzzy system is employed to adaptively adjust the parameters of PSO, such as the inertia weight and learning factors during the evolutionary process.

Cia *et al.*[158] have proposed a multi-objective chaotic particle swarm optimization (MOCPSO) method to solve the environmental/economic dispatch (EED) problems considering both economic and environmental issues. The proposed MOCPSO method has been applied in two test power systems. Compared with the conventional multi-objective particle swarm optimization (MOPSO) method for compromising minimum fuel cost per hour and pollutant emission per hour, MOCPSO method has reduced these two factors. The MOCPSO method also results in higher quality solutions in both of the test power systems

10.6. Issues and Sub-Issues of PSO and MOPSO

In order to apply the PSO strategy for solving MOPs, it is obvious that the original scheme needs to be modified. In multi-objective optimization, since solution set does not consist of a single solution but a set of different solutions (the so-called Pareto optimal set), following three main goals need to be achieved:

- Maximize the number of elements of the Pareto optimal set found.
- Minimize the distance of the Pareto front produced by the algorithm with respect to the true (global) Pareto front (assuming its location is known empirically).
- Maximize the spread of solutions found so that the distribution can be as smooth and uniform as possible.

Given the population-based nature of PSO and the characteristics of PSO algorithm, it is desirable to produce several (different) non-dominated solutions in a single run. As a result, following issues arise when extending PSO to multi-objective optimization:

(1) Selection of particles (to be used as leaders) must be done in such a way that preference will be given to non-dominated solutions over those that are dominated.
(2) These particles (leaders) must be retained in the external archive from one iteration to another.
(3) At the end of stopping criterion, these non-dominated solutions to be reported as final solutions must be non-dominated not only with

respect to the current population but also with respect to all the past populations.

(4) The diversity of the swarm must be maintained in order to avoid premature convergence to a single local optimal solution. Premature convergence is caused by the rapid loss of diversity within the swarm. Hence, suitable mechanism to promote diversity in each generation must be adopted. These mechanisms may be either updating of positions of particles through Eqs. (10.3) and (10.4) or using mutation (turbulence) operator or any other means. The final solutions in this way can be maintained well spread along the Pareto front.

(5) Decision to select a single solution out of current set of non-dominated solutions which are all equally good must be taken either in a random way or through the use of any additional criterion (quality measure). Most authors are adopting additional information to select leaders (e.g., information provided by a type of quality measure known as density measure[24]) in order to avoid a random selection of a leader from the current set of non-dominated solutions.

Density measure indicates the closeness of the particles within the swarm. This measure promotes diversity. Two of the most important density measures used in the area of multi-objective optimization are (1) Nearest neighbor density estimator[103] and (2) Kernel density estimator (or niche count).[122,159]

- Nearest neighbor density estimator gives an idea of how crowded are the closest neighbors of a given particle, in objective function space. This measure estimates the perimeter of the cuboid formed by using the nearest neighbors as the vertices. Particles with a larger value of this estimator are preferred.

- Kernel density estimator or niche count gives an idea of degradation of the fitness of particle which is sharing resources with other particles surrounding it within a certain perimeter. This amount of degradation is proportional to the number and closeness of surrounding particles. A neighborhood of a particle is defined in terms of a parameter called σ-share that indicates the radius of the neighborhood. Such neighborhoods are called niches. For each particle a niche is defined. Particles whose niches are less crowded are preferred.

Besides the above issues, there is one more very important issue, that is, the size of the external archive which retains the solutions that are non-dominated with respect to all the previous populations (or swarms) must be maintained in such a way that the time complexity of the updating process at each generation is not too high. This is possible if the size of the archive does not grow too much at each generation.

In the worst case, all members of the swarm may wish to enter into the archive at each generation. Thus, the corresponding updating process, at each generation, has a complexity of $O(kN^2)$, where N is the size of the swarm and k is the number of objectives. In this way, the time complexity of the updating process for the complete run is of $O(kMN^2)$, where M is the total number of iterations.

Considering above issue, most authors are maintaining the size of the external archive bounded in order to have "not very expensive" updating process computationally. Once the archive is full, they are adopting different techniques to prune the archive i.e., deciding which non-dominated solutions to retain and which to delete. Out of various techniques, ϵ-dominance technique[107] is the recent one and it is much faster than the previously used various clustering techniques[39,105] for fixing the archive size. The ϵ-dominance method can find solutions much faster than these clustering techniques with comparable (and even in better in some cases) convergence and diversity.

Connected with the main issue of maintaining diversity of the swarm by use of two well known main mechanisms, (1) updating of positions and (2) use of a mutation (or turbulence) operator, there are some sub-issues which need to be looked into.

- How to preserve diversity within the swarm a longer time without sacrificing speed of information transfer among the particles in the swarm? Currently this tradeoff can be handled either through proper selection of swarm topology (i.e., fully connected or local best or tree etc.)[24] or through proper selection of size of inertia weight (w in Eq. (10.3)).[160]
- How to escape from local optima and speed up the search when a swarm stagnates, that is, when the velocities of the particles are almost zero? Currently, such problem is handled either by using any one of different mutation operators to a single particle in the swarm or without using any kind of mutation operator. If mutation is applied, it can be either mutation of position or mutation of

velocity of the particle. Such mutation is other way known as 'craziness' or turbulence.

- How to ensure global convergence of the MOPSO algorithm to true Pareto front ? Currently, however, convergence of PSO algorithm to local or global optimum can be possible to ensure by correctly setting the parameters of the flight formula[161] and also if following two conditions are satisfied:

  (1) The $gbest_{t+1}$ solution can be no worse than the $gbest_t$ solution (monotonic condition).
  (2) The algorithm must be able to generate a solution in the neighborhood of the optimum with non-zero probability, from any solution of the search space.

However, for convergence of MOPSO algorithm, only correct setting of parameters in flight formula does not ensure the convergence to the true Pareto front. However, two conditions mentioned above need to be satisfied with modifications of condition 1 in the following way:

The solution contained in the external archive at iteration t+1 should be non-dominated with respect to the solutions generated in all iterations i, $0 \leq i \leq t + 1$, so far (monotonic conditions).

10.7. Summary and Future Scope

In this chapter we have shown most representative PSO developed for single and multi-objective optimization problems. The effect of topological structure for both static and dynamic environment has been illustrated. In the case of single objective PSO we have categorized into several groups by considering various parameters. We saw, most of these methods developed so far were either adopted or inspired by research reported in genetic algorithms literature. There are, however, other methods that are not directly derived from such literature, because they rely on hybridizations between PSO and another meta-heuristics, with the clear aim of benefitting from the advantages of both types of approaches.

References

1. J. Kennedy and R. C. Eberhart. Particle swarm optimization. In *Proceedings of the IEEE International Conference on Neural Networks*, pp. 1942–1948, Perth, Australia, (1995).

2. J. Moore, R. Chapman, and G. Dozier, Multiobjective particle swarm optimization, *ACM.* pp. 56–57, (2000).
3. K. Deb, *Optimization for Engineering Design: Algorithms and Examples.* (Prentice Hall of India, New Delhi, 1995).
4. S. S. Rao, *Optimization: Theory and Applications.* (Wiley, New York, 1984).
5. G. V. Rekiaitis, A. Ravindran, and K. M. Ragsdell, *Engineering Optimization Methods and Applications.* (Wiley, New York, 1983).
6. V. Chankong and Y. Y. Haimes, *Multi-Objective Decision Making Theory and Methedology.* (North-Holland, New York, 1983).
7. M. Ehrgott, *Multi-Criteria Optimization.* (Springer, Berlin, 2000).
8. K. Miettinen, *Nonlinear Multi-Objective Optimization.* (Kluwer, Boston, 1999).
9. Y. Y. Haimes, L. S. Lasdon, and D. A. Wismer, On a bicriterion formulation of the problems of integrated system identification and system optimization, *IEEE Transactions on Systems, Man, and Cybernetics.* **1**(3), 296–297, (1971).
10. R. L. Kenney, *Decisions with Multiple Objectives: Preferences and Value Tradeoffs.* (Wiely, New York, 1976).
11. W. Charnes, W. Coope, and R. Ferguson, Optimal estimation of executive compensation by linear programming, *Management Sciences.* **1**(2), 138–151, (1955).
12. J. P. Ignizio, *Goal Programming and Extensions.* (Lexington Books, Lexington, MA, 1976).
13. J. P. Ignizio, A review of goal programming: A tool for multi-objective analysis, *Journal of Operations Research Society.* **29**(11), 1109–1119, (1978).
14. S. M. Lee, *Goal Programming for Decision Analysis.* (Auerbach Publishers, Philadelphia, 1972).
15. C. Romero, *Handbook of Critical Issues in Goal Programming.* (Pergamon Press, Oxford, U.K., 1991).
16. E. R. Clayton, W. Weber, and B. W. Taylor, A goal programming approach to the optimization of multi-response simulation models, *IEE Transactions.* **14**(4), 282–287, (1982).
17. M. H. Sayyouth, Goal programming : A new tool for optimizations in petroleum reservoir history matching, *Applied Mathematics Modelling.* **5** (4), 223–226, (1981).
18. R. C. Eberhart, P. Simpson, and R. Dobbins, *Computational Intelligent PC Tools.* (A. P. Professional, San Diego, California, USA, 1996).
19. J. Kennedy and R. C. Eberhart, *Swarm Intelligence.* (Morgan Kaufmann Publishers, San Francisco, California, 2001).
20. A. P. Engelbrecht, *Computational Intelligence: An Introduction.* (John Wiley & sons, England, 2005).
21. J. Kennedy and R. C. Eberhart. A discrete binary version of the particle swarm algorithm. In *Proceedings of the IEEE Conference on Systems, Man and Cybernetics*, pp. 1764–1771, Canberra, Australia (December, 2003).
22. X. Hu, R. C. Eberhart, and Y. Shi. Swarm intelligence for permutation optimization: A case study on n-queen problem. In *Proceedings of the IEEE*

Swarm Intelligence Symposium, pp. 243–246, Indianpolis, Indiana, USA, (2003).

23. P. J. Angeline. Evolutionary optimization versus particle swarm optimization: Philosophy and performance differences. In *Proceedings of the 7th International Conference on Evolutionary Programming*, pp. 601–610, San Diego, California, USA, (1998).

24. M. R. Sierra and C. C. A. C., Multi-objective particle swarm optimizers: A survey of the state-of-the-art, *International Journal of Computational Intelligence Research.* **2**(3), 287–308, (2006).

25. A. P. Engelbrecht, *Fundamentals of Computational Swarm Intelligence.* (John Wiley & sons, England, 2002).

26. R. Hassan, B. Cohanim, O. de Weck, and G. Venter. A comparision of particle swarm optimization and genetic algorithm. In *46th AIAA/ASME/ASCE/AHS/ASC Structures, Structural Dynamic and Materials Conference*, pp. 1897–1909, Austin, Texas (April, 2005).

27. C. A. C. Coello and M. Lechuga. Mopso: A proposal for multi-objective particle swarm optimization. In *Proceedings of the 9th IEEE World Congress on Computational Intelligence*, pp. 1051–1056, Honolulu, Hawaii, USA, (2002).

28. J. Kennedy. Small words and mega minds: Effect of neighborhood topologies on particle swarm optimization. In *In proceedings of the 1999 IEEE congress on Evolutionary Computation(CEC 1999)*, pp. 1931–1938, (1999).

29. J. Kennedy and R. Mendes. Population structure and particle performance. In *In proceedings of the 2002 IEEE congress on Evolutionary Computation(CEC 2002)*, pp. 1671–1676, (2002).

30. C. Veenhuis. Advanced meta-pso. In *Proceedings of the Sixth International Conference on Hybrid Intelligent Systems*, pp. 54–59, (2006).

31. J. Li and X. Xiao. Multi-swarm and multi-best particle swarm optimization algorithm. In *Proceedings of 7th World Congress on Intelligent Control and Automation*, pp. 6281–6286, (2008).

32. Y. X. Wang and Q. L. Xiang. Particle swarms with dynamic ring topology. In *IEEE Congress in Evolutionary Computation*, pp. 419–423, (2008).

33. J. E. Fieldsend and S. Singh. A multi-objective algorithm based upon particle swarm optimization, an efficient data structure and turbulence. In *Proceedings of the Workshop on Computational Intelligence*, pp. 37–44, Brimingham, UK (Sept, 2002).

34. X. Hu and R. C. Eberhart. Multi-objective optimization using dynamic neighborhood particle swarm optimization. In *Proceedings of the IEEE World Congress on Evolutionary Computation*, pp. 1677–1681, Honolulu, Hawaii, USA, (2002).

35. K. E. Parsopoulos and M. N. Vrahatis. Particle swarm optimization method in multi-objective problem. In *Proceedings of the ACM Symposium on Applied Computing*, pp. 603–607, Madrid, Spain, (2002).

36. S. Mostaghim and J. Teich. Strategies for finding good local guides in multi-objective particle swarm optimization mopso. In *Proceedings of the IEEE Symposium on Swarm Intelligence*, pp. 26–33, (2003).

37. C. A. C. Coello, G. T. Pulido, and M. S. Lechuga, Handling multiple objectives with particle swarm optimization, *IEEE Transactions on Evolutionary Computation.* **8**(3), 256–279, (2004).

38. T. Ray and K. M. Liew, A swarm metaphor for multi-objective design optimization, *Engineering Optimization.* **34**(2), 141–153, (2002).

39. Y. Shi and R. C. Eberhart. Parameter selection in particle swarm optimization. In *7th Annual Conference on Evolutionary Programming*, pp. 591–600, SanDiego,USA, (1998).

40. E. Ozcan and C. K. Mohan. Particle swarm optimization: Surfing the waves. In *Proceedings of the 1999 Congress on Evolutionary Computation*, pp. 1939–1944, (1999).

41. P. N. Suganthan. Particle swarm optimizer and neighborhood operator. In *Proceedings of the 1999 Congress on Evolutionary Computation*, pp. 1958–1962, (1999).

42. S. Yang, M. Wang, and L.Jiao. A quantum particle swarm optimization. In *IEEE Proceedings of Congress on Evolutionary Computation*, vol. 1, pp. 320–324, (2004).

43. C. J. Liao, C. T. Tseng, and P. Luarn, A discrete version of particle swarm optimization for flowshop scheduling problems, *Journal of Computer and Operation Research.* **34**, 3099–3111, (2005).

44. E. S. Correa, A. A. Freitas, and C. G. Johnson. A new discrete particle swarm algorithm applied to attribute selection in a bioinformatics data set. In *Proceedings of the 8th Annual Conference on Genetic and Evolutionary Computation*, pp. 35–42, (2006).

45. J. Wang. A novel discrete particle swarm optimization based on estimation of distribution. In *International Proceedings on Intelligent Computing*, pp. 791–802, (2007).

46. M. Sarfraz and A. Al-Awami. Arabic character recognition using particle swarm optimization with selected and weighted moment invariants. In *Proceedings of 9th International Symposium on Signal Processing and its Application*, pp. 1–4, Sharjah (february, 2007).

47. B. Jarboui, N. Damak, P. Siarry, and A. Rebai, A combinatorial particle swarm optimization for solving multi-mode resource-constrained project scheduling problems, *Journal of Applied Mathematics and Computation.* **195**, 299–308, (2007).

48. Z. Zhen, L. Wang, and Z. Huang. Probability-based binary particle swarm optimization algorithm and its application to wfgd control. In *Proceedings of International Conference on Computer Science and Software Engineering*, vol. 1, pp. 443–446, (2008).

49. M. F. Tasgetiren. A discrete particle swarm optimization algorithm for the generalized traveling salesman problem. In *Proceedings of the 9th Annual Conference on Genetic and Evolutionary Computation*, pp. 158–165, (2007).

50. M. C. Lee, K. H. Tsai, and T. I. Wang. A discrete particle swarm optimization based approach for review course composition. In *Proceedings of Third International Conference on Convergence and Hybrid Information Technology*, pp. 639–644, (2008).

51. Y. Shi and R. C. Eberhart. Empirical study of particle swarm optimization. In *Proceedings of the 1999 Congress on Evolutionary Computation*, pp. 1945–1950, (1999).

52. A. Carlisle and G. Dozier. Adapting particle swarm optimization to dynamic environments. In *Proceedings of International Conference on Artificial Intelligence*, pp. 1958–1962, (2000).

53. Y. Shi and R. C. Eberhart. A modified particle swarm optimizer. In *Proceedings of IEEE International Conference on Evolutionary Computation*, pp. 69–73, Piscataway, NJ, (1998).

54. X. Hu and R. C. Eberhart. Adaptive particle swarm optimization: Detection and response to dynamic system. In *Proceedings of IEEE Congress on Evolutionary Computation*, pp. 1666–1670, (2002).

55. X. F. Xie, W. J. Zhang, and Z. L. Yang. Adaptive particle swarm optimization on individual level. In *Proceedings of 6th International Conference on Signal Processing*, pp. 1215–1218, (2002).

56. W. Zhang and Y. Liu. Adaptive particle swarm optimization for reative power and voltage control in power systems. In *Proceedings of International Conference in Natural Computation, LNCS 3612*, pp. 449–452, (2005).

57. Z. Zhen, Z. Wang, and Y. Liu. An adaptive particle swarm optimization for global optimization. In *Proceedings of Third International Conference on Natural Computation*, vol. 4, pp. 8–12, (2007).

58. F. Chunxia and W. Youhong. An adaptive simple particle swarm optimization algorithm. In *Proceedings of Chinese Control and Decision ConferenceInternational Conference on Computer Science and Software Engineering*, pp. 3067–3072, (2008).

59. Z. H. Zhan, J. Zhang, Y. Li, and H. S. H. Chung, Adaptive particle swarm optimization, *IEEE Transactions on Systems, Man and Cybernetics.* **39**(6), 1362–1381, (2009).

60. L. Hongwu. An adaptive chaotic particle swarm optimization. In *Proceedings of International Colloquium on Computing, Communication, Control, and Management*, vol. 2, pp. 254–257, (2009).

61. F. V. D. Bergh and A. P. Engelbrecht, A cooperative approach to particle swarm optimization, *IEEE Transaction on Evolutionary Computation.* **8**(3), 225–239, (2004).

62. A. Ahmadi, F. Karray, and M. Kamel. Multiple cooperating swarms for data clustering. In *Proceedings of the IEEE Swarm Intelligence Symposium*, pp. 206–212, (2005).

63. M. Inthachot and S. Supratid. A multi-subpopulation particle swarm optimization: A hybrid intelligent computing for function optimization. In *Proceedings of Third International Conference on Natural Computation*, vol. 5, pp. 679–684, (2007).

64. L. Jian, L. Zhiming, and C. Peng. Solving constrained optimization via dual particle swarm optimization with stochastic ranking. In *Proceedings of International Conference on Computer Science and Software Engineering*, vol. 1, pp. 1215–1218, (2008).

65. A. Ratnaweera, S. K. Halgamuge, and H. C. Watson, Self-organizing hierarchical particle swarm optimizer with time-varying acceleration coefficients, *IEEE Transaction on Evolutionary Computation.* **8**(3), (2004).

66. C. A. Ling, Y. Gen-ke, and W. Zhi-ming, Hybrid discrete particle swarm optimization algorithm for capacitated vehicle routing problem, *Journal of Zhejiang University.* **7**(4), 607–614, (2005).

67. T. M. Blackwell, Particle swarms and population diversity, *Journal of Soft-Computing - A Fusion of Foundations, Methodologies and Applications.* **9** (11), 793–802, (2005).

68. M. S. Voss. Principal component of particle swarm optimization. In *IEEE Proceedings of Swarm Intelligence Symposium*, pp. 401–404, (2005).

69. I. H. Grundy and A. Stacey, Particle swarm optimization with combined mutation and hill climbing, *Journal of Complexity International.* **12**, 1–10, (2005).

70. G. Pan, Q. Duo, and X. Liu. Performance of two improved particle swarm optimization in dynamic optimization environments. In *Proceedings of Sixth International Conference on Intelligent Systems Design and Applications*, vol. 2, pp. 1024–1028, (2006).

71. M. Meissner, M. Schmuker, and G. Schneider, Optimized particle swarm optimization (opsp) and its application to artificial neural network training, *Journal of BMC Bioinformatics.* **7**, (2006).

72. M. A. W. and N. C. Sahoo, Efficient computation of shortest paths in networks using particle swarm optimization and noising metaheuristics, *Journal of Discrete Dynamics in Nature and Society.* **2007**, 1–25, (2007).

73. D. P. Tian and N. Q. Li. Fuzzy particle swarm optimization algorithm. In *Proceedings of International Joint Conference on Artificial Intelligence*, pp. 263–267, (2009).

74. R. I. Lung and D. Dumitrescu, Evolutionary swarm cooperative optimization in dynamic environments, *Natural Computing.* (2009).

75. W. C. Yeh, W. W. Chang, and Y. Y. Chung, A new hybrid approach for mining breast pancer pattern using discrete particle swarm optimization and statistical method, *Journal of Expert Systems with Application.* **36**(4), 8204–8211, (2008).

76. U. Natarajan, V. M. Periaswamy, and R. Saravanan, Application of particle swarm optimization in artificial neural network for the application of tool life, *International Journal of Manufacturing Technology.* **31**, 871–876, (2006).

77. X. F. Xie, W. J. Zhang, and Z. L. Yang. A dissipative particle swarm optimization. In *Proceedings of 2002 Congress on Evolutionary Computation*, pp. 1456–1461, (2002).

78. K. E. Parsopoulos and M. N. Vrahatis. Initializing the particle swarm optimization using the non-linear simplex method. In *Advances in Intelligent Systems, Fuzzy Systems, Evolutionary Computation*, pp. 216–221, (2002).

79. J. Kennedy. Bare bones particle swarms. In *Proceedings of IEEE Swarm Intelligence Symposium*, pp. 82–87, Indianapolis, Indiana, USA, (2003).

80. G. DePuy, T. Hardin, and J. Greenwell. Statistical evaluation of particle swarm optimization algorithm. In *Experimental Design, Industrial Engineering 563*, University of Louisville, (2003).

81. C. Yang and D. Simon. A new particle swarm optimization technique. In *Proceedings of 18th International Conference on Systems Engineering*, pp. 164–169, (2005).

82. B. Koh, A. D. George, R. T. Haftka, and B. J. Fregly, Parallel asynchronous particle swarm optimization, *International Journal of Numerical Methods In Engineering.* **67**, 578–595, (2006).

83. A. Khosla and K. K. Aggarwal, Identification of strategy parameters for particle swarm optimizer through taguchi method, *Journal of Zhejiang University.* **7**(12), 1989–1994, (2006).

84. K. Chandramouli and E. Izquierdo. Image classification using chaotic particle swarm optimization. In *IEEE Proceedings of International Conference on Image Processing*, pp. 3001–3004, (2006).

85. V. Kadirkamanathan, K. Selvarrajan, and P. J. Fleming, Stability analysis of the particle dynamics in particle swarm optimizer, *IEEE Transaction on Evolutionary Computation.* **10**(3), 245–255, (2006).

86. Y. Tan and Z. M. Xiao. Clonal particle swarm optimization and its applications. In *Proceedings of IEEE Congress on Evolutionary Computation*, pp. 2303–2309, (2007).

87. T. Xiang, K. W. Wong, and L. X., A novel particle swarm optimizer with time-delay, *Journal of Applied Mathematics and Computation.* **186**, 789–793, (2007).

88. M. S. Arumugam, M. V. C. Rao, and A. W. C. Tan, A novel and effective particle swarm optimization like algorithm with extrapolation technique, *Journal of Applied Soft Computing.* pp. 308–320, (2008).

89. E. Liu, Y. Dong, J. song X. Hou, and N. Li. A modified particle swarm optimization algorithm. In *International Workshop on Education Technology and Training & International Workshop on Geoscience and Remote Sensing*, vol. 2, pp. 666–669, (2008).

90. R. Poli, Dynamics and stability of the sampling distribution of particle swarm optimizers via moment analysis, *Journal of Applied Evolution and Applications.* **2008**, 1–10, (2008).

91. B. K. Panigrahi, V. R. Pandi, and S. Das, Adaptive particle swarm optimization approach for sanctioned dynamic economic load dispatch, *Journal of Energy Conversion and Management.* **49**(6), 1407–1415, (2008).

92. H. Zhang and M. Ishikawa, Characterization of particle swarm optimization with diversive curiosity, *Neural Computing and Applications.* **18**, 409–415, (2009).

93. S. T. Hsieh, T. Y. Sun, C. C. Liu, and S. J. Tsai, Efficient population utilization strategy for particle swarm optimizer, *IEEE Transaction on Systems, Man, and Cybernetics.* **39**(2), (2009).

94. H. Zhu, C. P. Eguchiand, and K. J. Gu. Euclidean particle swarm optimization. In *Proceedings of Second International Conference on Intelligent Networks Intelligent Systems*, pp. 669–672, (2009).

95. L. Hong. A particle swarm optimization based on immune mechanism. In *Proceedings of International Joint Conference on Computational Sciences and Optimization*, vol. 1, pp. 670–673, (2009).

96. K. Winner, D. Miner, and M. Desjardins. Controlling particle swarm optimization with learned parameters. In *Proceedings of Third IEEE International Conference on Self-Adaptive and Self-Organizing Systems*, pp. 288–290, (2009).

97. F. Arifia, M. Messaoud, and M. Abid, Nonlinear adaptive filters based on particle swarm optimization, *Leonardo Journal of Sciences.* **14**, 244–251, (2009).

98. A. A. dePina, C. H. Albreoht, B. S. L. P. deLima, and B. P. Jacob. Tailoring the particle swarm optimization algorithm for the design of offsore oil production rises. In *Optimisation and Engineering*, (2009).

99. J. Lis and a. Eiben. A multi sexual genetic algorithm for multi-objective optimization. In *Proceedings of the International Conference on Evolutionary Computation*, pp. 59–64, Indianapols, Indiana, (1997).

100. K. Deb, Multi-objective genetic algorithms: Problem difficulties and construction of test problems, *IEEE Transactions on Evolutionary Computation.* **7**(2), 205–230, (1999).

101. J. D. Schaffer. Multi-objective optimization with vector evaluated genetic algorithms. In *Proceedings of the First International Conference on Genetic Algorithms*, pp. 93–100, (1985).

102. J. D. Knowles and D. W. Corne, Approximating the non-dominated front using pareto archived evolution strategy, *IEEE Transactions on Evolutionary Computation.* **8**(2), 149–172, (2000).

103. K. Deb, A. Pratap, S. Agrawal, and T. Meyarivan, A fast and elitist multi-objective genetic algorithm: Nsga-ii, *IEEE Transactions on Evolutionary Computation.* **6**(2), 182–197, (2002).

104. M. Laumanns, E. Zitzler, and L. Thiele. A unified model for multi-objective evolutionary algorithm with elitism. In *Proceedings of the IEEE World Congress on Evolutionary Computation*, pp. 46–53, Piscataway, NJ, (2000).

105. E. Zitzler, M. Laumanns, and L. Thiele. SPEA2: Improving the Strength Pareto Evolutionary Algorithm. Technical Report 103, Computer Engineering and Networks Laboratory(TIK), Swiss Federal Institute of Tehnology(ETH), Zurich, Gloriastrasse 35, CH-8092 (may, 2001).

106. S. Mostaghim, M. Hoffmann, P. H. Konig, T. Frauenheim, and J. Teich. Molecular force field parameterization using multi-objective evolutionary algorithms. In *IEEE World Congress on Evolutionary Computation*, vol. 1, pp. 212–219, Portland, Organ, USA (June, 2004).

107. S. Mostaghim and J. Teich. The role of ϵ-dominance in multi-objective particle swarm optimization methods). In *Proceedings of the IEEE World Congress on Evolutionary Computation*, pp. 1764–1771, Canberra, Australia (December, 2003).

108. C. A. C. Coello and G. T. Pulido. Multi-objective optimization using a micro genetic algorithm. In *Proceedings of the Genetic and Evolutionary Computation Conference*, pp. 274–282, San Francisco, California, USA, (2001).

109. J. E. Fieldsend. Multi-objective Particle swarm Optimization Methods. Technical Report 419, Department of Computer Science, University of Exter (march, 2004).

110. S. Janson and D. Merkle. A new multi-objective particle swarm optimization algorithms using clustering applied to automated docking. In *Hybrid Meta-heuristics Second International Workshop*, pp. 128–142, Barcelona, Spain, (2005).

111. G. T. Pulido and C. A. C. Coello. Using clustering techniques to improve the performance of a particle swarm optimizer. In *Proceedings of the Genetic and Evolutionary Comptation Conference*, pp. 225–237, Seattle, Washington, USA (June, 2004).

112. S. Mostaghim and J. Teich. Particle swarm inspired evolutionary algorithm *psea* for multi-objective optimization problem. In *Proceedings of the IEEE World Congress on Evolutionary Computation*, pp. 2292–2297, Canberra, Australia (December, 2003).

113. S. Mostaghim and J. Teich. Covering pareto optimal fronts by sub-swarms in multi-objective particle swarm optimization. In *Proceedings of the IEEE World Congress on Evolutionary Computation*, pp. 1404–1411, Portland, Oregon, USA (June, 2004).

114. T. Bartz-Beielstein, P. Limbaurg, J. M. K. E. Parsopoulos, M. N. Vrahatis, and K. Schmitt. Particle swarm optimizers for pareto optimization with enhanced archiving techniques. In *Proceedings of the IEEE World Congress on Evolutionary Computation*, pp. 1780–1787, Canberra, Australia (December, 2003).

115. X. Li. A non-dominated sorting particle swarm optimizer for multi-objective optimization. In *Proceedings of the Genetic and Evolutionary Computation Conference*, pp. 37–38 (July, 2003).

116. M. R. Sierra and C. A. C. Coella. Improving pso-based multi-objective optimization using crowing, mutation and ϵ-dominance. In *Proceedings of the 3rd International Conference on Evolutionary Multi-criterion Optimization*, pp. 505–519, Guanajuato, Mexico, (2005).

117. J. E. Alvarez-Bevifez, R. m. Everson, and J. E. Fieldsend. A mopso algorithm based exclusively on pareto dominance concepts. In *Proceedings of the 3rd International Conference on Evolutionary Multi-criterion Optimization*, pp. 459–473, Guanajuato, Mexico, (2005).

118. S. L. Ho, E. W. C. Shiyou, Y. andLo, and H. C. Wong, A particle swarm optimization-based method for multi-objective design optimization, *IEEE Transactions on Magnetics.* **41**(5), 1756–1759, (2005).

119. M. A. Villalobos-Anias, G. T. Pulido, and C. A. C. Coello. A proposal to use stripes to maintain diversity in a multi-objective particle swarm optimizer. In *Proceedings of IEEE Swarm Intelligence Symposium*, pp. 22–29, Pasadena, California, USA, (2005).

120. M. Laumanns, L. Thiele, K. Deb, and E. Zitzler, Combining convergence and diversity in evolutionary multi-objective optimization, *IEEE Transactions on Evolutionary Computation.* **10**(3), 263–282, (2002).

121. M. Salazar-Lechuga and J. Rowe. Particle swarm optimization problems.

In *Proceedings of IEEE World Congress on Evolutionary Computation*, pp. 1204–1211, Edinburgh, Scotland, (2005).

122. D. E. Goldberg and J. Richardson. Genetic algorithms with sharing for multimdal function optimization. In *Proceedings of the Second International Conference on Genetic Algorithms*, pp. 41–49, (1987).

123. C. R. Raquel and P. C. N. Jr. An effective use of crowding distance in multi-objective particle swarm optimization. In *Proceedings of the Genetic and Evolutionary Computation Conference*, pp. 257–264, Washington DC, USA, (2005).

124. B. Zhao and Y. J. Cao, Multiple objective particle swarm optimization technique for economic load dispatch, *Journal of Zhejjang University Science*. **6A**(5), 420–427, (2005).

125. L. V. Santana-Quintero, A. G. H.-D. C. A. C. Coello, and J. M. O. Velazquez. Surrogate-based multi-objective particle swarm optimization. In *Proceedings of IEEE Swarm Intelligence Symposium*, pp. 1–8, St.Louis, MO, USA (september, 2008).

126. A. Lewis. The effect of population density on the performance of a spatial social network algorithm for multi-objective optimization. In *Proceedings of IEEE International Symposium on Parallel & Ditributed Processing*, pp. 1–6, Rome, Italy (may, 2009).

127. W. F. Leong and G. G. Yen. Impact of tuning parameters on dynamic swarms in pso-based multi-objective optimization. In *Proceedings of IEEE Congress on Evolutionary Computation*, pp. 1317–1324, Hong Kong (june, 2008).

128. A. Lewis and D. Ireland. Automated solution selection in multi-objective optimization. In *Proceedings of IEEE Congress on Evolutionary Computation*, pp. 2163–2169 (june, 2008).

129. L. Cagnina, S. Esquivel, and C. A. C. Coello, A particle swarm optimizer for multi-objective optimization, *JCS&T*. **5**(4), (2005).

130. E. F. G. Goldbarg, G. R. deSouza, and M. C. Goldbarg. Particle swarm optimization for the bi-objective degree constrained minimum spanning tree. In *IEEE Congress on Evolutionary Computation*, pp. 16–21, Sheraton Vancouver Wall Centre Hotel, Vancouver, BC, Canada (july, 2006).

131. S. Ho, W. Ku, J. Jou, M. Hung, and S. Ho. Intelligent particle swarm optimization in multi-objective problems. In *PAKDD 2006, LNAI 3918*, pp. 790–800, Springer-Verlag Berlin Heidelberg, (2006).

132. D. Ireland, A. Lewis, S. Mostaghim, and J. W. Lu. Hybrid particle guide selection methods in multi-objective particle swarm optimization. In *Proceedings of second IEEE International conference on e-Science and Grid Computing*, p. 116, (2006).

133. M. K. Oppena and C. Veenhuis, Multi-objective particle swarm optimization by fuzzy-pareto-dominance meta-heuristic, *International Journal of Hybrid Intelligent Systems*. **3**, 179–186, (2006).

134. S. Chiu, T. Sun, and S. Hsieh, Cross-searching strategy for multi-objective particle swarm optimization, *Expert System Application*. **37**(8), 5872–5886, (2010).

135. W. Leong and G. G. Yen, Pso-based multiobjective optimization with dynamic population size and adaptive local archives, *IEEE Transactions on Systems, Man, and Cybernetics.* **38**(5), 1270–1293, (2008).

136. WeiPeng and Q. Zhang. A decomposition-based multi-objective particle swarm optimization algorithm for continuous optimization problems. In *IEEE International Conference on Granular Computing*, pp. 534–537, Hangzhou (october, 2008).

137. N. Padhye and S. Branke, J.and Mostaghim. Empirical comparison of mopso methods guide selection and diversity preservation. In *IEEE Congress on Evolutionary Computation, CEC '09*, pp. 2516–2523, Trondheim, (2009).

138. Q. Jiang, M. Huang, and C. Wang. A novel method for finding good local guides in multi-objective particle swarm optimization. In *Third International Conference on Natural Computation, 2007*, pp. 737–741, Haikou (november, 2009).

139. C. Tsou, H. Fang, H. Chang, and C. Kao. An improved particle swarm pareto optimizer with local search and clustering. In *Proceedings of Simulated Evolution and Learning, LNCS 4247*, pp. 400–407, Springer Berlin (october, 2006).

140. J. C. F. Cabrera and C. A. C. Coello. Micro-mopso: A multi-objective particle swarm optimizer that uses a very small population size. In *Proceddings of Studies in Computational Intelligence, Vol . 261*, pp. 83–104, Springer Berlin (november, 2010).

141. C. K. Goh, K. C. B. Tan, D. S. B. Liu, and S. C. Chiamb, A competitive and cooperative co-evolutionary approach to multi-objective particle swarm optimization algorithm design, *European Journal of Operational Research,.* **202**, 42–52, (2009).

142. S. Tsai, T. Sun, C. Liu, S. Hsieh, W. Wu, and S. Chiu, An improved multi-objective particle swarm optimizer for multi-objective problems, *Expert Systems with Applications.* **37**(8), 5872–5886, (2010).

143. X. Zheng and H. Liu, A hybrid vertical mutation and self-adaptation based mopso, *Computers and Mathematics with Applications.* **57**, 2030–2038, (2009).

144. Y. Wang and Y. Yang, Particle swarm optimization with preference order ranking for multi-objective optimization, *Information Sciences.* **179**(12), 1944–1959, (2009).

145. Y. Bazi and F. Melgani. A multi-objective pso inflation methedology for svm regression with limited training samples. In *Proceedings of IEEE International Conference on Geo-Science and Remote Sensing Symposium*, pp. 4360–4363, Barcelona, (2007).

146. N. C. Chauhan, M. V. Kartikeyan, and A. Mittal. Design of rf window using multi-objective particle swarm optimization. In *Proceedings of International Conference on Recent Advances in Microwave Theory and Applications*, pp. 34–37, Jaipur, India, (2008).

147. R. Falcon, B. Depaire, K. Vanhoof, and A. Abraham. Towards a suitable reconciliation of the findings in collaborative fuzzy clustering. In *Proceed-*

ings of Eighth International Conference on Intelligent Systems Design and Applications, vol. 3, pp. 652–657, (2008).

148. A. B. de Carvalho, A. Pozo, S. Vergilio, and A. Lenz. Predicting fault proneness of classes through a multi-objective particle swarm optimization algorithm. In *Proceedings of 20th IEEE International Conference on Tools with Artificial Intelligene*, vol. 2, pp. 387–394, (2008).

149. J. E. Martin, M. F. Pantoja, A. R. Bretones, S. G. Garcia, C. M. de Jong van Coevorden, and R. G. Martin. Exploration of multi-objective particle swarm optimization on the design of uwb antennas. In *Proceedings of 3rd European Conference on Antennas and Propagation*, pp. 561–565, Berlin, Germany, (2009).

150. H. Pang and F. Chen. An optimization approach for intersection signal timing based on multi-objective particle swarm optimization. In *Proceedings of IEEE Coference on Cybernetics and Intelligent Systems*, pp. 771–775, Chegdu,China, (2008).

151. J. Hazra and A. K. Sinha, Congestion management using multi-objective particle swarm optimization, *IEEE Transactions on Power Systems*. **22**(4), 1726–1734, (2007).

152. S. N. Qasem and S. M. Shamsuddin, Radial basis function network based on time variant multi-objective particle swarm optimization for medical diseases diagnosis, *journal of Applied Soft Computing*. (2010).

153. M. N. Pindoriya, S. N. Singh, and S. K. Singh, Multi-objective mean-variance-skewness model for generation portfolio allocation in electricity markets, *Electric Power Systems Research*. **80**(10), 1314–1321, (2010).

154. D. Y. Sha and H. Lin, A multi-objective pso for job-shop scheduling problems, *Expert Systems with Applications*. **37**(2), 1065–1070, (2010).

155. I. Montalvo, J. Izquierdo, S. Schwarze, and R. Prez-Garca. Multi-objective particle swarm optimization applied to water distribution systems design: An approach with human interaction. In *Mathematical and Computer Modelling* (february, 2010).

156. Y. Liu, Automatic calibration of a rainfall-runoff model using a fast and elitist multi-objective particle swarm algorithm, *Expert Systems with Applications*. **36**(5), 9533–9538, (2009).

157. W. Zhang and Y. Liu, Multi-objective reactive power and voltage control based on fuzzy optimization strategy and fuzzy adaptive particle swarm, *International Journal of Electrical Power & Energy Systems*. **30**(9), 525–532, (2008).

158. J. Cai, X. Ma, Q. Li, L. Li, and H. Peng, A multi-objective chaotic particle swarm optimization for environmental/economic dispatch, *Energy Conversion and Management*. **50**(5), 1318–1235, (2009).

159. K. Deb and D. E. Goldberg. An investigation of niche and species formation in genetic function optimization. In *Proceedings of the Third International Conference on Genetic Algorithms*, pp. 42–50, San Mateo, California, USA (june, 1989).

160. Y. L. Zhen, L. H. Ma, L. Y. Zhang, and J. X. Qian. On the convergence analysis and parameter selection in particle swarm optimization. In *Proceedings of the Second International Conference on Machine Learning and Cybernetics*, pp. 1802–1807, (2003).

161. F. V. den Bergh. *An Analysis of Particle swarm Optimizer*. PhD thesis, Department of Computer Science, University of Pretoria, Pretoria, South Africa, (2002).

Chapter 11

Coherent Biclusters of Microarray Data by Imitating the Ecosystem: An Ant Colony Algorithmic Approach

D. Mishra[1,*], A. K. Rath[2,†] and M. Acharya[1]

[1]*Department of Computer Applications,*
Institute of Technical Education and Research,
Siksha 'O' Anusandhan University, Bhubaneswar, Orissa, India
* debahuti@iter.ac.in

[2]*Department of computer Science and Engineering,*
College of Engineering, Bhubaneswar, Orissa, India
[†] *amiyaamiya@rediffmail.com*

Microarray technology is a powerful tool for geneticists to monitor interactions among tens of thousands of genes simultaneously. There has been extensive research on coherent subspace clustering of gene expressions measured under consistent experimental settings. However, these methods assume that all experiments are run using the same batch of microarray chips with similar characteristics of noise. Algorithms developed under this assumption may not be applicable for analyzing data collected from heterogeneous settings, where the set of genes being monitored may be different and expression levels may not be directly comparable even for the same gene. In this chapter, we propose a biclustering model that imitates the ecosystem taking into account the features of biological data for mining subspace coherent patterns from heterogeneous gene expression data. We have implemented the system using an Ant colony algorithm (ACO). The algorithm decides the number of bi-clusters automatically. This processes the input biological data, runs the ACO, assigns biclusters to the genes and conditions simultaneously and displays the output.

11.1. Introduction to Microarray

Several companies have recently announced the availability of products that enable scientists to probe gene expression from the entire human genome on a single DNA microarray. The methodologies which are employed to

305

create such microarrays and the implications of the whole human genome microarray for future biological studies have made the achievement possible. The use of DNA arrays for monitoring gene expression has become a tool for molecular biologists. The ability of arrays to monitor thousands of separate but inter-related events simultaneously has captured the imagination of scientists. Most early microarrays for gene expression studies required cloning from complementary DNA (cDNA) sequences copied from isolated messenger RNAs (mRNAs), followed by the subsequent implication and deposition of PCR products from each clone.[1,2] Microarrays are also known as Biochips. Biochips are similar to semiconductors, except that instead of having electronic circuits, they have Biological material, DNA or RNA or protein, attached to the surface of a 'chip', which can be glass, plastic or silicon.

11.1.1. *Preparation of Microarrays*

Deoxyribonucleic acid, DNA, contains all necessary instructions for all living organisms. DNA is a double stranded and contains a Nucleoside base (Adenine, Cytosine, Guanine or Thymine) (See Fig. 11.1), a ribose sugar and a Phosphate group. DNA genes are differentially expressed, meaning that of the time the majority of the DNA is silent or unexpressed by being tightly twisted together.

But when a particular section of DNA is needed, it is expressed, or unwound so that the DNA can be read and the appropriate proteins are made. This is done by copying the DNA's instructions to RNA, specifically messenger RNA (mRNA). This process is called Transcription. The mRNA then translated the DNA 'message' into Protein.

11.1.2. *Design of Microarray or DNA Chip*

DNA chip or microarray in computer technology is a small piece of semiconducting material containing an electronic circuit. Such chips are usually less than five centimeters per slide. Their small size helps to make modern computers fast, compact and relatively inexpensive. There are three different methods for creating the microarray:

- Spotting long DNA fragments
- Array of prefabricated Oligonucleotides
- In-situ synthesis of Oligonucleotides

11.1.3. *How to Perform an Array Experiment*

The process can be described as follows: (See Fig. 11.2)

- A large array of cDNA or Oligonucleotide DNA sequences is fixed on a glass, nylon or quartz wafer. (Affymatrix Inc.).
- This array is then reacted generally with two series of mRNA probes (samples) that are labeled with two different colors of florescent probes.
- After hybridization of probes, the microarray is scanned using generally a laser beam to generate an image of all the spots. The intensity of the florescent signals at each spot is taken as a measure of the labels of the mRNA associated with specific sequence at that spot.
- The image of all the spots is analyzed using sophisticated software linked with information about the sequence of the DNA at each spot.

This then generates a general profile of gene expression level for the selected experimental and control conditions. Thus, in brief, a microarray experiments includes the following steps:

- Microarray Preparation
- Probe Preparation, Hybridization
- Low Level Information Analysis
- High Level Information Analysis

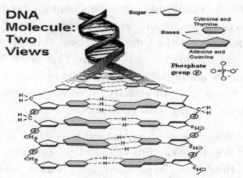

Fig. 11.1.: Structure of DNA.

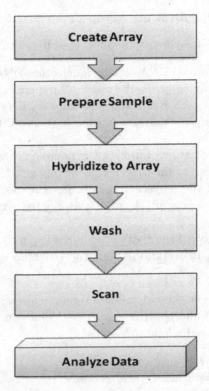

Fig. 11.2.: How to perform Array experiment.

11.1.4. *Importance of Gene Expression or Microarray Data*

Most cells in eukaryotic organisms contain genes that make up the entire Genome of the organism. Yet, these genes are selectively expressed in each cell depending on the type of cell and tissue and general conditions both within and outside the cell. Figure 11.3 given below shows a microarray. Thus,understanding of how expression of genes is selectively controlled has become the domain of activity in biological research. Some of the question arises when dealing with gene expression:

- How does gene expression level differ in various cell types?
- What are the functional roles of different genes?
- How their expression level varies in response to physiological changes within the cellular environment?
- How is gene expression affected by various diseases?

Fig. 11.3.: Microarray.

- Which genes are responsible for specific hereditary diseases?
- Which genes are affected by the treatment with pharmological agents such as drugs?

11.1.5. *Applications of Gene Expression or Microarray Data*

Gene expression or Microarray is generally applied for the following purposes:

- Tissue/cell phenotype analysis by multiple gene expression experiments.
- Identification of pathogens.
- Genome wide mutation survey and diagnosis of genotype polymorphisms and mutation.
- Elucidating the molecular and genetic bases of normal and abnormal conditions.
- Clinical tools for diagnosis, prognosis and treatment of diseases.
- Adaptation of microarray technology makes it possible for investigators to analyze the genetic pathways and dynamic interactions of genes in various diseases.

11.1.6. *Advantages of Gene Expression or Microarray Data*

Advantages of whole genomic Gene Expression data are:

- *Lower Cost*: Prior to the whole human genome microarray, most commercial offerings utilized are at least two microarrays to provide coverage of most of the human genome. Now, not only will one Microarray be less expensive than two, but processing will inherently require smaller amounts of costly reagents. Further, less time will be required to use detection instruments and associated software.

- *Streamlined Workflow:* Since researchers will need only to prepare and process one microarray instead of two or more per sample, they will have fewer steps in processing and tracking the sample and there will be less work associated with creating and analyzing the associated data derived from the microarray.
- *Greater Reproducibility:* Since, all the data for the entire genome will be generated from one microarray; there will be as a result, less variability in the data. When two or more microarrays associated with the same sample are processed separately, they are always questions of variability of the experimental conditions used to process each microarray.
- *Less sample use:* A smaller quantity of sample material is needed to perform an experiment.

11.2. Towards Biclustering of Microarray Data

DNA microarray technology has now made it possible to simultaneously monitor the expression levels of thousands of genes during important biological processes and across collections of related samples. Elucidating the patterns hidden in gene expression data offers a tremendous opportunity for an enhanced understanding of functional genomics. However, the large number of genes and the complexity of biological networks greatly increase the challenges of comprehending and interpreting the resulting mass of data, which often consists of millions of measurements. A first step towards addressing this challenge is the use of biclustering techniques, which is essential in the data mining process to reveal natural structures and identify interesting patterns in underlying data.

Biclustering is the most popular approach of analyzing gene expression data and has proven successful in many applications, such as discovering gene path way, gene classification and function prediction. There is a very large body of literature on biclustering in general and on applying biclustering techniques in microarray data in particular. Several representative algorithmic techniques have been developed and experimented in clustering gene expression data, which include but are not limited to hierarchical clustering,[3] self organizing maps,[4] and graph theoretic approaches e.g., CLICK.[5]

Recent studies have focused on the problem of discovering hidden module structures in large expression matrices. This involves simultaneous clustering of genes and conditions and is thus an instance of biclustering. Using that terminology, the modules we seek can be referred to as biclusters. The aim of biclustering is to identify subset pairs (each pair consisting of a subset of genes and a subset of conditions) by clustering both the rows and the columns of an expression matrix.

11.2.0.1. *What is Biclustering?*

Biclustering is simultaneous clustering of both rows and columns of a data matrix.Here, we consider a r by c data matrix, A$=(x, y)$, where

$$X = (X_1,\ldots,X_r) = \text{set of } r \text{ rows and}$$
$$Y = (Y_1,\ldots,Y_c) = \text{set of } c \text{ columns,}$$

a_{ij} = numeric value (discrete or real) representing the relation between row i and column j.

Table 11.1.: Gene Expression Matrix.

Gene	Condition 1	...	Condition j	...	Condition c
Gene$_1$	a_{11}	...	a_{1j}	...	a_{1c}
Gene...	...	...	...	...	...
Gene$_i$	a_{i1}	...	a_{ij}	...	a_{ic}
Gene...	...	...	...	...	...
Gene$_r$	a_{r1}	...	a_{rj}	...	a_{rc}

In case of gene expression matrices

$$X = \text{set of genes and}$$
$$Y = \text{set of conditions}$$

a_{ij} = expression level of gene i under condition j (real value). Here, a_{11} to arc are expression levels. For exmple,$a_1 1$ is the expression level of gene G_1 under condition 1 and $a_1 2$ is the expression level of gene G_2 under condition 2. Gene expression matrices have been extensively analyzed in two dimensions:

- Gene dimension
- Condition dimension

11.2.0.2. *Example of Biclustering*

In the example:
$X = (G_1, G_2, G_3, G_4, G_5, G_6)$
$Y = (C_1, C_2, C_3, C_4, C_5, C_6, C_7, C_8, C_9, C_10)$
$I = (G_2, G_3, G_4)$
$J = (C_4, C_5, C_6)$
Cluster of columns $(X, J) = (C_4, C_5, C_6)$ and Cluster of Rows $(I, Y) = (G_2, G_3, G_4)$, Bicluster $(I, J) = [(G_2, G_3, G_4), (C_4, C_5, C_6)]$, where, $A(X, Y) =$ a matrix; $I =$ subset of rows and $J =$ subset of columns. $(I, Y) =$ a subset of rows that exhibit similar behavior across the set of all columns $=$ cluster of rows. $(X, Y) =$ a subset of columns that exhibit similar behavior across set of all rows $=$ cluster of columns. $(I, J) =$ is a bicluster i.e., subset of genes and subset of conditions, where the genes exhibit similar behavior across the conditions and vice-versa.

11.2.1. *Goal of Biclustering*

- Identify subgroups of genes and subgroup of conditions, where the genes exhibit highly correlated activities for every condition[9]
- Identify sub-matrices with interesting properties.
- Perform simultaneous clustering on the rows and column dimensions of the gene expression matrix.

11.2.1.1. *Why Biclustering?*

- Clustering can be applied to either the rows or the columns of the data matrix, separately.
- Biclustering methods are applied to perform clustering in the two dimensions simultaneously.
- Clustering methods derive a global model where as bi-clustering methods derive a local model.
- As global model use all the conditions and all the genes to define each gene and each condition respectively.
- Local model uses a specific subset to bicluster the genes and conditions.

11.2.1.2. *Problem Formulation*

Here we want to identify a set of biclusters $B_K = (I_K, J_K)$ each bicluster B_K must satisfy some specific characteristics of homogeneity.

11.2.2. *Types of Bicluster*

A biclustering algorithm concerns the identification of the type of bicluster the algorithm is able to find. They are:

- Bicluster with constant values.
- Bicluster with constant values on rows or columns.
- Bicluster with coherent values.
- Bicluster with coherent evolutions.

11.2.2.1. *Bicluster with Constant Values*

A perfect constant bicluster is a submatrix (I, J) where all the values within the bicluster are equal for all i *in* I *and* j *in* J. $a_{ij} = \mu$

1.0	1.0	1.0	1.0
1.0	1.0	1.0	1.0
1.0	1.0	1.0	1.0
1.0	1.0	1.0	1.0

Fig. 11.4.: Bicluster with constant values.

11.2.2.2. *Bicluster with Constant Values on Rows and Columns*

A bicluster with constant row is a sub-matrix (I, J), where all the values with in the bicluster can be obtained using one of the following expression.

$$a_{i}j = \mu + \alpha_i$$
$$a_{i}j = \mu * \alpha_i$$

Where, μ is the typical value within the bicluster and α_i is the adjustment for row i *in* I. A bicluster with constant columns is a sub matrix (I, J) where all the values within the bicluster can be obtained using one of the following expressions.

$$a_{i}j = \mu + \beta_i$$
$$a_{i}j = \mu * \beta_i$$

Where μ is the typical value within the bicluster and β_j is the adjustment for column $j \in J$.

1.0	1.0	1.0	1.0
2.0	2.0	2.0	2.0
3.0	3.0	3.0	3.0
4.0	4.0	4.0	4.0

1.0	2.0	3.0	4.0
1.0	2.0	3.0	4.0
1.0	2.0	3.0	4.0
1.0	2.0	3.0	4.0

(a) Bicluster with constant rows   (b) Bicluster with constant columns

Fig. 11.5.: Bicluster with constant values on rows and columns.

11.2.2.3. *Bicluster with Coherent Values*

- These approaches view the elements of the matrix as symbolic values and try to discover subsets of rows and subjects of columns with coherent behaviors regardless of the exact numeric values in the data matrix.
- A bicluster (I, J), defined using an additive model is a subset of rows and a subset of columns, where values are predicted using the following expression:

$$a_i j = \mu + \alpha_i + \beta_j$$

- Where μ is the typical value within the bicluster, i is the adjustment for row *iin* I and j is the adjustment for row j *in* J.
- A bicluster (I, J) defined using a multiplicative model is a subset of rows and a subset of columns, whose values a_{ij} are predicted using the following expression:

$$a_i j = \mu * \alpha_i * \beta_j$$

1.0	2.0	5.0	0.0
2.0	3.0	6.0	1.0
4.0	5.0	8.0	3.0
5.0	6.0	9.0	4.0

1.0	2.0	0.5	1.5
2.0	4.0	1.0	3.0
4.0	8.0	2.0	6.0
3.0	6.0	1.5	4.5

(a) Additive model   (b) Multiplicative model

Fig. 11.6.: Bicluster with coherent values.

S1	S1	S1	S1
S1	S1	S1	S1
S1	S1	S1	S1
S1	S1	S1	S1

S1	S1	S1	S1
S2	S2	S2	S2
S3	S3	S3	S3
S4	S4	S4	S4

(a) Overall coherent evolutions   (b) Coherent evolutions on rows

Fig. 11.7.: Bicluster with coherent evolutions.

11.2.2.4. *Bicluster with Coherent Evolutions*

- These approaches view the elements of the matrix as symbolic values and try to discover subsets of rows and subsets of columns with coherent behaviors regardless of the exact numeric values in the data matrix.
- The co-evolution property can be observed as:
  - On the entire bicluster.
  - On the rows of the bicluster.
  - On the columns of the bicluster.

11.3. Bicluster Structure

Biclustering algorithm, assumes the existence of several biclusters in the data matrix, the following bicluster structures can be obtained.

- Single bicluster
- Exclusive row and column biclusters (rectangular diagonal blocks after row and column recorded).
- Non-overlapping biclusters with checkerboard structure.
- Exclusive-rows biclusters.
- Exclusive-column biclusters.
- Non-overlapping biclusters with tree structure.
- Non-overlapping non-exclusive biclusters.
- Overlapping biclusters with hierarchical structure.
- Arbitrarily positioned overlapping biclusters.

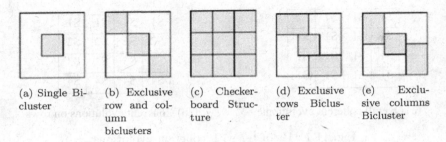

(a) Single Bicluster   (b) Exclusive row and column biclusters   (c) Checkerboard Structure   (d) Exclusive rows Bicluster   (e) Exclusive columns Bicluster

Fig. 11.8.: Bicluster Structure(1).

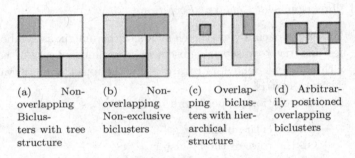

(a) Non-overlapping Biclusters with tree structure   (b) Non-overlapping Non-exclusive biclusters   (c) Overlapping biclusters with hierarchical structure   (d) Arbitrarily positioned overlapping biclusters

Fig. 11.9.: Bicluster Structure(2).

11.3.1. *Biclustering Algorithms*

- Biclustering algorithms may have two different objectives: to identify one or a given number of biclusters.
- Some approaches attempt to identify one bicluster at a time.
- Ex: Cheng and Church[10] and Sheng *et al.*[11] identify a bicluster at a time, mask it with random numbers, and repeat the procedure in order to eventually find other bicluster.
- Other bicluster approaches discover one set of biclusters at a time.
- The complexity of the bicluster algorithm may depend on:
  - The exact problem formulation.
  - The merit function used to evaluate the quality of a given bicluster.
- In simplest form the data matrix A is a binary matrix.
- Every element a_{ij} is either 0 or 1.

- It can be modeled as a bipartite graph. When this case is found out, a bicluster corresponds to a biclique.
- Finding maximum size bicluster is therefore equivalent to finding the maximum edge biclique in a bipartite graph, which is a NP-complete problem.[12]
- A number of different heuristic approaches have been used to identify biclusters. They can be divided into five classes:

  - Iterative row and column clustering combination.
  - Divide and conquer
  - Greedy iterative search
  - Exhaustive bicluster enumeration.
  - Distribution parameter identification.

1. Iterative Row and Column Clustering Combination

- Apply clustering algorithms to the rows and columns of the data matrix, separately.
- Combine the results using some sort of iterative procedure to combine the two cluster arrangements.

2. Divide And Conquer

- Break the problem into several sub problems that are similar to the original problems but smaller in size.
- Solve the problem recursively.
- Combine the intermediate solutions to create a solution to the original problem.
- Usually break the matrix into sub matrices (biclusters) based on a certain criterion and then continue the biclustering process on the new sub matrices.

3. Greedy Iterative Search

- Always make a locally optimal choice in the hope that this choice will lead to globally good solution.
- Usually perform greedy row/column addition/removal.

4. Exhaustive Bicluster Enumeration

- A number of methods have been used to speed up exhaustive search
- In some cases the algorithms assume restrictions on the size of the bicluster that should be listed

5. Distribution Parameter Identification

- Distribution parameter identification biclustering approaches assume a given statistical model and try to identify the distribution parameters used to generate the data by minimizing a certain criterion through an iterative approach.

11.3.2. *Applications of Biclustering*

- Even though most recent applications of biclustering are in biological data analysis, there exists much other possible application in very different application domains. Examples of these application areas are: information retrieval and text mining, collaborative filtering, recommendation systems, target marketing, database research, data mining and even analysis of electoral data.
- Biclustering techniques can be used in collaborative filtering to identify subgroups of customers with similar preferences or behaviors towards a subset of products with the goal of performing target marketing or use the information provided by the bicluster in recommendation systems.
- Recommendation systems and target marketing are important application in the E-commerce area. In these applications the goal is thus to identify sets of customers with similar behavior so that we can predict the customer's interest and make proper recommendations.
- In information retrieval and text mining, biclustering can be applied to identify subgroups of documents with similar properties relatively to subgroups of attributes, such as words or images. This information can be very important in query and indexing in the domain of search engines.
- The application of biclustering use data matrices with electoral data and try to identify biclusters to discover subgroup of rows with the same political ideas and electoral behaviors among a subject of the attributes considered.

11.3.2.1. *Biological Applications*

- Cheng and Church[10] applied bicluster to two gene expression data matrices, specifically to the yeast saccharomyces cerevisae cell cycle data with 2884 genes and 17 conditions.

- The human B-cells expression data 4026 genes and 96 conditions.[10]
- GetZ et al.,[13] applied biclustering to two gene expression data matrices containing cancer data.
- Bun-Dor et al.,[14] used a breast tumor dataset with gene expression data from 3226 genes under 22 experimental conditions.

11.3.3. *Procedures for Biclustering Analysis*

The procedures for bicluster analysis include:

- Feature selection
- Bicluster algorithm selection
- Bicluster validation
- Result interpretation

The intimately connected steps of cluster analysis with feedback pathway[9] are shown in the following Fig. 11.10.

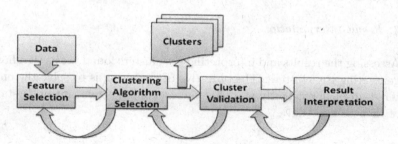

Fig. 11.10.: Procedures for biclustering analysis.

1. *Feature Selection*

Microarray experiments provide expression information of large number of genes (from 103 to 104 or 105). It is essential to consider which feature (gene) subset will be employed in biclustering analysis, by eliminating the least interesting and highlight the most interesting genes. Distinctive features from a set of candidates are neatly selected, while feature extraction exploits some alteration to produce useful and novel features from the original ones which are very essential to the efficiency of biclustering purpose.[9]

2. Bicluster Algorithm Selection

Different biclustering algorithms and methods have been developed to improve the preceding ones.[15] There is no absolute clustering method that can be universally used to solve all problems. So in order to select or generate a suitable clustering strategy, it is vital to investigate the features of the problem. As in Ref. 9 revealed the step is usually combined with the selection of a corresponding proximity measure and the construction of a criterion function. Patterns are grouped according to whether they resemble each other. Once a proximity measure is chosen, the construction of a biclustering criterion function makes the partition of biclusters an optimizing problem.

3. Bicluster Validation

Finding the number of biclusters in a dataset and many of this method have been proposed some of which are in Ref. 12 for gene expression data which evaluates the partitions generated using biclustering algorithm and find the preeminent partition based on intracluster and intercluster distance.

4. Result Interpretation

Assessing the results and interpreting the clusters found are as significant as generating the clusters. The objective of clustering is to solve the encountered problem efficiently and offer the users with significant understanding of their original data.

11.3.4. Related Research

Because of the characteristics of the DNA chip data such as the high complexity,large in amount, and variable properties, we propose a biclustering model which uses an algorithm that imitates the ecosystem. There are currently many algorithms that imitate the ecosystem. The Genetic algorithm, Neural Network algorithm, Particle Swarm algorithm and Ant Colony algorithm are the most popular algorithms. The genetic algorithm (GA) is a search technique used in computer science to find approximate solutions to optimization and search problems. Specifically it falls into the category of local search techniques and is therefore generally an incomplete search. Genetic algorithms are a particular class of evolutionary algorithms that use techniques inspired by evolutionary biology such as inheritance, mutation, selection, and crossover (also called recombination).[12] Neural Networks (NNs) is an information processing paradigm that is inspired by the way biological nervous systems, such as the brain process information. The key element of this paradigm is the novel structure of the information processing system. It is composed of a large number of highly interconnected processing elements (neurons) working in union to solve specific problems. An ANN is configured for a specific application, such as pattern recognition or data classification, through a learning process. Learning in biological systems involves adjustments to the syntactic connections that exist between the neurons.[16]

Particle Swarm Optimization (PSO) is a recently proposed algorithm by Kennedy and Eberhart in 1995, motivated by social behavior of organisms such as bird flocking and fish schooling. PSO is an optimization tool, provides a population-based search procedure in which individuals called particles change their position (state) with time. In a PSO system, particles fly around in a multidimensional search space. During flight, each particle adjusts its position according to its own experience, and according to the experience of a neighboring particle, making use of the best position encountered by itself and its neighbor.[17] The Ant Colony Optimization algorithm that our system uses is a probabilistic technique for solving computational problems which can be reduced to finding good paths through graphs.[18] There are some works on clustering algorithms based on the ACO. Yuqing *et al.* proposed the K-means clustering algorithm based on Ant Colony[19] and Xiao *et al.*, proposed gene clustering using self-organizing maps and particle swarm optimization.[20] Handl *et al.* proposed Ant-based clustering.[21] Cluster analysis using ants and also algorithms for clustering using ants has been reviewed in.[26–28] Our model extends this work to adopt Ant Colony Optimization (ACO) for Microarray data analysis.

11.3.5. *Goal*

Objective is to cluster objects that exhibit similar patterns on a subset of dimensions both column and row wise.

- We propose a new biclustering model, to capture not only the closeness of objects but also the similarity of patterns exhibited by the objects.
- Our model uses ACO for designing Meta heuristic algorithms for combinatorial optimization problem.

- Our model is a generalized subspace clustering. However, it finds much broader range of applications, including DNA array analysis and collaborative filtering, where pattern similarity among a set of objects carry significant meanings.
- Our method is deterministic in that it discovers all qualified clusters, while the bicluster approach is a random algorithm that provides only an approximate answer.

11.3.6. *Problem Definition*

Our research is highly related to pattern based clustering. Cheng and Church[22] introduced *bicluster* model. Given a subset of objects I and a subset of attributes J, the coherence of the sub matrix (I,J) is measured by the *mean squared residue score(H)*.

$$H(I,J) = \frac{1}{|I||J|} \sum_{i \in I, j \in J} (d_{ij} - d_{iJ} - d_{Ij} + d_{IJ})^2 \qquad (11.1)$$

Where a_{ij} is the value of object i on j, a_{iJ} is the average value of row i, a_{Ij} is the average value of column j, and a_{IJ} is the average value of the sub matrix (I, J). The problem of biclustering is to mine sub matrices with low mean squared residue scores.

If the bicluster B (I, J) is perfect, the H (I, J) equals to zero. However, there always exists noise. In order to find the bicluster with noise, according to Cheng and Church,[13] given a matrix X (G, C) and a threshold δ, find the largest sub matrix $B(I, J)$ from the matrix which satisfies:

$$H(I,J) < \delta \qquad (11.2)$$

As a typical application, a microarray data set can be modeled as a numerical data matrix recording the expression levels of genes on samples. An important task of analyzing microarray data is to find co-expressed genes and phenotypes. A group of co-expressed genes are the ones that demonstrate similar expression patterns over a substantial subset of samples, and the subset of samples may correspond to some *phenotype*.

Moreover, given a microarray data set, a gene can belong to more than one co-expressed gene group. Since, it may correlate to more than one phenotype or a sample can manifest more than one phenotype, such as tumor vs. normal tissues and male vs. female samples. To address the novel requirements, recently, a new theme of coherent pattern based biclustering, is being proposed.

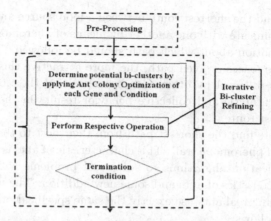

Fig. 11.11.: Model for finding coherent biclusters using Ant Colony Optimization Algorithm.

In this section we propose a model of mining coherent pattern based biclusters, based on a user specified quality/utilization function. Our model uses ACO algorithms to show the potentiality of using artificial pheromone and artificial ants to derive the search of always better solutions for complex optimization problem.

11.3.7. *Proposed Model*

In this section, the Fig. 11.11 shows the proposed model for finding the biclusters by applying the ACO algorithm in both gene and condition dimension. In the first step, the preocessing techniques such as, selecting relvant conditions, reducing the attributes, applying normalization to scale the data as per our requirement, discretizing the attribute values to find the regular intervals for further processing. In the second step, we want to find the potential biclusters by using the ACO algorithm in both the gene and condition direction. In the third step, the respective operation is carried out iteratively refining the found bicllusters if the constraint is satisfied, else the total process is terminated.

11.3.8. *Ant colony optimization*

Ant colony optimization is Meta heuristic, a novel population-based approach was recently proposed in 1992 by Marco Dorigo *et al.* to solve several discrete optimization problems.[23] The ACO mimics the way real ants find the shortest route between a food source and their nest. The ants communicate with one another by means of pheromone trails and exchange information about which path should be followed. The more the number of ants traces a given path, the more attractive this path (trail) becomes and is followed by other ants by depositing their own pheromone. This auto catalytic and collective behavior results in the establishment of the shortest route.

Ants find the shortest path from their nest to the food source with the help of pheromone trail. This characteristic of ants is adapted on ant colony optimization algorithms to solve real problems with using exactly some characteristics of ants and some new addition. The method is improved by modeling real ants use exactly the same specifications taken from real ants is as follows:

- The communication established with ants through pheromone trail
- Paths deposited more pheromone preferred previously
- Pheromone trail on short paths increase more rapidly

Addition of new specifications to this new technique is as follows:

- They live in an environment where time is discrete
- They will not be completely blind, they will reach the details about the problem
- They will keep information formed for the solution of the problem which has some memory.

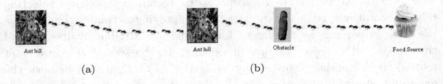

Ant hill   Ant hill   Obstacle   Food Source

(a)   (b)

Fig. 11.12.: Behaviors of real ants between their nest and food source(1).

As shown in Fig. 11.12(a), ants start from their nest and goes along a linear path through the food source.

Figures 11.12 and 11.13 shows behaviors of real ants between their nest and food source.

(Fig. 11.12a) Ants following a path between their nest and food source

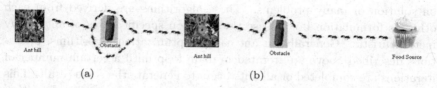

(a)   (b)

Fig. 11.13.: Behaviors of real ants between their nest and food source(2).

(Fig. 11.12b) Encountering an obstacle of ants
(Fig. 11.13a) Selection of ants
(Fig. 11.13b) Finding the shortest path of ants

Actually, if there exists a difficulty on the path while going to the food source (Fig. 11.12(b)), ant lying in front of this difficulty can not continue and has to account a preference for the new outgoing path. In the present case, selection probability of the new direction alternatives of ants is equal. In other words, if ant can select anyone of the right and left directions, the selection chance of these directions is equal (Fig. 11.13(a)). Namely, two ants start from their nest in the search of food source at the same time to these two directions. One of them chooses the path that turns out to be shorter while the other takes the longer path. But it is observed that following ants mostly select the shorter path because of the pheromone concentration deposited mostly on the shorter one.

The ant moving in the shorter path returns to the nest earlier and the pheromone deposited in this path is obviously more than what is deposited in the longer path. Other ants in the nest thus have high probability of following the shorter route. These ants also deposit their own pheromone on this path. More and more ants are soon attracted to this path and hence the optimal route from the nest to the food source and back is very quickly established. Such a pheromone-meditated cooperative search process leads to the intelligent swarm behavior.

The instrument of ants uses to find the shortest path is pheromone. Pheromone is a chemical secretion used by some animals to affect their own species. Ant deposit some pheromone while moving, they deposit some amount of pheromone and they prefer the way deposited more pheromone than the other one with a method based on probability. Ants leave the pheromone on the selected path while going to the food source, so they help following ants on the selection of the path (Fig. 11.13(b)). There are many algorithms derived from ant colony meta-heuristic and they are used

on solution of many problems. These algorithms are derived from each
other as formulation but all use the common specifications of ant colony
meta-heuristic. Generally, in ant colony optimization algorithms, operations described above are iterated in main loop until a certain number of
iterations are completed or all ants begin to generate the same result. This
situation is named as stagnation behavior, because after a point, algorithm
finishes to generate alternative solutions. The reason of this situation is,
after a certain number of iterations, ants generate continuously the same
solutions because pheromone amount intensifies in some points and the
difference between pheromone concentrations on paths become very huge.
Most ant colony optimization use this algorithmic diagram demonstrated
as follows: Initiation of the parameters which determines the pheromone
trail

While (until result conditions supplied) *do*
Generate Solutions
Apply Local Search
Update Pheromone Trail
End

11.3.9. *Biclustering with Ant Colony Optimization*

In this section, ant colony optimization algorithm has been used aiming to
solve the data biclustering problem and proposed a new technique and is
explained in detail and the solutions are compared.

We use an ACO algorithm for data clustering, in which a set of con-
current distributed agents collectively discover a sensible organization of
objects for a given dataset.[24] In the algorithm used in this study, each
agent discovers a possible partition of objects in a given dataset and the
level of partitioning is measured subject to some metric like Euclidean distance. Information associated with an agent about clustering of objects is
accumulated in the global information hub (pheromone trail matrix) and
is used by the other agents to construct possible clustering solutions and
iteratively improve them. The algorithm works for a given maximum number of iterations and the best solution found with respect to a given metric
represents an optimal or near-optimal partitioning of objects into subsets
in a given dataset.

The aim of data clustering is to obtain optimal assignment of N objects
in one of the K clusters where N is the number of objects and K is the
number of clusters. Artificial ants used in algorithm are named as software

Table 11.2.: Illustrative dataset to explain ACO algorithm for clustering with $N=10$ and $n=4$ (N: number of agents, n: number of attributes).

Gene	S_0	S_1	S_2	S_3	S_4	S_5	Cluster
G_0	0.15	-0.07	-0.25	-0.3	-1.12	-0.67	1
G_1	0.21	0.03	0.18	-0.27	-0.32	0.62	2
G_2	-0.03	-0.07	0.28	0.32	-0.27	-0.36	3
G_3	-0.25	0.58	0.77	0.28	0.32	0.65	1
G_4	0.11	0.04	0.75	0.82	0.21	-0.2	1
G_5	0.24	0.31	0.95	0.12	0.18	0.69	2
G_6	-0.3	0.22	0.02	-0.64	0.06	-0.04	2
G_7	-0.15	-0.25	0.18	0.06	-0.15	-0.17	2
G_8	0	-0.74	-0.38	0.87	-0.34	0.12	2
G_9	-0.15	0.2	0.31	0.15	0.04	-0.22	2

ants or agent and number of agents expressed with R. Ants start with empty solution strings and in the first iteration the elements of the pheromone matrix are initialized to the same values. With the progress of iterations, the pheromone matrix is updated depending upon the quality of solutions produced. To describe the algorithm in detail, a data set with 10 test data is formed. The data of this test data set are obtained from UCI's machine learning repository.[25] Test data are shown in Table 11.2 and in real data set, data are divided into 3 subsets, so $K=3$.

Each agent used in this model is defined with solution strings expressed with S of length N and at start of the algorithm solution string of each element is empty. Each element of string corresponds to one of the test samples and its value includes which bicluster it will be assigned. For example, a representative solution string, S_4 in Table 11.4 constructed for $N=10$ and $K=3$ is given as follows:

$$S_4 : [\, 2\ 1\ 3\ 2\ 2\ 3\ 2\ 1\ 2\ 3\,]$$

This means, the first element of the above test string S_4 is assigned to second cluster; the second element is assigned to first cluster and so on.

To construct a solution, the agent uses the pheromone trail information to allocate each element of string S to an appropriate cluster label. At the start of the algorithm, each agent or software ant starts with empty solution string and the pheromone matrix keeping each element is assigned to which cluster is initialized to some small value 0. Hence, at first iteration, each element of solution string S of each agent is assigned randomly to one of the K clusters.

The trail value, τ_{ij} at location (i, j) represents the pheromone concentration of sample i associated to the cluster j. So, for the problem of separating N samples into K clusters the size pheromone matrix is $N * K$. Thus, each sample is associated with K pheromone concentrations. The pheromone trail matrix evolves as we iterate. At any iteration level, each agent or software ants will develop solutions showing the probability of each ant belonging to which cluster using this pheromone matrix.

After generating the solutions of R agents, a local search is performed to further improve fitness of these solutions. The pheromone matrix is then updated depending on the quality of solutions produced by the agents. Then, the agents build improved solutions depending on the pheromone matrix and the above steps are repeated for certain number of iterations. At the end of any iteration level each agent generates the solution using the information derived from updated pheromone matrix. The pheromone matrix at any iteration level for test dataset is shown in Table 11.3.

The pheromone concentrations for the first sample as shown in Table 11.3 are: $\tau_{11} = (0.014756)$, $\tau_{12} = (0.015274)$ and $\tau_{13} = (0.009900)$. It indicates that at the current iteration, sample number 1 has the highest probability of belonging to cluster number 2, because τ_{12} is the highest.

Each agent selects a cluster number with a probability value for each element of S string to form its own solution string S. The quality of constructed solution string S is measured in terms of the value of objective function for a given data-clustering problem. This objective function is defined as the sum of squared Euclidian distances between each object and the center of belonging cluster. Then, the elements of the population, namely agents are sorted increasingly by the objective function values. Because, the lower objective function value, the higher fitness to the real solution, namely, lower objective function values are more approximated to real solution values. Table 11.4 shows the solution string values of ten agents in the test data set and the fitness values of each agent sorted decreasingly.

Most of existing ant colony optimization algorithms uses some local search procedures to develop the generated solutions discovered by software ants. Local search helps to generate better solutions, if the heuristic information can not be discovered easily. Local search is applied on all generated solutions or on a few percent R. In this chapter, local search is performed on 20% of the total solutions. So in the test data set of 10 data, local search is applied on the top 2 solutions as shown in Table 11.4. In the local search procedure, the objective function values of top 2 agents are computed again. These solutions can be accepted only if there is an

Table 11.3.: Pheromone trail matrix generated at any iteration level of the ACO algorithm for test dataset.

N Sample No/ (K Cluster No)	1	2	3
1	0.014756	0.015274	0.009900
2	0.015274	0.009900	0.014756
3	0.015274	0.014756	0.009900
4	0.009900	0.015274	0.014756
5	0.014756	0.015274	0.009900
6	0.009900	0.014756	0.015274
7	0.009900	0.020131	0.009900
8	0.015274	0.014756	0.009900
9	0.009900	0.015274	0.014756
10	0.014756	0.015274	0.009900

Table 11.4.: For data-clustering problem generated solutions sorted decreasingly.

S Solution String/ N Sample No	1	2	3	4	5	6	7	8	9	10	F(Fitness)
1	2	1	1	2	2	3	3	1	2	2	4.003931
2	2	3	1	2	2	3	2	3	2	2	7.172357
3	2	1	1	2	2	3	2	1	2	3	7.864054
4	2	1	3	2	2	3	2	1	2	3	8.455329
5	2	2	1	2	2	3	2	1	2	2	10.36714
6	2	1	1	2	3	3	2	1	1	3	10.92255
7	1	1	1	2	2	3	2	1	2	3	11.94087
8	2	1	1	2	1	3	2	1	1	1	12.00959
9	1	1	2	2	2	3	1	1	2	2	13.26286
10	1	1	2	2	2	3	3	1	2	3	13.33634

improvement on the fitness, namely, if the newly computed objective function value is lower than the first computed value, newly generated solution replaces the old one.

After the local search procedure, the pheromone trail matrix is updated. Such a pheromone updating process reflects the usefulness of dynamic information provided by software ants. The pheromone matrix used in ant colony optimization algorithm is a kind of adaptive memory that contains information provided by the previously found superior solutions and is up-

dated at the end of the iteration. The pheromone updating process used
in this algorithm includes best L solutions discovered by R agents at iteration level t. This L agent mimics the real ants' pheromone deposition by
assigning the values of solutions. The trail information is updated using
the following rule as

$$\tau_{ij}(t+1) = (1 - \rho)\tau_{ij}(t) + \sum_{l=1}^{L} \Delta\tau_{ij}^l \qquad (11.3)$$

where $i=1,\ldots,N$ and $j=1,\ldots,K$, where ρ is a persistence or trail and lies
between $[0, 1]$ and $(1-\rho)$ is the evaporation rate. Higher value of ρ suggests
that the information gathered in the past iterations is forgotten faster. The
amount of $\Delta\tau_{ij}^l$ is equal to $1/F_l$, if cluster j is assigned to i^{th} element of
the solution constructed by ant l and zero otherwise.

An optimal solution is that solution which minimizes the objective function value. If the value of best solution in memory is updated with the best
solution value of the current iteration if it has a lower objective function
value than that of the best solution in memory, otherwise the best solution
is kept in memory. This process explains that an iteration of the algorithm
is finished. Algorithm iterates these steps repeatedly until a certain number of iterations and solution having lowest function value represents the
optimal partitioning of objects of a given dataset into several groups.

Ants follow the path between their nest and the food source according to
the pheromone amount deposited on the path. Following ants decides which
path to go depending on the pheromone concentrations on the path. After a
number of iterations ants follow continuously the same path because of the
enormous pheromone concentration present, than the not used paths. This
behavior of ants is called stagnation behavior as explained before. To avoid
from this disadvantage, reference algorithm is improved with the addition
of two new techniques and the solutions are compared with each other.

For the purpose of increasing the working performance of the model
developed to cluster data with ant colony optimization technique, proposed
model brings the pheromone amount to initial values every 50 iteration to
avoid from stagnation behavior. In other words, clustering is performed in
a progressive technique.

Aiming minimize the stagnation behavior of ants, the proposed model
follows the pheromone amounts of ants and if there is no change on the
pheromone concentration of every path after last 10 iterations, it brings
the pheromone amount to initial values. In other words, to improve the
solution, a feedback technique is applied on the algorithm.

With the aim of generating the optimal solutions of the presented SOMs and ACO algorithm developed for solving data biclustering problem and added two new techniques, an application program is written with MAT-LAB 7.0 and the program is applied on the iris database existing in the data warehouse of UCI.[25] The iris database consists of 150 data and it is stored in a text file. Our aim is to compare the actual cluster values and the new values produced by the ACO algorithm. The main application program has number of iterations, biclusters, agents, local search agents and initial pheromone values evaporation rate of pheromone and some values needed for the algorithm are specified. Program runs the algorithm until a number of iterations supplied by user.

Table 11.5.: Parameters taken into consideration.

Attribute Name	No.of Clusters	No.of Agents	No.of Clusters	Q(Thres- -hold Value)	Evaporation Rate Value	Initial Pheromone
S1,S2,S3, S4,S5,S6	3	10	2	100 (User Supplied)	0.98	0.01

Table 11.6 shows the statistical result values worked on the application program with 100 iterations with two solutions. 'Solution1' represents initial biclusters in SPECT heart database and its performance is 52% and 'Solution2' represents our proposed model and its performance is 80%.

Figure 11.14, shows the graph screen of these two methods worked on the application program with 100 iterations and the given criterion (see Table 11.5) Curve specifying the 'Solution1' shows the initial biclusters in iris database and its working performance is 52% and curve specifying the 'Solution2' shows the proposed technique's results and its working performance is 80% (see Fig. 11.14).

11.4. Conclusion

In this chapter, ant colony optimization algorithm developed for data biclustering problem is verified on an application program and to increase the working performance of the ACO algorithm, the new techniques are proposed and these new techniques are also verified on the application program. With the comparison of these two methods, it is shown that the

Table 11.6.: Statistical results values of the ACO methods worked with the parameters of Table 11.5.

Iteration No	Fitness Value	Real Fitness Value	Solution1	Solution2
98	135.72	10051	603, 21012, 40	373, 14272, 77
			622, 20776, 41	374, 14239, 77
			651, 20458, 43	415, 14235, 77
			673, 20280, 44	586, 13876, 79
			675, 20221, 44	627, 13870, 79
			687, 20198, 44	648, 13764, 79
			851, 20185, 44	659, 13754, 79
			852, 19901, 46	670, 13709, 80
			854, 19818, 46	731, 13709, 80
			858, 19547, 48	733, 13672, 80

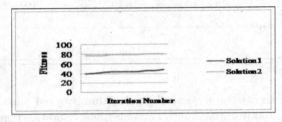

Fig. 11.14.: Graph screen showing the result values of the ACO methods worked with the criterion specified in Table 11.5 and the real solution.

proposed model increase the performance of the reference ACO algorithm and the best results are derived. Consequently, our proposed two model increased the success of the ACO algorithm developed for solving the data biclustering problem applying simultaneously both on row and column profile.

References

1. Kerr, M. K. and Churchill, G. A. (2001).Experimental Design for Gene Expression Microarrays. *Biostatistics*,2:183-201.
2. Churchill, G. A. (2002).Fundamentals of Experimental Design for cDNA Microarrays. *Nature Genet.*,32:S490-S495.
3. Liping Ji, Kenneth Wei-Liang Mock, Kian-Lee Tan (2006).Quick Hierarchical Bi-clustering on Microarray Gene Expression Data, *Proceedings of the Sixth*

IEEE Symposium on Bio-Informatics and Bio-Engineering Pages, pp. 110-120.

4. S. Kaski, J. Nikkil, and G. Wong(2003). Analysis And Visualization Of Gene Expression Data Using Self- Organizing Maps, Proceedings of NSIP-01, *IEEEEURASIP Workshop on Nonlinear Signal and Image*.

5. R. Sharan and R. Shamir(2000). Click: A clustering algorithm with applications to gene expression analysis. In *ISMB*, pages 307-216.

6. a b Mirkin, Boris (1996) 'Mathematical' classification and clustering, *Kluwer Academic Publishers*, ISBN 0792341597.

7. Hartigan J.A. (1972). Direct clustering of a data matrix *Journal of the American Statistical Association 67* (337) : 123-9.

8. Van Mechelen I. Bock HH, De Boeck P (2004). Two-mode clustering methods: a structured overview. *Statistical methods in Medical research 13* (5) : 363-94.

9. Sara C. Madeira and Arlindo L. Oliveira (2004). Biclustering algorithms for biological data analysis: A Survey.

10. Yizong cheng and George M. Church. Biclustering of expression data. *In proceedings of the 8th International conference on intelligent systems for molecular Biology (ISMB '00')*, pages 93-103, 2000.

11. Qizheng sheng, yves Moreau and Bart De Moor (2003). Biclustering microarray data by gibbs sampling. In *Bioinformatics, volume 19* (suppl. 2), pages ii 196- ii 205.

12. Ren Peeters(2003). The maximum edge biclique problem in NP-complete Discrete applied Mathematics 131 (3) : 651-654.

13. 13. G GetZ, E. Levine and E. Domany (2000). Coupled two-way clustering analysis of gene microarray data. *In proceedings of the Natural Academy of Sciences* USA, pages 120 79-12084.

14. Amcr Ben-Dor, Benny Chor, Richard karp and Zohar yakhini (2002). Discovering local structure in gene expression data. The order-preserving sub matrix problem. *In proceedings of the 6th International computational Biology (RECOMB' 02)* Page 49-57.

15. Special Issue on Bioinformatics (July 2002), *IEEE computer*, Vol-35.

16. Aleksander I. and Morton H. An introduction to Neural Computing, 2nd Edition.

17. Particle Swarm Optimization Homepage, http://www.cis.syr.edu/ mohan/pso/.

18. WIKIPEDIA, http://en.wikipedia.org/wiki/Ant Clony Optimization.

19. Peng Yuqing, Hou Xiangdan, Liu Shang (2003). The K-means Clustering Algorithm based on Density and Ant colony, *IEEE Int. Conf. Neural Networks and Signal Processing Nanjing*, China, December 14-17.

20. Xiang Xiao, Ernst R. Dow, Russell Eberhart, Zina Ben Miled, Robert J. Oppelt (2003). Gene Clustering Using Self-Organizing Maps and Particle Swarm Optimization, *IEEE International Workshop On High Performance Computational Biology*.

21. Julia Handl, Joshua Knowles, Marco Dorigo (2003). Ant-Based Clustering: A Comparative Study of its relative performance with respect to k-means, average link and 1D-SOM, *IRIDIA-Technical Report Series*.

22. Cheng, Y. and Church, G.M.(2000). Biclustering of Expression Data. *ISMB'00*

23. DI CARO, G., DORIGO, M.,(1998). Extending AntNet for Best-effort Quality-of-Services Routing, Ant Workshop on Ant Colony Optimization, htpp://iridia.ulb.ac.be/ants98/ants98.html, 15-16,

24. Shelokar, V.K., Jayaraman, V.K., Kulkarni, B.D.(2004). An Ant Colony Approach for Clustering, *Analytica Chimica Acta 509*, 187-195.

25. UCI Repository for Machine Learning Databases retrieved from the World Wide Web: http://www.ics.uci.edu

26. Urszula Boryczka (2009), Finding groups in data: Cluster analysis with ants, *Applied Soft Computing*, Volume 9, Issue 1, Pages 61-70.

27. Benlian Xu, Qinglan Chen, Xiaoying Wang, Jihong Zhu, (2009),A novel estimator with moving ants, *Simulation Modelling Practice and Theory*,Volume 17, Issue 10, Pages 1663-1677.

28. R.J. Mullen, D. Monekosso, S. Barman, P. Remagnino (2009),A review of ant algorithms, *Expert Systems with Applications*, Volume 36, Issue 6, Pages 9608-9617.

Author Index

Subject Index